The Puzzle Universe

The Puzzle Universe

A History of Mathematics in 315 Puzzles

IVAN MOSCOVICH

FIREFLY BOOKS

A FIREFLY BOOK

Published by Firefly Books Ltd. 2015

First printing

Publisher Cataloging-in-Publication Data (U.S.)

Moscovich, Ivan.
    The puzzle universe : a history of mathematics in 315 puzzles / Ivan Moscovich.
[396] pages : photographs (some color) ; cm.
Summary: "An accessible and engaging guide to puzzles invented by history's greatest puzzle and game masterminds, Moscovich has assembled 350 puzzles and games that he professes can change the way we think and make us more human through creative problem solving." — from Publisher.
ISBN-13: 978-1-77085-475-8
1.Mathematical recreations. 2. I. Title.
793.74 dc23   QA95.M583 2015

Library and Archives Canada Cataloguing in Publication

Moscovich, Ivan, author
    The puzzle universe : a history of mathematics in 315 puzzles / Ivan Moscovich.
Includes index.
ISBN 978-1-77085-475-8 (bound)
    1. Mathematical recreations. I. Title.
QA95.M68 2015   793.74   C2015-901630-4

Published in the United States by
Firefly Books (U.S.) Inc.
P.O. Box 1338, Ellicott Station
Buffalo, New York 14205

Published in Canada by
Firefly Books Ltd.
50 Staples Avenue, Unit 1
Richmond Hill, Ontario L4B 0A7

Design: Studio Lannoo

Printed in China

"If you always do what interests you, at least one person is pleased."

— Katherine Hepburn

Image Credits

(Key: a-above; b-below; l-left; r-right; c-center)
All images are part of the collection of Ivan Moscovich or are part of the public domain, apart from the images on the following pages:
Shutterstock 12, 17, 24, 30, 31, 32 (a.r. and b.l.), 33, 35 (l.), 38, 39 (b.), 46 (r.), 47 (a.), 55 (a.), 60 (c.), 71 (a.), 78 (a.l. en a.r.), 88, 90 (b.), 98 (a.), 101 (a.l.), 105, 108 (a.), 121 (b.), 126 (b.), 162 (b.), 163 (l.), 165 (b.), 169 (b.), 171, 173 (a. and c.), 181 (b.); 191 (b.), 193 (l. and r.), 204, 242, 246, 250 (a.), 304 (a., b.l., b.r.), 308 (l.), 328 (b.l.), 350); Tatjana Matysik: 27 (b.r.), 40, 43 (b.r.), 58 (a.r.), 64 (a.r.), 106 (c.r.), 170 (c.), 269, 276 (a.), 324, 339 (a.); RobAid 32 (a.l.); Royal Belgian Institute of Natural Sciences 36 (b.); Norman Rockwell 43 (c.); Andy Dingley 44 (a.); Getty Images 46 (l.); Carole Raddato 55 (a.); Ernst Wallis 60 (a.); Tate, London 89 (r.); Taty2007 102 (l.); Luc Viatour 113 (b.), National Gallery, London 112 (a.), 114 (b.); National Maritime Museum, London 125 (a.); JarektBot 127 (b.); Gauss-Gesellschaft Göttingen e.v./A. Wittmann 172 (l.); Rigmor Mydtskov 253 (l.); Piet Hein 253 (r.); Topsy Kretts 254 (a.); Konrad Jacobs 254 (a.); Anna Frodesiak 259 (a.); Dan Lindsay 300 (l.); Ed Keath 307 (b.); J. Jacob 308 (b.r.); Rinus Roelofs 319; Scott Kim 325; Erik Nygren 328; Jeremiah Farrell 344 (l.); Nick Baxter 346; Nick Koudis 349; Bruce Whitehill 352 (a.); Michael Taylor 352 (b.); Oscar van Deventer 357 (r.), 358 (r.); José Remmerswaal 358 (l.); Antonia Petikov 358, 359; Teja Krasek 361.

CONTENTS

This book is a labor of love. I dedicate it warmly to

my dear wife Anitta, with love and gratitude for her infinite patience, valuable judgment and assistance;

my dear daughter Hila, whom I love and admire for her inspiring new insights and creativity;

my adorable granddaughter Emilia, who is well on her way with her wonderful life journey

… and to all those who like beauty, puzzles and mathematics.

FOREWORD BY HAL ROBINSON

Over 350 stimulating thought-provoking puzzles and problems from the history of puzzles, mathematics and brain games, selected and presented (and sometimes invented) by the recognized mastermind of the world of puzzles, games and recreational mathematics.

Each of the puzzles in the book appears in newly visualized original layout and artwork created by the author.

The text does much more than just set the challenges; it also explains in layman's terms the puzzle's significance in the history of puzzles and mathematics, and indicates its mathematical and educational significance, too, especially in light of the revelations of recent studies.

The historical dimension sets *The Puzzle Universe* apart from other books in this field, none of which create the interactive visual experience that is Ivan Moscovich's recognized specialty and trademark. It is the stamp of his genius that takes the experience and the learning and the fun of puzzles to a whole new level — as so many of Ivan Moscovich's thousands of loyal followers and supporters online will attest. The author is celebrated worldwide as one of the leading inventors and presenters of visual games and puzzles, with more than 40 illustrated books to his credit — one of which, *The Big Book of Brain Games,* has sold well over a million copies and has been translated into more than 20 languages. Ivan Moscovich is also the inventor of more than 100 successful games, puzzles and toys, commercially produced by many of the major companies in the Toy Business, demonstrating the highly original character of his inventions, which is reflected in his approach to books as well.

Ivan Moscovich writes: "We are born to play and get to know our world. One of the reasons I'm so passionate about puzzles and games is that I believe they can change the way people think and improve the quality of life. They can make us more inventive, more creative, more artistic. They can allow us to see the world in new ways. They can inspire us to tackle the unknowable. They can remind us to have fun. They can make us more healthy and even prolong our lives."

Humans derive great satisfaction from finding patterns, but we gain even greater pleasure from understanding what lies behind the patterns. The discovery of an unexpected connection, of some hidden magical regularity, provokes a delightful combination of surprise and intellectual satisfaction. We may feel a sense of awe at the beauty of what we have uncovered...

This book is for all those who like beauty, surprise, challenges, mathematics — and puzzles.

The Puzzle Universe is the story of ideas to entertain, teach and play, regardless of age.

— *Hal Robinson*
Executive Director of the British Interactive Media Association, Creator of brain games and other puzzles

INTRODUCTION

The history of ideas and recreational mathematics is full of puzzles. A puzzle is fun, a mental challenge that requires a solution.

I love games and puzzles. Over the last 60 years or so, I have collected, designed, researched and invented thousands of them, as well as hands-on interactive exhibits, toys, books and much more.

One of the reasons I am so passionate about games and puzzles is that I believe that they can change the way people think. They can make us more inventive, more creative, more artistic and even more human. They can allow us to see the world in new ways. They can remind us to have fun, make us healthier and even prolong our lives.

Child psychologists have long known that children learn about the world through games. We can understand the most abstract and difficult concepts if we allow ourselves the luxury of approaching them not as work, but as fun — and as a form of exploration. G.C. Lichtenberg,

the 18th century German physicist, famous for his wit and quotations, noted: "What you have been obliged to discover by yourself leaves a path in your mind which you can use again when the need arises."

The Puzzle Universe is intended for general readers of all ages. It is about puzzles, mathematics and its latent beauty. It is full of challenging historical facts, thinking puzzles, paradoxes, illusions and problem solving. But it is much more.

The puzzles in this book, which combine entertainment and brain teasing, expand on that idea and apply it to concepts common to art, science and mathematics. Well-placed quotations, historical anecdotes, biographical notes and in-depth explanations of the solutions all try to create a pleasant and enjoyable environment for creative discovery, problem solving, fun and enjoyment.

For these reasons I have given them a new name: Playthinks. A Playthink may be a visual challenge, riddle or

puzzle; it may be a toy, game or illusion; it may be an art object, a conversation piece or a three-dimensional structure.

There are many discoveries and problems in mathematics that do not depend on specialist knowledge. They may only involve basics, common sense and a bit of intuition. Some of the Playthinks are original, while others are novel adaptations or visualizations of classic and modern challenges. Whatever its form, a Playthink will ideally transfer you to a state of mind where thinking, pure play and problem solving coexist for the betterment of your brain.

Because playing and experimenting with Playthinks stimulates creative thinking, you may find the book slyly educational.
I certainly hope it is! My goal is for you to play, think about the problems, maybe even solve some of them, and go away satisfied, more curious and more creative.

— *Ivan Moscovich*

THINKING ABOUT PLAYTHINKS AND YOUR BRAIN

CREATIVITY AND INTELLIGENCE

Creativity and creative people have been regarded with wonder and admiration throughout human history. They seem to be able to remain connected with that childlike state of wonder, and use it for fun, enjoyment, creativity and much more.

How do they do it? How can we learn to be more creative? Creative people themselves can be a resource in answering this question.

Great scientists, artists and thinkers are highly motivated, challenge assumptions, recognize hidden patterns, see in new ways, make new connections and take advantage of chance.

Without creativity, human beings would have remained in a Paleolithic existence. Creativity is the most powerful mode of human thought and advancement. It is the resource that we all have to draw upon to enjoy and understand our lives and build our world.

There is no such thing as a recipe for creativity. It is also very difficult to define. It is more than just the process by which new ideas are generated. Creativity is really a different way of thinking, one that is preoccupied with fundamental relationships, arrangements and connections.

And, indeed, the more connections a creative mind can make, the more paths open up to finding a unique and satisfying answer to a problem. Psychologist Edward de Bono called this mindset lateral thinking, and it is commonly found not only in pioneering scientific minds, but also in artists and other visionaries.

The ability to "think outside the box," to think differently, in a novel and unconventional way, is highly sought after today. We are entering an age where creativity is becoming increasingly important.

Yet these creative people are not endowed with any special gift. For the first five years of life, every child is a creative thinker, with insatiable curiosity. Later, as we get older, we acquire mental blocks that obscure the essence of problems and often lead us away from even the most obvious solutions. We all have the potential to be creative; we just don't think creatively most of the time.

Eureka!

Innovation begins with creative ideas, requiring expertise. "Chance favors only the prepared mind," observed Louis Pasteur. The second component is imaginative thinking skills. In moments of creativity we see things in new ways, recognize patterns, make new connections — The Eureka Moment. A venturesome personality, continuously seeking new experience is the third component of innovation, followed by intrinsic motivation and the will to push through. Ideally, all of this is encouraged by a creative environment.

Most of us grew up with a concept of intelligence that is driven by tests: the person who can answer the most questions is thought to be the most intelligent. But imagining that intelligence can be boiled down to a single number — the IQ — is an obsolete notion. There have been a number of attempts to develop a Creativity Quotient for an individual, along the same lines as the Intelligence quotient (IQ), but they have been unsuccessful. Another flaw in early attitudes was the idea that intelligence is fixed at birth. Many late researches have shown that IQ scores can be significantly raised through appropriate training. According to Bernard Devlin, genes account for no more than 48 percent of a person's IQ, while 52 percent is a result of prenatal care, environment and education.

If you find yourself having difficulty with some of the puzzles, don't worry that you are not "smart" enough to do puzzles. It is all a matter of freeing up your latent creativity. With the proper mindset, anyone can do these puzzles.

And if you find the puzzles easy, congratulations!

> **"Creative thinking — in terms of idea creativity — is not a mystical talent. It is a skill that can be practised and nurtured."**
>
> — *Edward de Bono, psychologist (Malta, 1933)*

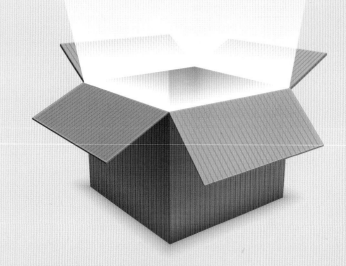

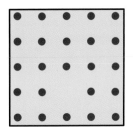

PEGBOARD SQUARES

How many different sizes of squares can you get by stretching rubber bands along four pegs of the pegboard?

1

CHALLENGE ● ● ○ ○ ○ ○
REQUIRES 🧠 ✏️
COMPLETED ○

BLIND DATE PROBLEM

The computer dating website has set you up on a blind date. She will be waiting for you holding this book. Approaching the place you see Julia Roberts with the book. What's the problem?

2

CHALLENGE ● ○ ○ ○ ○ ○
REQUIRES 🧠
COMPLETED ○

CREATIVITY AND PROBLEM SOLVING

Problem solving can teach us a little about the way in which our brains function. Thinking can be hard work, hence the natural human tendency to do as little of it as possible. This is visible in the hit-and-run approach to problem solving: pick the first solution that comes to mind and run with it.

Your subconscious mind stores all of your previous life experiences, your beliefs, your memories, your skills, all situations you've been through and all images you've ever seen. It's time to remove consciousness from its privileged place in problem solving. The real power of creativity lies in the subconscious. As modern cognitive science has revealed and confrmed, our brains operate with a vast unconscious mind that even Freud never suspected. Much of our information processing occurs below the level of conscious awareness — offstage and out of sight. Our unconscious mind feeds our insight, creativity and intuition. It is fast, automatic, and effortless, while our conscious mind is deliberate, sequential and rational, requiring an effort to employ. Theories of Amos Tversky and Daniel Kahnemann indicate that humans have developed mental shortcuts, called heuristics, which enable quick and efficient judgments and actions. Experience teaches our intuition and many judgments become automatic (driving a car; chess masters who may have 50,000 chess patterns stored in their memories can play "blitz chess" picking the right moves intuitively, etc.).

> **"Creativity comes from the abrasive juxtaposition of life experiences."**
>
> — *Mario Capecchi, Nobel Prize winner for Medicine. His mother was taken to Dachau concentration camp. Mario, age 4, landed on the streets for six years, until they were reunited.*

INTRODUCING YOUR SUBCONSCIOUS

"In the course of our lives, the subconscious handles over 90 percent of our problem solving. Complex decisions are best left to your unconscious mind to work out," according to recent research carried out by Prof. Ap Dijksterhuis at the University of Nijmegen. "Overthinking a problem could lead to expensive mistakes."

It suggests that the conscious mind should be trusted with simple decisions only.

Thinking hard about complex decisions based on multiple factors appears to confuse the conscious mind resulting in people concentrating only on a subset of information, often resulting in unsatisfactory decisions. In contrast, the unconscious mind appears to be more capable of considering the full information and therefore produce a more satisfactory decision.

Experiments demonstrate that we are all

capable of such "nonconscious learning," as we shall experience from the two counting puzzles on pages 14 and 15. Faced with complex decisions involving many factors, the best advice may be to take time out — to "sleep on it" — and to await the intuitive result of our unconscious processing. Today's cognitive science enhances our appreciation for intuition but also reminds us to always carefully check it against reality.

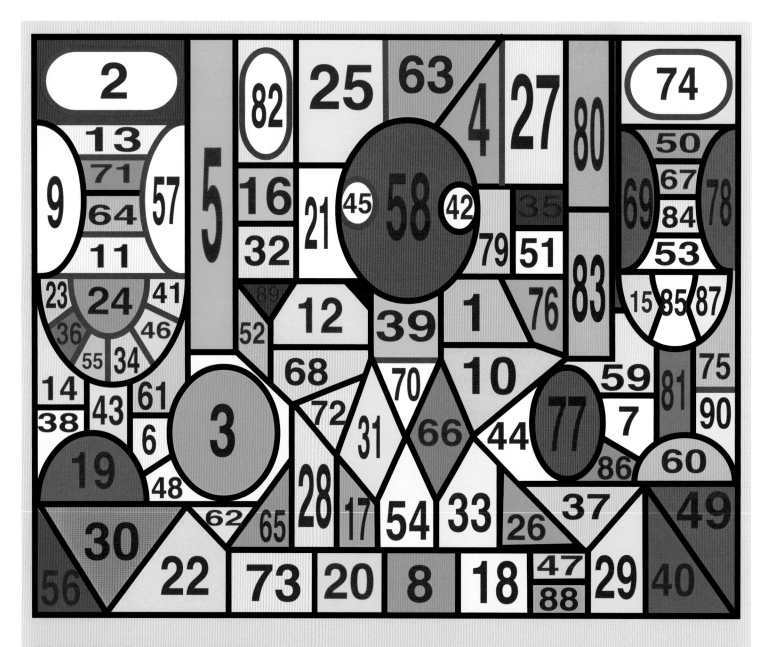

COUNTING FUN — TEST 1

Two surprising visual tests may reveal how your subconscious works and solves problems, in this instance just by counting from 1 to 90.

Counting is the oldest mathematical activity of humankind and also one of the most powerful and fundamental ideas ever conceived by humans.

The idea that for each natural number, there is a next number, offered great mathematical advances.

This test and the test on the following page will give counting a new twist. The object in both is simply to find how long it will take you to find the consecutive numbers from 1 to 90 in succession, just by looking, without skipping any (consecutive) number or numbers, and also without marking the page as you search for the next consecutive number.

To complete the two tests, you must start by finding 1, then 2, 3, 4 and up to 90.

Cheating is of course not allowed and it would spoil the fun, believe me.

Repeat each test two or three times and mark your times in minutes for each count in the chart on the next page.

Your first surprise will be that the tests will take much longer than expected. You will also note that as you repeat the tests your times may improve. But the results of the second test will be even more surprising.

3   CHALLENGE ●●
REQUIRES
COMPLETED

COUNTING FUN — TEST 2

Repeat the two tests two or three times and mark your scores in the tables on the right.

It may surprise you that in test 2 there may be a very significant improvement in your scores compared to your results in test 1.

If this happens, can you find an explanation why such unexpected improvements happened?

YOUR SCORES

TEST 1	Score in minutes
first	
second	
third	

TEST 2	Score in minutes
first	
second	
third	

4

CHALLENGE ● ● ○ ○ ○ ○
REQUIRES 🧠
COMPLETED ○

PUZZLES AND YOUR BRAIN

Puzzles are more than just an entertaining pastime

Solving a variety of different types of puzzles will improve your brain and ward off mental deterioration and age-related illnesses. The brain is a very complex machine that is constantly creating and reinforcing connections between its 100 billion cells. Exercising the brain, for example by solving puzzles, helps to create new connections and enhance mental performance in the long run. Memories are formed by these connections between brain cells, and by the capacity of each neuron to chemically signal the 10,000 or so other cells it connects to. Puzzles can support essential brain functions such as memory retrieval and the ability to process new information by strengthening the connections between brain cells. Engaging in problem solving exercises stimulates the creation of new connections in the brain whilst strengthening old ones.

Use it or lose it

If you don't give your brain a solid workout, your mental strength starts to deteriorate. As you age, it is important to keep your brain fit by playing a diversity of different types or puzzles.

Problems and puzzles can be simplistically divided into ones that require insight (a mental leap) and ones that are solved through a more systematic analysis. What is more common in problem solving: insight or analysis? Like most of these kinds of debates, it appears that both play an important role. The ability to switch between these two brain states is important, and if you try to solve problems that ask

for both deep analysis as well as out-of-the box insight (as many types of puzzles do), this would add to your brain plasticity and health during the aging process.

Nowadays, brain scientists are asked whether doing puzzles helps stave off

attention
memory
speed
flexibility
problem solving

BRAIN AREAS

Alzheimer's and other forms of mental deterioration. Alzheimer's is a brain disease that causes increasing loss of memory and other mental abilities. It is the most common cause of severe memory loss in older adults. It attacks very few people before the age of 60, becoming increasingly common thereafter. In the not-too-distant past, older adults suffering from severe memory loss were often labeled "senile." But it is recognized today that they were probably suffering from Alzheimer's.

As a brain disease, Alzheimer's raises the question of whether or not puzzle activities can be used in helping to pre-

vent it. Many studies strongly suggest that puzzle activities may indeed prevent mental deterioration and many organizations worldwide endorse puzzles as part of a preventive strategy.

Puzzle instinct

However, the literature on the correlation between puzzles and brain fitness is not extensive and, when looked at with a critical eye, really does not establish a definite correlation. The reason is that doing a certain genre of puzzle over and over, as most people are inclined to do, does not provide enough diversity for the brain. The brain does seem to need many different types of stimulating input in order to keep functioning. A great variety is needed.

In general, one can argue that the areas of the brain activated by puzzles of particular kinds might be larger than normal. This seems to be a logical assumption, but more studies are needed.

Studies by Marcel Danesi at the University of Toronto revealed that we do indeed all have different puzzle preferences and problem solving abilities. Some people only love crossword puzzles, others just like logic ones (like Sudoku). A few enjoy a mixture of them. In some of Danesi's experiments, puzzle materials that fell into "un-preferred" categories were given to groups of students. After a period of eight months, a significant number (around 74 percent) claimed that they started to like the puzzle genre they once disliked. By simply doing puzzles, our "puzzle instinct" seems to kick in and allow us to enjoy all genres.

PUZZLE BASICS

We continue with a random sample of 24 typical classic thinking puzzles from my collection of about 5000 classic puzzles to warm up your puzzle-solving ability. Thinking puzzles require a variety of strategies to be solved, this includes mathematical (though a particular knack for math is not imperative), but also logical and the application of basic principles.

5	CHALLENGE	● ● ● ○ ○
	REQUIRES	🧠✏️
	COMPLETED	○

SNAIL'S PACE

A little snail climbs up a window 90 cm high. If every day it climbed up 11 cm, and every night it dropped back 6 cm, how many days without stopping did the little snail travel to reach the top?

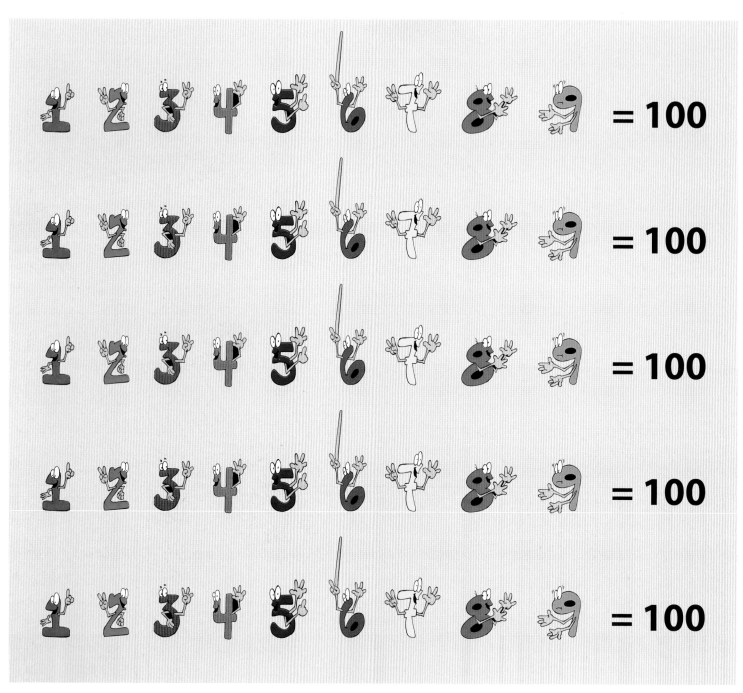

1 2 3 4 5 6 7 8 9 = 100

1 2 3 4 5 6 7 8 9 = 100

1 2 3 4 5 6 7 8 9 = 100

1 2 3 4 5 6 7 8 9 = 100

1 2 3 4 5 6 7 8 9 = 100

CHALLENGE ● ● ● ○ ○

REQUIRES 🧠 ✏️

COMPLETED ○

A HUNDRED TOTAL

A very old classic arithmetic problem is the one using consecutive digits from 1 to 9 and inserting mathematical symbols between them to form the sum total of 100. There are endless variations of the problem, including the one in which only plus or minus signs are allowed. Martin Gardner demonstrated the solution using the maximum number of plus signs and the minimum number of minus signs.

Can you insert plus or minus signs in the above number sequence to achieve such a solution? Hint: two of the consecutive digits will form a two-digit number. There may be other solutions using more than just one two-digit number, and/or more than just minus and plus signs. How many solutions can you find for this problem?

7 CHALLENGE ● ● ○ ○ ○ ○
REQUIRES 🧠
COMPLETED ○

HIDING REGULAR POLYGONS AND A STAR

How long will it take you to find the outlines of seven regular polygons and a ten-pointed regular star?

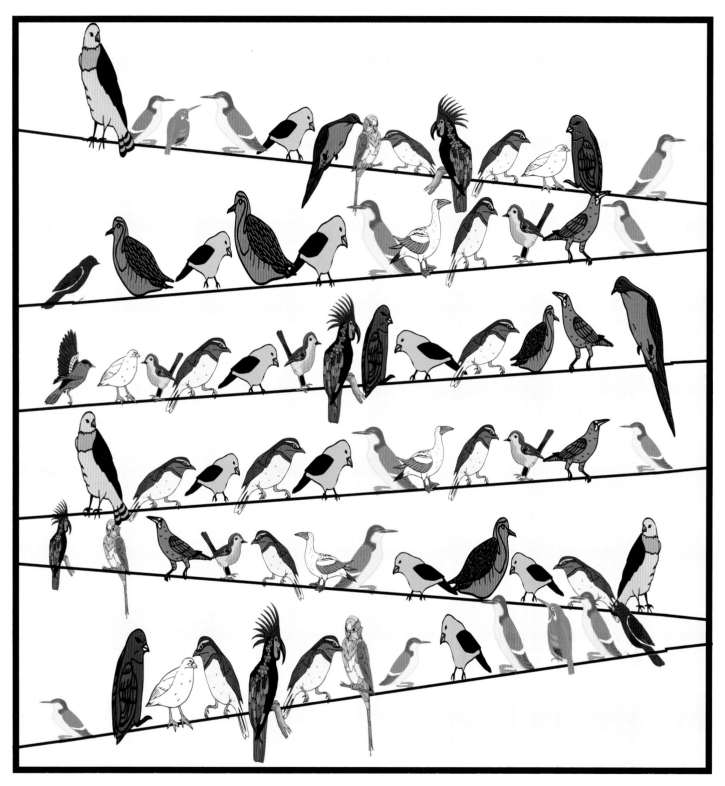

CHALLENGE ● ● ● ● ○
REQUIRES 🧠 ✏️
COMPLETED

BIRDS ON A WIRE

Imagine a very long wire on which a multitude of birds are randomly distributed looking one way or another along the wire at their nearest neighbor. Can you guess how many birds will be observed by one, two, or none of their neighbors if there is an infinite number of birds on the wire? In the example above only 72 birds were randomly distributed, so your guess will be an approximation.

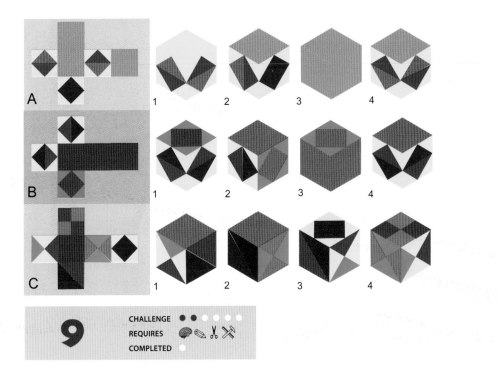

ANGLES IN A TRIANGLE

Euclid proved that the three angles of a triangle add up to a straight angle (180 degrees). The beauty of mathematics is that often, amateurs with a bit of insight can make new discoveries and come up with new evidence.

Luther Washington, today a mathematician at Stanford, as a very young student had an idea to prove it in a simpler way, using only a pencil. Can you work out how he did it?

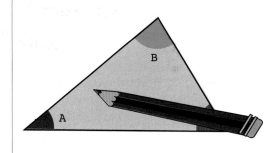

CUBE FOLDS

On the left are unfolded nets of cubes with designs on their faces (A-B-C). On the right are four isometric drawings of cubes. The object in each case is to match up each unfolded cube with the correct isometric cube.

ELEVATOR GOING UP AND DOWN

In an 18-floor building there is only one strange elevator, which only has two buttons: one up and the other down. The up button takes you up 7 floors (or doesn't move at all if there are no floors available), and the down button takes you down 9 floors (or doesn't move if there are no floors available). Is it possible to get from the ground floor to any floor by taking the elevator? How many times will the maintenance man have to push the buttons to get from the ground floor to all the other floors and in what sequence will he visit the floors? The first three trips are demonstrated.

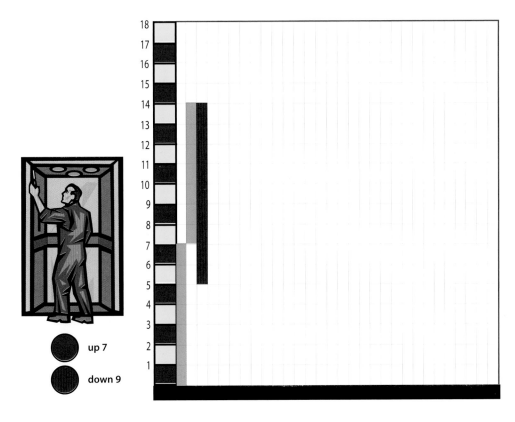

ENCLOSURE

Tim found 14 bricks to make a fence in the garden for his new pet, a small turtle. Since his turtle has grown over the years, Tim wants to enlarge the enclosure as much as possible, using the same number of bricks. How can he do this?

12	CHALLENGE	● ● ○ ○ ○ ○
	REQUIRES	🧠 ✏️
	COMPLETED	○

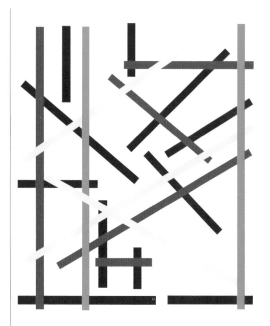

PLAYGROUND

You are looking from above on a playground on which planks are piled one on top of another. Can you identify the highest point?

13	CHALLENGE	● ● ○ ○ ○ ○
	REQUIRES	🧠
	COMPLETED	

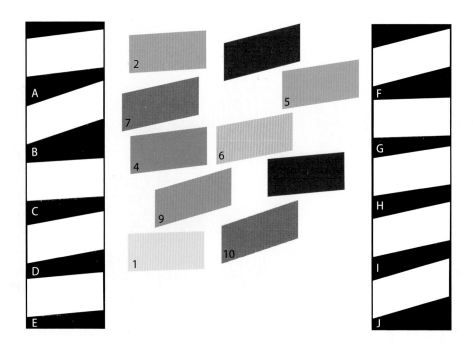

A	
B	
C	
D	
E	
F	
G	
H	
I	
J	

FILLING GAPS

Just by looking, fit the numbered colored shapes into their corresponding gaps and pair them with the lettered gaps in the accompanying chart. How many mistakes did you make? You may be surprised!

14	CHALLENGE	● ● ○ ○ ○ ○
	REQUIRES	🧠 ✏️
	COMPLETED	

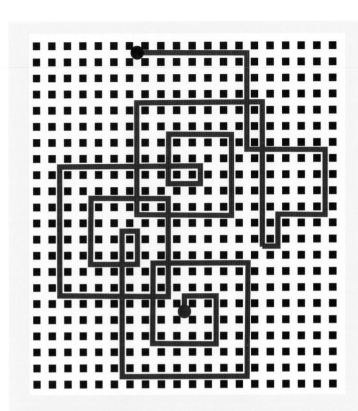

MOLE WALKS

The mole starts from the red point. The red line shows its path until it ends at the blue point. Can you work out the logic of its path up till the point where it was forced to change the rule? At which point did this happen?

LINE UP

How many lines will you be able to trace through just by looking, before you lose track? Stability of attention is the ability to direct it towards something for a long period of time.

PIPES

Nine pipes are tightly tied together by a red metal band. How long is the red band?

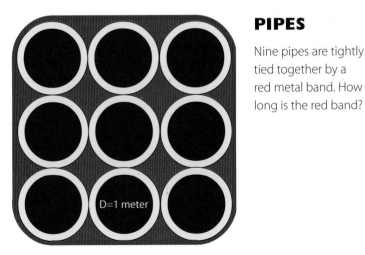

SWEETS

A piece of cake and an ice-cream cost two and a half dollars between them, but the cake costs a dollar more than the ice-cream. How much does each cost?

PATTERNS OF MIND

The ancient Greek "deiknymi," or thought experiment, was the most ancient pattern of mathematical proof. Thought experiments are ideas of imagination used to investigate the nature of the unknown, often resembling riddles. The common features of thought experiments include visualization of a hypothetical scenario, experimentation and conceptualization of what is happening.

The idea and simple reasoning behind thought experiments is that just by the power of thinking we can discover new things about the world, which early on created great interest in philosophy. Many famous thought experiments played an enormous role in advancing mathematics and science, like Einstein's elevator, Newton's apple, Schrödinger's cat, Maxwell's demon, Newton's satellite principle, Galileo's balls and many others.

The infinite universe

One of the most beautiful early-thought experiments in its simplicity and elegance is the "Infinite Space" thought experiment of Archytas and Epicurus. Archytas believed that the universe is infinite and has no limits. In his thought experiment he used the idea that someone is at the border of the universe and stretches his hands beyond this limit. About a hundred years later, Epicurus, in a similar thought experiment, imagined an arrow flying endlessly through space without encountering any obstacle, thus proving that space is infinite. On the other hand, if it bounces back encountering an obstacle like a wall, at the "end" of space, it would again prove that space is infinite, since there must be something behind the wall.

The ideas of Archytas and Epicurus are beautifully visualized in the Flammarion engraving from 1888, believed to have been originally created in the 16th century.

Plato and Aristotle did not accept the idea of an infinite universe because they had problems accepting the idea of the infinite in general. Up to medieval times, Aristotelian cosmology was accepted in terms of which space was finite and had a definite edge.

EPICURUS (341-270 BC)

The philosophy of Epicurus was to make a happy life for everyone with more pleasure and less sadness. According to him, the best way to achieve this objective is not to want anything. Epicurus's philosophy had an important influence on early Christianity. His Infinite Space thought experiment is a beautiful and important classic.

ARCHYTAS (428-347 BC)

Ancient Greek mathematician, scientist of the Pythagorean school and one of the founders of mathematical mechanics. Among his many achievements he constructed a wooden pigeon, the earliest self-propelled flying device (probably by steam) capable of flying some 200 meters.

SALE

The saleslady sold two sofas at the bargain price of 1200 dollars apiece. She made a 25 percent profit on the first sofa and a 20 percent loss on the other. She assumed that she still made a profit on the combined sale. Was she right in this assumption?

19	CHALLENGE	● ● ○ ○ ○ ○
	REQUIRES	🧠
	COMPLETED	○

SEATING ARRANGEMENT

In how many different ways can you seat men and women in a row of four chairs so that no two women are sitting next to each other? Can you find a general solution to this problem for n numbers of chairs and work out the answers for any number of chairs?

20	CHALLENGE	● ● ● ○ ○ ○
	REQUIRES	🧠
	COMPLETED	○

22	CHALLENGE	● ● ● ● ○ ○
	REQUIRES	🧠
	COMPLETED	○

SQUARES IN AN EQUILATERAL TRIANGLE

Three identical squares set in an equilateral triangle dissect it into 22 regions. Using three different identical squares, can you do better?

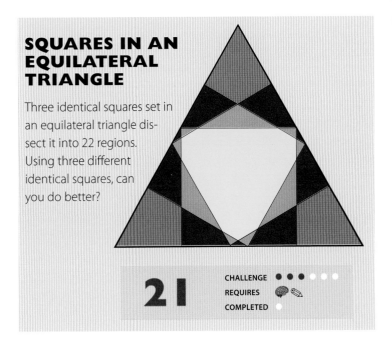

21	CHALLENGE	● ● ● ○ ○ ○
	REQUIRES	🧠 ✏️
	COMPLETED	○

LINES THROUGH 16 POINTS

Six straight lines are needed to create a continuous line through 16 points. How many solutions can you find with the smallest number of intersections? And how many solutions can you find forming symmetrical patterns?

TRIANGULATING POLYGONS

Draw a polygon of any number of sides. Place a colored dot at each of the corners, and then place any number of dots at random on the inside. Then, using the dots as corners, divide the polygon into non-overlapping triangles and label their corners using three colors, red, blue and yellow. Coloring the triangles this way there can be 10 different types of triangles as shown on the right.

A triangle having all three colors at its vertices is a complete triangle.

In my example I have only colored the border dots. Can you color the inside dots so as to create two complete triangles and no more?

Leaving the border as it is, you can subdivide the inside differently as you wish and try again.

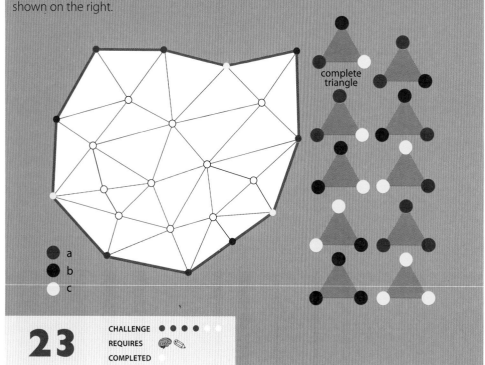

complete triangle

a
b
c

23

CHALLENGE ● ● ● ● ○ ○
REQUIRES
COMPLETED

ROLLING A DICE

Mutually exclusive events are related in such a way that only one of them can possibly happen. In such a situation, the probability that any of them will occur is the sum of their individual probabilities. Rolling a four with one dice is 1/6. So is the chance of rolling a six. What is the probability of rolling either a four or a six on a given toss?

24

CHALLENGE ● ● ○ ○ ○ ○
REQUIRES
COMPLETED

TOP VIEW

The top view of a building is shown from above. Can you envisage the three-dimensional form of the top of the building as seen from above?

25

CHALLENGE ● ● ○ ○ ○ ○
REQUIRES
COMPLETED

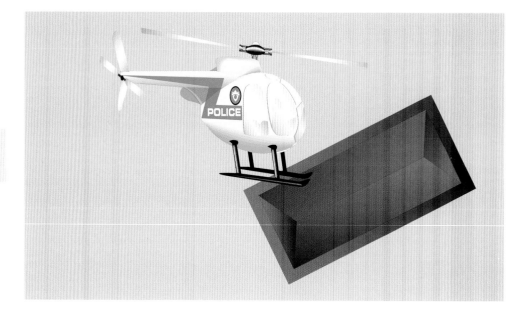

HOW MANY SQUARES? (1)

A distinct group of recreational math puzzles deals with the problem of "How Many?," including "How Many Squares?".

Just by looking at the three patterns, how many squares of different sizes can you count?

The beautiful third pattern was created by Cliff Pickover and he is challenging puzzlists for the right answer, since he is getting different answers from his colleagues. Can you provide it and finally settle the issue?

26 CHALLENGE ● ● ○ ○ ○ ○
REQUIRES 🧠 ✏️
COMPLETED ○

HOW MANY SQUARES? (2)

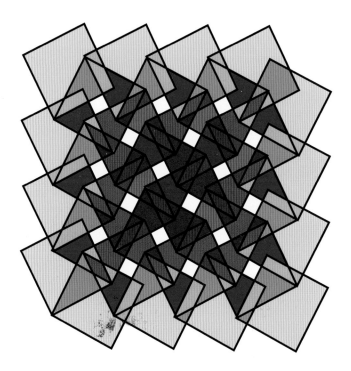

27 CHALLENGE ● ● ○ ○ ○ ○
REQUIRES 🧠 ✏️
COMPLETED ○

HOW MANY SQUARES? (3)

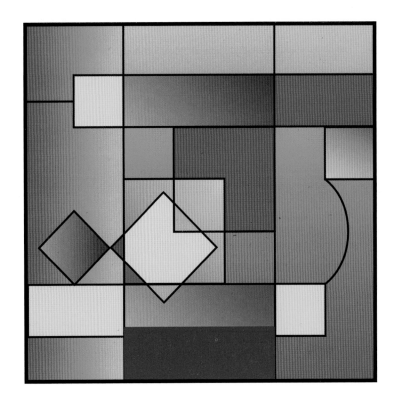

28 CHALLENGE ● ● ○ ○ ○ ○
REQUIRES 🧠 ✏️
COMPLETED ○

CHAPTER

2

BEGINNINGS, GREEK MATH, GEOMETRY AND AHMES' PUZZLE

MATHEMATICS AS A SCIENCE OF PATTERNS

For the ancient Greeks, mathematics was the science of numbers. But this definition of mathematics has been incomplete for hundreds of years.

In the middle of the 17th century, Isaac Newton in England and Gottfried von Leibniz in Germany independently invented calculus, the study of motion and change. Contemporary mathematics comprises 80 distinct disciplines, some of which are still being split into subcategories.

So today, rather than focusing on numbers, mathematicians think that their field is better defined as the science of patterns. As the science of patterns, mathematics affects every aspect of our lives; abstract patterns are the basis of thinking, communication, computation, society and even life itself.

Patterns are everywhere and everyone recognizes them, but mathematicians are able to see patterns within the patterns. Despite the somewhat imposing language used to describe their work, the goal of most mathematicians is to find the simplest explanations for the most complex patterns. Part of the magic of mathematics is how a simple, amusing problem or puzzle can often lead to far-reaching insights.

It gives us great joy to discover patterns, but even greater joy to understand what lies behind them. The discovery of an unexpected connection, some hidden magical regularity, provokes pleasurable mixtures of beauty, awe and surprise. This is what this book hopes to show you! Or as prof. E.D. Bergman said: "Is the inherent beauty of a mathematical theorem less appealing than the beauty of a painting? Is the elegance of a physical apparatus inferior to that of a beautiful poem or a great work of literature? Is the history of scientific thought less inspiring than the history of religions? Or, is the fight against hunger and disease less heroic than the war of conquest or even liberation?"

"Never before in my life have I troubled myself over anything so much, and I have gained enormous respect for mathematics, whose more subtle parts I considered until now, in my ignorance, as pure luxury!"

— *Albert Einstein, in a letter to a friend, 1916*

"Is the inherent beauty of a mathematical theorem less appealing than the beauty of a painting? Is the elegance of a physical apparatus inferior to that of a beautiful poem or a great work of literature? Is the history of scientific thought less inspiring than the history of religions? Or, is the fight against hunger and disease less heroic than the war of conquest or even liberation?"

— *Prof. E.D. Bergman*

> **"We all live in a gigantic mathematical object."**
>
> — *Max Tegmark, Swedish cosmologist*

OUR MATHEMATICAL UNIVERSE — 13,8 BILLION YEARS AGO

Our universe is mathematical. Nature is a master builder playing endless variations with a number of basic shapes. The circle, the square, the triangle and the spiral may be compared to the letters of an alphabet, which can be used in combinations to form more elaborate shapes with new and unique properties.

The idea of a mathematical universe goes all the way back to the philosophy of Ancient Greece. Today, some scientists, noteworthy among them Max Tegmark, an unconventional Swedish cosmologist, push the concept to the extreme by arguing that the universe is not just described by mathematics. They claim it is mathematics!

Tegmark's fascinating Mathematical Universe Hypothesis (MUH) is based on the following premise: "All structures that exist mathematically also exist physically. Mathematical patterns and formulas create reality." He argues that with a sufficiently broad definition of mathematics, our physical world is an abstract mathematical structure. Or as Tegmark says: "We don't invent mathematical structures — we discover them, and invent only the notation for describing them."

One of the delights of exploring the world around us through mathematical eyes is that we are able to perceive patterns that would otherwise be concealed from us. As with other theories of the universe, Tegmark's "MUH" is forcefully criticised by some scientists, mathematicians, and philosophers. Tegmark's response to these criticisms involves another hypothesis, the External Reality Hypothesis (ERH), which states that an external physical reality exists independently of humans, implying the existence of MUH and also the concept of many parallel universes (a compelling conversation piece).

POWERED FLIGHT — 410 MILLION YEARS AGO

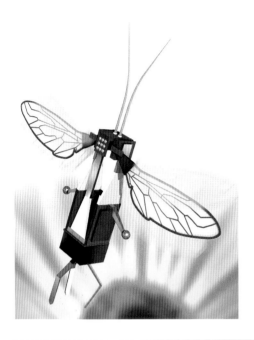

The first flying creatures were dragonfly-like insects, which evolved around 410 million years ago. Even today, the complex aerodynamics of how insects fly with flexible flapping wings is not yet fully understood. The engineering of extremely small micro air vehicles (MAVs) or nano air vehicles (NAVs) as small as a few centimeters or even smaller is at a very early stage.

The robotics research team at Harvard University has a head start. In 2007, inspired by the biology of flying insects, they built a life-sized tiny robotic fly, the RoboBee, capable of tethered flight, the culmination of 12 years of research. They successfully created artificial muscles capable of beating the wings 120 times per second.

The aim of the RoboBee project is to develop a fully autonomous swarm of flying robots, with applications ranging from search and rescue to artificial pollination. The power supply and decision-making functions are still reliant on thin cables tethered to the robot, so to achieve true autonomy researchers on a way to integrate them within the frame. With a wingspan of just 3 centimeters, the RoboBee is the smallest man-made device modeled on an insect to achieve flight.

WHY IS THE TOP OF THE WING OF AN AIRPLANE CURVED?

The wings of an airplane are designed in a way that ensures that air rushes across their upper surfaces faster than it rushes past the lower. For this reason the top surface of the wings is made longer than the bottom. According to Bernouilli's Principle, this lowers the pressure more above the wings than below, producing the force called lift — the force that keeps the plane in the air when it moves forward. When an airplane is in the air, the combined weight of the plane, fuel, passengers and cargo all exert a heavy pull downward. However, this total weight is overcome by the lift, allowing the airplane to fly.

Daniel Bernoulli (1700–1782)

Daniel Bernoulli, one of the many prominent mathematicians in the Bernouilli-family. He is most famous for his Bernouilli principle.

PRIME NUMBERS AND CICADAS — MILLIONS OF YEARS AGO

There are about 3,500 species of cicadas (Latin for "tree cricket"), harmless flying insects, all over the world.

Most cicadas have a life cycle from two to five years, but some species have a much longer, curious and strange life cycle.

The Magicicadas, or periodical cicadas of North America, have a life cycle lasting 13 or 17 years. Apart from their life stories, these have attracted enormous interest from biologists and mathematicians because 13 and 17 are prime numbers, playing a major role in mathematics as well as life.

Paul Erdös, the legendary mathematician, in a moment of despair declared: "It will be another million years until primes are understood." The significance of 13 and 17 in the lives of cicadas, is still not convincingly explained.

Periodical cicadas live underground as nymphs for 13 or 17 years. But like clockwork, after these periods, they build an exit tunnel to the surface and emerge in millions.

What is it that makes cicadas come out during these years? How do cicadas know prime numbers? There could be no coincidence.

Stephen Jay Gould advanced a theory that they were driven below ground for such long periods to evade shorter lived predators.

Then they shed their skin and become flying insects. The males constantly noisy, the females silent. They don't eat and their only purpose in their short life of just a few weeks is to mate and preserve their species.

No wonder the beautiful Mexican song "La Cigarra" romanticizes cicadas as a creature that sings until it dies. And indeed, two months later, they are all dead, leaving behind millions of eggs that will hatch, with their nymphs rushing underground for the next 13 or 17 years until their life cycle starts all over again.

The life of a cicadas is impressive proof of the mathematical brilliance of nature. Nature's innate knowledge of prime numbers provides cicadas with a valuable survival skill.

Evolution is a long-term game. It made a compromise by choosing a relatively large prime number (e.g. 13 or 17) for cicadas to avoid predators. For example, if the cicada has a life cycle of 17 years and its predator has a life cycle of 5 years, then they will only meet every (17x 5) = 85 years.

WHEN DID CREATIVITY START?

Isn't it thought provoking that more than 200,000 years ago, our forefathers had flashes of creativity and innovation, far earlier than previously thought — even before the emergence of Homo sapiens, as indicated by the latest research. This chapter takes you on a journey exploring these highlights from around the world. The wheel (in all its variations), the first dice, board games in ancient Egypt and Pythagoras' theory: the knowledge humankind has gathered in the past centuries is really breathtaking!

ACHEULEAN HAND AXE — TWO MILLION YEARS AGO

Human lineage emerged in Africa around six million years ago. We have no visible record of innovation for nearly four million years.

Then, at some point, nomadic humans discovered fire and started flaking stones to produce cutting tools, perfecting the technique for the Acheulean hand axe shown, dated about two million years ago.

You could say that mathematics, art and technology began when early man chipped a stone with symmetrical perfection to produce a point and very sharp edges. It was a truly unique creation.

ISHANGO BONE — 16,000 BC

The Ishango bone was discovered at Ishango in 1960 by the Belgian geologist Jean de Heinzelin de Braucourt and is now displayed at the Royal Institute for Natural Sciences of Belgium in Brussels. It is a small tool comprising a bone handle with a piece of quartz set into its end, and with a series of notches carved in three rows (shown below).

The Ishango bone is dated about 16,000 BC. The earliest records of counting were found dating from 35,000 BC, but the three rows of notches on the Ishango bone indicate knowledge of mathematics — astonishing, given the time. It is therefore considered one of the earliest mathematical artifacts of substantial mathematics.

Although it was initially believed that these notches were similar to the early tally marks used to record counts and encountered throughout the world, there appears to be much more to the Ishango bone than a simple tallying tool. Let's look more closely at each row.

In the first row, apart from the last pair at the left, all pairs indicate multiplication by two. In the second row, the difference between the two pairs is 10.

The third row is the most astonishing. It lists the prime numbers between 10 and 20, in order.

Are the notches on the Ishango bone an intentional list of primes? Probably not. It is more likely that the notches formed a primitive calendar system.

However, let's say we select four positive integers at random, all below 30. Ten primes fall within this range, so the probability of all four being primes is 1/81. Astonishing, isn't it!

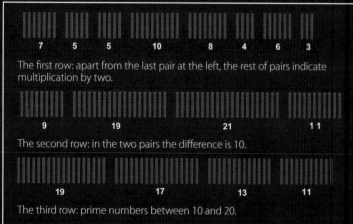

The first row: apart from the last pair at the left, the rest of pairs indicate multiplication by two.

The second row: in the two pairs the difference is 10.

The third row: prime numbers between 10 and 20.

ROLLING CIRCLES AND ROLLERS — BEFORE 6000 BC

It is reasonable to suppose that man discovered the roller before the wheel. The difference between the two is important. Unlike the wheel, the roller is independent of the vehicle that it transports. As the load on top of the roller is propelled forward, it moves the roller beneath it. The result is that both the load and the roller move forward.

Early civilizations around the world independently discovered that rollers can greatly facilitate the transportation of heavy loads. Without this discovery, building pyramids, temples and giant stone monuments would have been impossible.

The circumference of each of the two rollers is one meter. If the rollers make a full revolution, how far will the weight be carried forward?

29

CHALLENGE ●● ○ ○ ○ ○

REQUIRES 🧠✏️

COMPLETED ○

THE WHEEL — ROTARY MOTION — 6000 BC

The wheel is probably the most important mechanical invention of all time.

Aristotle imprinted his image of perfect symmetry onto the motions of all heavenly bodies, proclaiming that they could only orbit in circles. This "knowledge" was uncritically accepted for more than 2,000 years, even by Copernicus.

The introduction of the wheel, or, in abstract form, rotary motion, represented an event of enormous importance in history.

It took thousands of years to conceive the idea of a form of motion unprecedented in man's immediate surroundings. The discovery of the wheel required a capacity for abstract thinking and the ability to pass from the object itself to the idea of it — from the phenomenon to the theory. Once this problem was solved, the development of the wheel has not subsequently seen any outstanding advances, as is the case with many other really great inventions. There are only differences in accessories between the first wheel of Ur, Mesopotamia, and the 20th century wheel with its pneumatic tires.

Cow-on-wheels — 4000 BC
Ceramic toy (Romania)

LABYRINTHS AND MAZES

Mazes are ancient structures. One of the first recorded mazes is the Egyptian Labyrinth (c. 900 BC). Herodotus, a Greek traveler and writer, visited the Egyptian Labyrinth in the fifth century BC, and recorded: "The Pyramids surpass description, but the Labyrinth surpasses the Pyramids." Little remains of this once-impressive structure today.

Legend has it that the first maze was built by Daedalus to house the Minotaur, half bull, half monster, of King Minos of Crete. Theseus found his way back out of this maze by using a ball of golden thread.

From the mathematical standpoint a maze is a problem in topology. A maze can be solved quickly on paper by shading all the blind alleys until only the right route remains. But when you do not possess a map of the maze and you are inside it, the maze can be solved by placing your hand against the right (or left) wall and keeping it there all the time as you walk. You are sure to reach the exit, though your route may not be the shortest one. This method does not work with mazes in which the goal is within the labyrinth and surrounded by closed circuits. Mazes that contain no closed circuits are called "simply connected," i.e. they have no detached walls. Mazes with detached walls are sure to contain closed circuits, and are called "multiply connected."

Top: a simply connected maze
Bottom: a multiply connected maze

ADRIAN FISHER'S MAZES

Adrian Fisher is internationally recognized as the world's leading maze designer, having created more than 500 full-sized mazes across 30 countries.

Working with his wife Marie from his family estate in Dorset, England, Adrian has created half of the world's ingenious mirror mazes, the world's first cornfield maize maze and water mazes, plus he holds several world records in the *Guinness Book of World Records*.

Adrian is also an original puzzle inventor of over 400 puzzles published internationally in leading magazines and on television, and has written over a dozen excellent books on mazes.

Adrian Fisher

In a mirror maze created by Adrian Fisher

DUDENEY'S MAZE

Henry Ernest Dudeney (1857–1930) was an English author and mathematician who specialized in logic puzzles and mathematical games. He is known as one of England's foremost creators of puzzles.

His initial efforts to engage the general public led him to submit puzzles to newspapers and magazines, frequently using the pseudonym "Sphinx." He produced much of his earlier works with American puzzlist Sam Loyd, such as a series of articles in the English penny weekly *Tit-Bits,* published in 1890. Their collaboration was broken off abruptly when Dudeney accused Loyd of stealing his puzzles and publishing them under his own name.

His maze is essentially not much different from the ancient maze designs of about 6000 years ago found all around the world, but it is a challenging puzzle. Can you solve it? Dudeney found 600 ways to solve his maze without taking the same route twice…

DICE — 5000 BC

Dice have been used throughout Asia since before recorded history. The oldest known dice were excavated as part of a 5,000-year-old backgammon set at the Burnt City, an archeological site in south-eastern Iran. The Bible contains numerous references to the practice of "casting lots." This suggests that this form of gambling was probably quite common in the region at the time of King David. In one derative form of knucklebones, a skill game played by women and children, all four sides of the bone were allocated different values, similar to modern dice. A popular form of entertainment among the Greek upper classes in particular was to use two or three dice in a gambling game; this was a familiar diversion during a symposium.

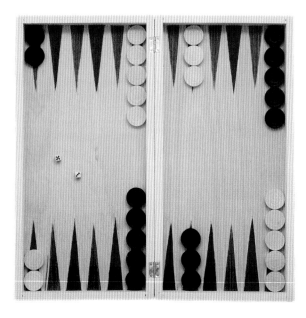

BACKGAMMON — 3000 BC

Backgammon is one of the oldest board games for two players. You move the playing pieces according to the roll of two dice, and win the game if you succeed in removing all your pieces from the board. There are numerous derivatives and variations of the game, the majority of which are closely related.

You need a portion of luck to win the game, but strategy is equally important. Each throw of the dice requires the player to select one of a number of possible options, at the same time anticipating the opponent's coming moves. Crucially, players may raise the stakes during the game, a move which may be rejected by the opponent conceding the game.

LIBRO DE LOS JUEGOS

Alfonso X (1221-1284) succeeded his father as King of Castile, Galicia and Leon in 1257. During his reign he commissioned numerous works in various fields, including the Libro de los Juegos, ("Book of Games"), which was completed in his scriptorium in Toledo in 1283. It is a wonderful example of Alfonso's literary legacy, consisting of 97 leaves of parchment, many with color illustrations, and containing 150 miniatures.

The text comprises an exposition of three games: a game of skill, or chess; a game of chance, or dice; and a third game, backgammon, combining elements of both skill and chance. The book contains the earliest known description of some of these games and hence is one of the most important documents for researching the history of board games. The library of the monastery of San Lorenzo del Escorial near Madrid in Spain contains the only known original of this work.

DICE GAMES AND PUZZLES AND SENET — 3000 BC

Dice games are games that use or incorporate dice as their sole or central component, usually as a randomizing device.

Archeological findings show evidence of pharaohs playing dice games with their ladies or even the spirits from the underworld, like the game Senet. Egyptian tombs, dating back to 2000 BC, have been found containing dice reputedly dating back to 6000 BC. Senet is the oldest board game, comparable to backgammon in game play. Senet was essentially racing game, with each of the two players moving a team of five pieces across a board of 30 squares arranged into three parallel rows of 10 squares. The game was won by the first player to remove all five pieces from the board. Players were allowed to pass and block the opponent's pieces, as well as capture them and force them to restart from the beginning or mid-point of the board. Moves around the board were determined by throwing either four casting sticks or two knucklebones. Senet is not just a game of chance; an element of skill and strategy is also required.

Queen Nefertari playing Senet.
(Tomb painting from tomb of Nefertari, 1279-1213 BC)

DICE PROBLEMS — A PAIR OF DICE

In many games dice are thrown in pairs with the aim of achieving a desired total.

"Chance favors the prepared mind," said Louis Pasteur in 1854. That's true. When asked to calculate the probability of getting a particular sum when two dice are thrown, many adults are at a complete loss. Even the great mathematician and philosopher Gotfried Leibniz thought that the probabilities of throwing 11 and 12 with two dice were the same, because he thought there was only one way to throw each (a five and six for 11, and a pair of sixes for 12).

1. What was wrong with this reasoning?
2. Throw a pair of dice. What is the probability that an even number will come up as the total? Is an even or an odd total equally likely to come up?

What we know initially is that the possible totals can be from two to 12. In the diagram on the left the number of ways each of these totals can come up is visualized. The distribution graph of these outcomes approximates the famous "Normal Distribution" or "Gauss Curve."

We know that when the odds of something occuring are 50-50, then half of the time, on average, the event will happen. But fewer might realize that the average usually nears 50 percent only after a very large number of events.

How large does the number of events have to be to rely on probabilistic forecasts? You can check this yourself with a little experimentation. The grey diagram shows the probabilistic outcome of throwing a pair of dice 106 times (3 x 36). The red diagram is the result of the author's experiment. Try it yourself. You may be surprised that even such a relatively small number of events can give a surprisingly good approximation of the theory.

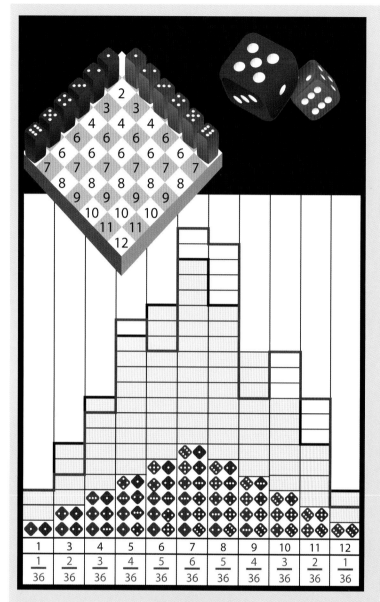

1	3	4	5	6	7	8	9	10	11	12
$\frac{1}{36}$	$\frac{2}{36}$	$\frac{3}{36}$	$\frac{4}{36}$	$\frac{5}{36}$	$\frac{6}{36}$	$\frac{5}{36}$	$\frac{4}{36}$	$\frac{3}{36}$	$\frac{2}{36}$	$\frac{1}{36}$

31 CHALLENGE ●●● ○ ○ ○
REQUIRES 🧠
COMPLETED ☐

THREE DICE

In how many ways can three dice be thrown? The number of pips on three dice can add up to the totals of three through 18. Can you work out the probability of getting a total of seven and 10 in a throw of three dice?

For centuries it was believed that there were only 56 possible ways of throwing three dice; people failed to recognize the difference between a set (combination) and a sequence (permutation). They counted sets only, when sequences should have been counted to get an accurate assessment of each roll's possibility. It was not until around 1250, when Richard de Fournival first described the true numbers of ways three dice can fall, that the correct possibilities were found.

32 CHALLENGE ●●● ○ ○ ○
REQUIRES 🧠 ✏️
COMPLETED ☐

THROWING A DICE

Your friend throws a dice, then you throw the same dice. What is the probability that you will throw a higher number than your friend?

33 CHALLENGE ● ● ● ● ○ ○
REQUIRES 🧠
COMPLETED ○

THROWING A SIX

What is the probability of throwing a six at least once if you throw one dice six times?

34 CHALLENGE ● ● ● ○ ○ ○
REQUIRES 🧠
COMPLETED ○

SIX TOSSES

If you toss a dice six times, what is the probability that each face will turn up exactly once in the six tosses?

35 CHALLENGE ● ● ● ● ○ ○
REQUIRES 🧠
COMPLETED ○

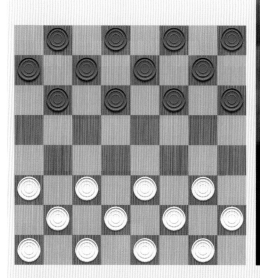

CHECKERS — DRAUGHTS — 3000 BC

The board game called checkers in North America and draughts in Europe is one of the oldest known games. The earliest form of the game was unearthed in the ancient city of Ur in Mesopotamia, modern Iraq, carbon dated at 3000 BC. In Ancient Egypt, from around 1400 BC, there was a game played in a similar manner; it was called Alquerque and used a 5x5 board.

In 1100 AD an innovative Frenchman adapted the game to be played on a chessboard increasing the amount of pieces to 12 for each player and calling the game "Le Jeu Plaisant De Dames," because it was considered a women's social game.

Today, computer programs playing checkers are defeating even the best players, and checkers is as popular as ever. It can be a good training exercise in logic and thinking.

Checkers is a game for two players, each player having 12 playing pieces that are placed on the black squares of a chessboard. In international draughts, players have 20 pieces and play on a 10x10 game board. Black moves first, having a slight advantage — very slight at beginner's level.

The object is to eliminate all opposing pieces or to create a situation in which it is impossible to make legal moves.

From the starting position, checkers may only move diagonally forward. The moves fall within two categories: capturing and non-capturing. Non-capturing moves are simply a diagonal move forward from one square to an unoccupied adjacent square. Capturing moves involve a player "jumping" opposing checkers. This is also done on the diagonal and can only happen when the square behind (on the same diagonal) is empty.

A piece may make a number of jumps in a capturing move. If a player makes one jump and is then in a position to make another, he is free to do so. This means that a player may make several jumps in succession, capturing several pieces during a continuous jump. Note that when a player is in a position to make a capturing move, he must do so.

When a checker reaches the opponent's edge of the board (called the "king's row"), it is crowned and becomes a powerful King. A King gains an added ability to move backward, and also to jump in either direction or even in both directions in one turn (during multiple jumps).

Something about strategy. First, always keep in mind the possibility of using the forced capture rule to maneuver your opponent into a position where he gives up

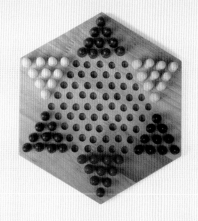

Chinese checkers dating back to 1892 is a race board game and not a variation of the classic checkers. The game was invented in Germany, as a variation of the older American game Halma.

two pieces for only one of your own. A one-piece advantage can often be significant in the end game. During the game, you should always attempt to block your opponent's access to your own King. The acquisition of a King by either player makes any uncrowned checker in the open highly vulnerable to attack.

Usually, the victory will be due to complete elimination, but sometimes a game can end in a situation in which it is impossible for your opponent to make any further moves.

CHECKERS GAME WITH CIRCUS CLOWN

Checkers has inspired many artists. This delightful painting by Norman Rockwell shows a clown involved in a game of checkers with the ringmaster and other circus performers.

CHECKERS SOLVED

Chinook, a computer program developed by Jonathan Schaeffer and colleagues in 2007, proved that checkers is a no-win game if you play it perfectly. As in Tic Tac Toe, when both players make no wrong moves the games end in a draw. Chinook played a series of draws against the world champion Marion Tinsley.

Grandmaster Marion Tinsley: since 2007 no longer the only checkers champion

FASCINATING HISTORY OF THE GEAR

CHINESE SOUTH POINTING CHARIOT

Gears are one of the oldest basic mechanisms of mankind. Their origin can be traced back to the Chinese South-Pointing Chariot of the 27th century BC — a vehicle built on two wheels that bore a movable indicator that always pointed south, irrespective of the direction in which chariot moved. The chariot, said to have been designed by mechanical engineer Ma Jun, had mechanically geared rotating wheels that kept the indicator toward the south.

The earliest description of gears was written in the fourth century BC by Aristotle. He wrote that the "direction of rotation is reversed when one gear wheel drives another gear wheel." In the third century BC, various Greek inventors used gears in water wheels and clocks. The 19th century saw the first use of form cutters and rotating cutters and in 1835 English inventor Whitworth patented the first gear hobbing process.

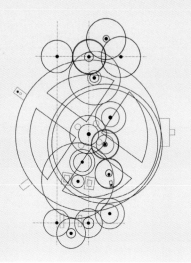

THE ANTIKYTHERA MECHANISM

The Antikythera mechanism is an ancient analog computer designed to calculate the movement and positions of astronomical bodies, resembling a classical clockwork. It was recovered in 1900–1901 from the Antikythera shipwreck, but its significance and complexity were not understood until a century later. New research suggests that the ancient mechanism was built in the second century BC.

It is an unbelievable achievement of the ingenuity of our forefathers. Such complex and intricate mechanisms were not to be seen for another millennium, with the construction of mechanical astronomical clocks in Western Europe.

Professor Michael Edmunds of Cardiff University, who led a 2006 study of the mechanism, said: "This device is just extraordinary, the only thing of its kind. The design is beautiful, the astronomy is exactly right. The way the mechanics are designed just makes your jaw drop. Whoever has done this has done it extremely carefully … in terms of historic and scarcity value, I have to regard this mechanism as being more valuable than the Mona Lisa," (November 30, 2006).

The original Antikythera mechanism can be seen at the National Archaeological Museum of Athens, together with a reconstruction made and donated to the museum by Professor Derek de Solla Price.

THE GENERAL PLAN OF THE ANTIKYTHERA GEARING

The discovery of the Antikythera gearing by Derek de Solla Price (American Philosophical Society) in 1974 initiated a complete reassessment of modern views on the achievements of ancient technology. The system contained 32 gears, combined in a mechanism that accurately reproduced the motion of the sun and the moon against the background of fixed stars, giving their relative position and the phases of the moon. When it was first published there were serious suggestions that it had been dropped into the wreck at a much later date or even that it was the work of aliens.

It was not possible to create a comprehensive plan of the gearing until after 1971, when what remained of the mechanism was examined with gamma rays on the initiative of De Solla, thus enabling the calcareous block covering the gears to be penetrated.

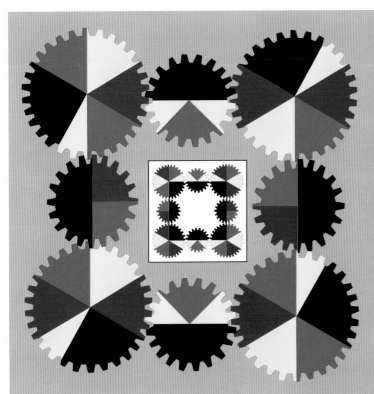

GEAR SQUARE

How many times would you have to turn the upper small gear to form the black square in the circle of 8 meshed gears as seen in the small center diagram? The small gears have 20 teeth, the large 30 teeth.

36
CHALLENGE ● ● ● ○ ○ ○
REQUIRES 🧠 ✏️
COMPLETED ○

GEARS UP OR DOWN?

If you turn the bottom red gear counter-clockwise, what will happen to the four numbered weights? Which will go up and which will go down?

37
CHALLENGE ● ● ● ○ ○ ○
REQUIRES 🧠 ✏️
COMPLETED ○

SPINNING TOPS — 3500 BC

The discovery of rotary motion introduced the invention of the top all over the world, independently from one another. Clay tops were found in the ancient city of Ur (Iraq) dating from 3500 BC.

A top (also known as a spinning top) is a toy designed to be spun on an axis, balancing on a point. This motion is produced in the simplest form of top by twirling the stem using the fingers. More sophisticated tops are spun by holding the axis and sharply pulling a string or twisting a stick.

The operation of the spinning top is based on complex mechanical principles explained only thousands of years after its invention.

The gyroscopic effect keeps the top upright and spinning. The top will initially wobble until it is forced upright by the shape of the tip and its interaction with the surface. After spinning upright for awhile, the angular momentum, and therefore the gyroscopic effect will gradually lessen, leading to ever increasing precession, finally causing the top to topple in a frequently violent last thrash.

Over the centuries countless designs and variations of the top appeared. Among these, the Tippe Top is an interesting kind of top. When it is spun at a high angular velocity, its handle slowly tilts downwards more and more until suddenly and astonishingly it lifts the body of the top off the ground with the stem pointing downward. As the top's spinning rate slows, it loses stability and eventually topples over.

It is easy at first sight to mistake the inversion of the top for an energy gain by the object. This is due to the object's center of mass being raised by the top's inversion, resulting in an increase in the potential energy. The inversion and the consequent increase in potential energy are actually the result of a torque caused by surface friction, with the result that the top's kinetic energy is also reduced, so the total energy does not actually increase.

YO-YO

The earliest surviving yo-yo dates back to 500 BC and was made using terra cotta skin disks. A Greek vase of this era has a painting of a boy playing with a yo-yo. Contemporary records contain descriptions of toys made from metal, wood and terra cotta (fired clay). The terra cotta yo-yos were offered to the gods in a coming-of-age ceremony, while the metal and wood yo-yos were intended for play.

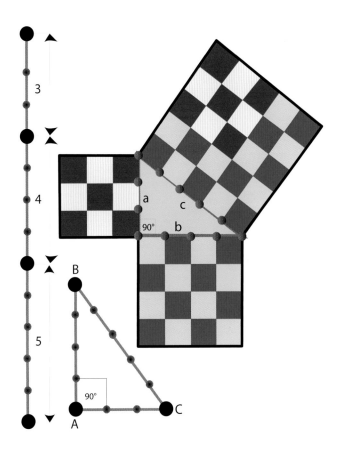

EGYPTIAN TRIANGLE — 2000 BC

By 2000 BC the ancient Egyptians had a primitive numeral system and some geometric ideas about triangles, pyramids and the like. There are unverified historical records about their ingenious method of creating right angles. Egyptian surveyors used a loop of rope 12 units in length, divided into 12 equal parts by knots. They used such a rope to form a triangle whose sides are in the ratio 3 : 4 : 5 and with an area of six units containing a right angle. This triangle is called the Egyptian Triangle, and is used to demonstrate the Pythagorean theorem in its simplest form. They fixed the rope along a line between A and B and pulled the remaining loop taut at point C.

The result was a right angle. A visual proof of the Pythagorean theorem for the Egyptian Triangle is shown on the following page. You can use a similar rope to create other shapes as well.

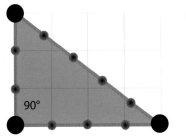

The Egyptian rope stretched into the Egyptian triangle of a six-unit area

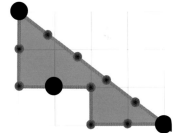

The Egyptian rope formed into a polygon of a four-unit area

38 CHALLENGE ● ● ● ○ ○
REQUIRES 🧠 ✏️
COMPLETED ○

1. Can you form polygons with an area of four units using such a rope, stretching it to form polygons with straight lines? One solution is shown above. Can you find others?
2. What is the largest area that can be encompassed by the Egyptian rope held straight between points?

PYTHAGOREAN TRIPLES — 2000 BC

The ancient Babylonians knew of Pythagorean triples thousands of years ago. George Plimpton found a clay tablet (the famous Plimpton 322) featuring triplets, including the Egyptian Triangle.

A Pythagorean triple consists of three positive integers a, b and c, such that $a^2 + b^2 = c^2$.

Such triples are commonly written as (a, b, c). The smallest Pythagorean triple is the set of numbers (3, 4, 5), which is the Egyptian triangle (see previous page).

The three sides of the Egyptian triangle are all integers (3,4,5). The Pythagoreans believed that the three sides of every right-angled triangle are also integers. They were awfully wrong. Can you find the smallest right-angled triangle of sides that are not all integers?

There is a simple formula (Euclid's formula) that gives Pythagorean triples.

Given a pair of positive integers m and n, with m greater than n, the formula states that the integers a, b and c:
$a = m^2 - n^2$; $b = 2mn$; $c = m^2 + n^2$, form a Pythagorean triple.

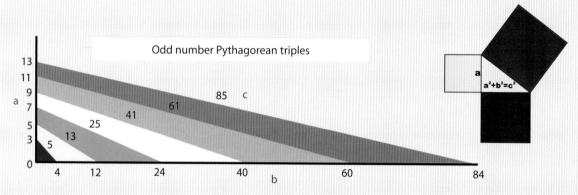

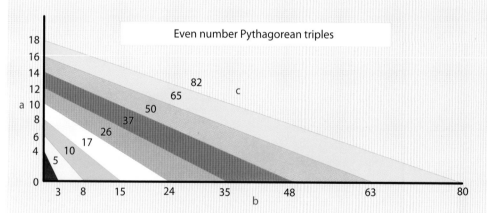

Above are the first six Pythagorean triples, when the "a" side is chosen as a consecutive sequence of odd numbers.

There is also an infinite number of Pythagorean triples in which the "a" side is even. Every even number can be the "a" side of a Pythagorean triple. The first eight even triples are shown (below side 100).

The famous Plimpton 322, ancient Babylonian cuneiform tablet, dated about 1800 years BC, listing the first 15 Pythagorean triples.

FERMAT'S LAST THEOREM

Fermat's famous last theorem involves triples. Fermat (1601–1665) stated that: $a^n + b^n = c^n$ has no solution in non-zero integers a, b and c, except for n=2, which is our Pythagorean theorem. In 1637 Fermat wrote on the margin of a book, the famous sentence that baffled mathematicians for nearly 400 years: "I have discovered a truly marvellous proof of this proposition which this margin is too small to contain."

It was the most famous unsolved problem in the history of mathematics until 1994, when Andrew Wiles in his Eureka moment solved the theorem.

SACRED GEOMETRY — 1800 BC

Sacred geometry is an ancient science based on the theory that all things are created and unified by energy patterns and explaining how the energy of Creation is organized. All natural patterns, movement and growth, irrespective of scale, necessarily conform to one or more geometric shape(s).

The concept of Sacred Geometry is beautifully visualized by the Sri Yantra. A yantra is the yogic equivalent of the Buddhist mandala. It is a geometric figure used to balance the mind and focus it on spiritual concepts. Some traditions claim it has magical benefits.

The Sri Yantra or the Yantra of Creation dates back to around 1800 BC. It is formed by nine interlocking triangles surrounding a central point, known as the "bindu." These nine triangles vary in size and intersect with one another. The bindu at the center is the power point, which represents the invisible focal point from which the complete figure, and indeed the cosmos itself, radiate. The nine interconnected triangles form a network of 43 smaller triangles that represents the cosmos or a womb symbolizing creation.

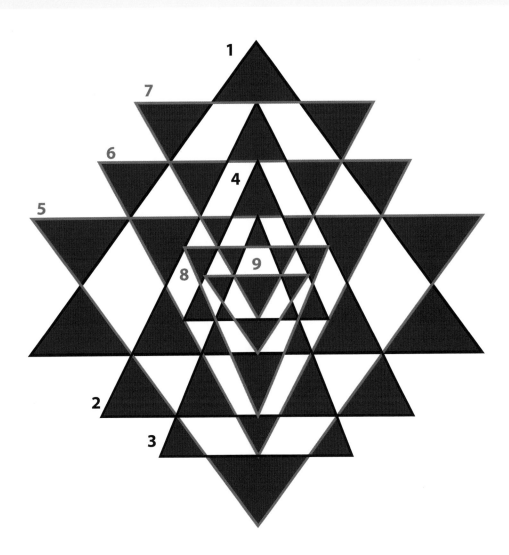

SRI YANTRA

The image of the Sri Yantra above consists of 43 red triangles. How many of the nine surrounding blue and green triangles must be removed to eliminate all red triangles?

39   CHALLENGE ● ● ● ● ○
REQUIRES 🧠✎
COMPLETED ○

HOUSE-CAT-MICE-WHEAT

Seven houses each have seven cats. Each cat kills seven mice. Each of the mice would have eaten seven ears of wheat. Each ear of wheat would have produced seven unit measures of flour. How many unit measures of flour were saved by the cats?

40 CHALLENGE ● ●
REQUIRES 🧠 ✏️
COMPLETED

AHMES' PUZZLE — 1650 BC

The Rhind papyrus, sometimes called the Ahmes papyrus is a scroll with a length of six meters, kept today at the British Museum. It is one of the oldest mathematical documents ever found. It is our main source of knowledge of Egyptian mathematics.

It contains a collection of 84 mathematical problems and their solutions, devoted to arithmetic, calculation of areas and the resolution of "linear equations."

It is also a convincing document indicating how much of early Egyptian mathematics were based on puzzle problems.

It's problem 79, the "House-Cats-Mice-Wheat" puzzle is a classic that became famous. It indicates some notion about the application of geometrical progressions, and is also probably one of the earliest puzzles connected with combinatorics (the first one may have been the Chinese I Ching.)

ST. IVES RIDDLE

Ahmes' puzzle has inspired many variations, including the St. Ives riddle. Leonardo of Pisa (also known as Fibonacci) published the riddle in his *Liber Abaci* in 1202. It is unclear how he got access to the Rhind papyrus at that time. "As I was going to St. Ives, I met a man with seven wives. Every wife had seven

41 CHALLENGE ● ● ○ ○ ○
REQUIRES 🧠
COMPLETED

bags. Every bag had seven cats. Every cat had seven kits. Kits, cats, bags and wives. How many were going to St. Ives?"

COMBINATORICS

Combinatorics is a branch of mathematics studying the formation of all possible complex systems by combination and permutation of elements, or, to put it more simply, the mathematical discipline that attempts to answer "how many" questions without actually having to count. More specifically, combinatorics studies finite collections of objects that satisfy specified criteria. More specifically, combinatorics studies finite collections of objects that satisfy specified criteria. "Enumerative combinatorics" focuses on the counting of objects in those collections, "extremal combinatorics" on deciding whether certain optimal objects exist and "algebraic combinatorics" on the algebraic structures these objects contain.

TIC-TAC-TOE — 1400 BC

Tic-tac-toe originated in ancient Egypt. It was probably the earliest game in the paper-and-pencil game category for two players, X and O, taking turns marking the spaces in a 3-by-3 square grid. The player who succeeds in placing three of his or her marks in a horizontal, vertical or diagonal row wins the game.

Players soon discover that the best play by both players leads to a draw. Computer programs play the game perfectly against human opponents. They can enumerate 765 essentially different game positions, as well as 26,830 possible games, a large number for such a simple game.

Despite its apparent simplicity, tic-tac-toe requires detailed analysis to determine even a few elementary combinatorial facts, like the number of possible games and the number of possible positions.

An early variant of the game was played in the Roman Empire in which each player had only three markers, so they had to move them around empty space in the play. In this form the game was probably the earliest form of the sliding block game category.

The first reference to "noughts and crosses," the British name of the game, appeared in 1864, while in the U.S. the game was renamed as tic-tac-toe in 1952.

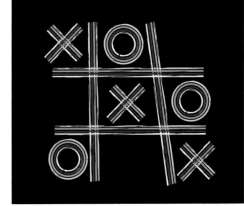

MILL — NINE MEN'S MORRIS — 1400 BC

42 CHALLENGE ●● ○○○○
REQUIRES 🪨✏️
COMPLETED ○

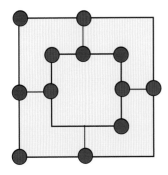

Six Men's Morris

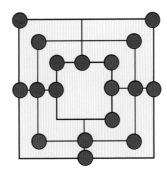

Nine Men's Morris

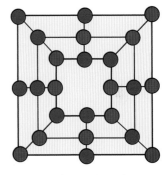

Twelve Men's Morris

Nine Men's Morris is a strategy board game for two players that emerged in Ancient Egypt around 1400 BC. The game is also known as Mills.

Each player has nine pieces, or "men," which move across a board with 24 spots. In order to win the game, you need to leave the opposing player with fewer than three pieces or unable to make a legal move. At the start of the game the board is empty. The players take turns placing their pieces on empty spots. When a player successfully forms a horizontal or vertical row of three pieces, he has a "mill," at which point he can remove one of his opponent's men from the board; once removed, these pieces may not return to the game. Players may not remove a piece from a mill until all other pieces have been removed from the board. Once all 18 pieces have been used, players take turns sliding one of their pieces to an adjacent empty spot. If this is not possible, the game is lost. In one common variation, once a player is reduced to three pieces, his pieces may "fly" to any empty spots, not only adjacent ones.

Many board games, such as Pente, Connect Four or Quarto, share the principle of trying to be the first to get a given number of pieces in a row. The object of the games is to leave the opponent with fewer than three pieces or unable to make a legal move. Strategically it is more important to place the pieces in specific locations than to create mills. In Nine Men's Morris such a situation is shown: even if it is red's turn, blue can easily win. How? In Twelve Men's Morris the board can be filled in the placement as shown. The game is a draw.

PYTHAGOREAN THEOREM PROOFS — 550 BC

The Pythagorean theorem is one of the most frequently used theorem in all of mathematics. The ingenious visual proofs of the Pythagorean Theorem give a maximum of intuitive insight (Martin Gardner called these "look-see-proofs"), are educationally meaningful and also beautiful.

A selection of such famous proofs is demonstrated below:

1. The first statement of the Pythagorean Theorem was discovered on a Babylonian tablet dated c. 1900 BC. Pythagoras is credited to be the first to provide a proof, which was probably a proof by dissection, similar to the beautiful proof in the Chou Pei Suan Ching, a Chinese manuscript ("Arithmetic Classic of the Gnomon and the Circular Paths of Heaven") whose origin is placed somewhere before 200 BC.

2. Leonardo's proof — Leonardo da Vinci, a Renaissance painter, engineer and inventor.

3. Baravalle's proof — Hermann Baravalle, a New York mathematician, published a five-step dynamic proof in 1945.

4. The simplest proof — by Stanley Jashemski, age 19, from Ohio, proof number 230, reinvented by Eli Maor as The Folding Bag.

5. Perigal's proof — The elegant proof of Henry Perigal, an amateur astronomer in 1830.

Can you understand and explain these proofs, just by looking?

43 CHALLENGE ● ● ○ ○ ○
REQUIRES 🧠
COMPLETED ○

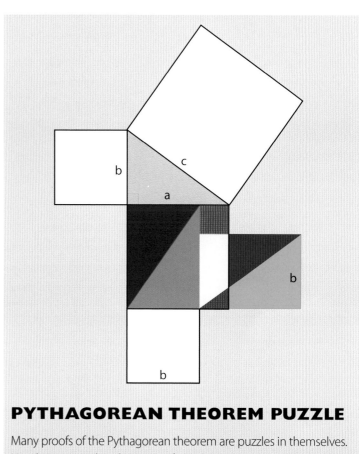

PYTHAGOREAN THEOREM PUZZLE

Many proofs of the Pythagorean theorem are puzzles in themselves. Use the seven colored pieces to form the square on the hypotenuse.

44 CHALLENGE ● ● ● ● ○
REQUIRES 🧠 ✏️
COMPLETED ○

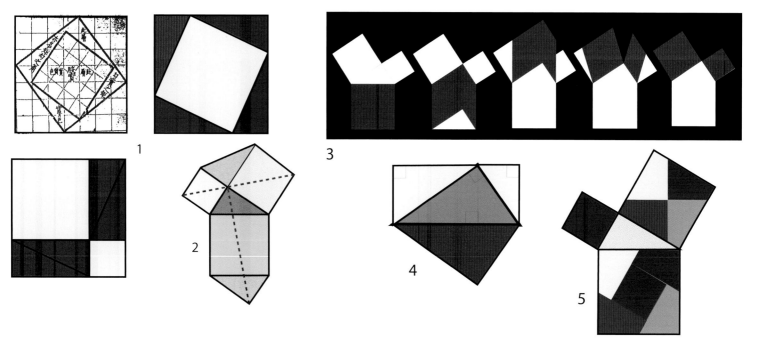

1

2

3

4

5

DYNAMIC DEMONSTRATION MODELS

Citing from Eli Maor's *Pythagorean Theorem — A 4,000 Year History,* the ultimate book on the Pythagorean theorem:

What is it that gives the Pythagorean theorem its universal appeal? Part of it, no doubt, has to do with the great number of proofs that have been proposed over the centuries. Elisha Scott Loomis (1852–1940), an eccentric mathematics teacher from Ohio, spent a lifetime collecting all known proofs — 371 of them — and writing them up in his "The Pythagorean Proposition" (1927).

Loomis claimed that in the Middle Ages, it was required that a student taking his Master's degree in mathematics offered a new and original proof of the Pythagoren theorem. Some of these proofs are based on the similarity of triangles, others on dissection, others still on algebraic formulas, and a few make use of vectors. There are even "proofs" ("demonstrations" would be a better word) based on physical devices: in a science museum in Tel Aviv, Israel, I saw a demonstration in which colored liquid flowed freely between the squares built on the hypotenuse and on the two sides of a rotating, plexiglass right triangle, showing that the volume of liquid in the first square equals the combined volume of the other two. (The Dynamic Pythagorean Demonstration models were invented by me, when I was director of the science museum, in 1960. They were shown at many exhibitions as teaching aids and kinetic art objects, and can still be found worldwide in many science museums and science centers.)

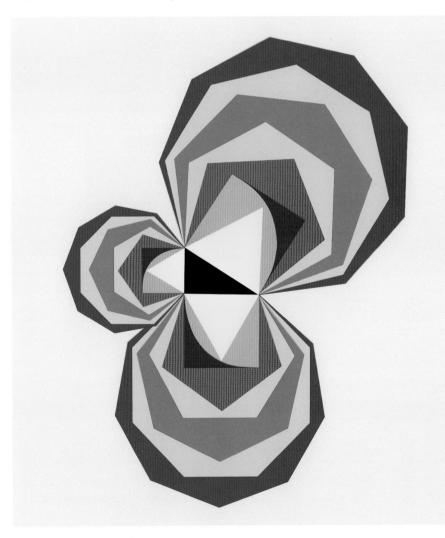

GENERALIZED FORMS

We know that the areas of the two squares on the sides of a right-angled triangle are equal to the area of the square on the hypotenuse.

But it is a lesser known fact that the validity of the Pythagorean relationship also holds for an endless number of other figures (as long as they are geometrically similar).

PYTHAGOREAN CURIOSITY

Prof. Elisha Scott Loomis (1852–1940), an American mathematician, wrote several books on geometry and the teaching of mathematics in high schools. He is probably best known for his work, *The Pythagorean Proposition,* a collection of more than 250 proofs of the famous theorem known in his time. The manuscript was prepared in 1907 and published in 1927. A second edition appeared in 1940, and this was reprinted by the National Council of Teachers of Mathematics in 1968 as part of its *Classics in Mathematics Education* series.

In his book, Loomis included a beautiful and ingenious composition based on the Pythagorean Theorem, which he called the Pythagorean Curiosity and which included a great number of fascinating mathematical relationships of lengths and areas, such as: the yellow triangles and the Pythagorean triangle are equal in area; the violet trapezoids are equal in area; the two red squares equal five blue squares in area, and more. Loomis traces the curiosity to John Waterhouse, a New York engineer.

RIDDLE OF THE SPHINX — 500 BC

The sphinx is one of the most famous mythological creatures of all times. The oldest known sphinx, thought to date back to 9500 BC, was found at Nevali Cori near Gobekli Tepe, Turkey. Throughout the ages, it has taken on different forms. The Egyptian sphinx for example usually combines a man's head with a lion's body, whereas the Greek sphinx has the haunches of a lion, the wings of a great bird, and the face of a woman.

Legend says that the Greek Sphinx guarded the entrance to the city of Thebes, requiring travelers to solve a riddle before they were allowed to enter. It strangled and devoured anyone unable to answer. The famous Greek hero Oedipus was the first to succeed.

Can you solve the riddle? It goes as follows: "Which creature walks on four legs in the morning, two legs in the afternoon, and three legs in the evening?"

The great sphinx at Giza (ca. 2500 BC). The Sphinx at Giza with a man's face and a lion's body, thought to have been modeled after Pharaoh Chephren, is by all standards a monumental work, carved out of rock.

FIGURATE NUMBERS — 500 BC

The mathematical study of figurate numbers originated with Pythagoras, possibly based on Babylonian or Egyptian precursors. It seems to be certain that the fourth triangular number of 10 objects, called tetraktys in Greek, was a central part of the Pythagorean religion.

The modern study of figurate numbers is derived from the Fermat polygonal number theorem. It became subject of particular significance to Euler, who made a made a number of discoveries related to figurate numbers, including an explicit formula for all triangular numbers that are perfect squares. (More on perfect squares: see Chapter 7.)

Figurate numbers have played a significant role in modern recreational mathematics.

The ancient Greeks liked to represent numbers as patterns composed of dots forming triangles, squares and other polygons from which they discovered meaningful relationships. If whole numbers are represented by dots arranged in certain geometric shapes, they can form groups or "series" called "polygonal" or "figurate" numbers. The geometric visual representation of figurate numbers in many cases is so simple and beautiful that the truth or a proof of a theorem is almost seen and understood without words ("look-and-see proofs").

PYTHAGORAS BY RAPHAEL

Pythagoras (570–495 BC) demonstrating his theory of numbers to his disciples and the triangular numbers which he considered the "best" numbers, and specifically the "best of all these," the "Divine Tetraktys" shown on the tablet at the bottom.

Detail of "The School of Athens" by Raphael, with Pythagoras (left) writing in his book.

DIVINE TETRAKTYS

The fourth triangular number is composed of 10 points, the sum of the first four consecutive integers, arranged in the shape of a pyramid, called the Tetraktys. It was devised by Pythagoras as a symbol of the creation of the known Universe, and used as a sacred basis of the Pythagoreans' oath.

The dots represent numbers from 1 to 10, the rows represent dimensions and the organization of space.

First row — a point — zero dimensions.

Second row — a line of two points — one dimension.

Third row — a plane defined by a triangle of three points — two dimensions.

Fourth row — a tetrahedron defined by four points — three dimensions.

The last row also symbolizes the four elements — Earth, Air, Fire, and Water. Tetraktys is a beautiful symbol representing the evolution of simplicity into complexity, and the abstract into concrete.

Can you make an intelligent guess of how many different ways there are to place the 10 numbers in the Tetraktys, not counting reflections and rotations?

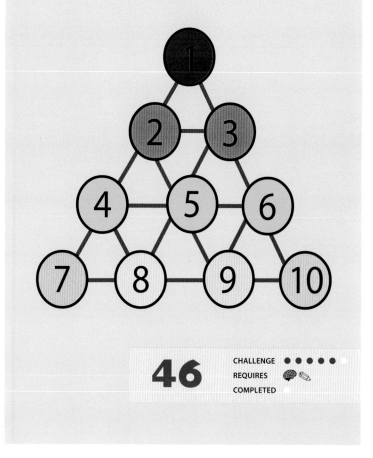

46

CHALLENGE ● ● ● ● ● ○

REQUIRES 🧠✏️

COMPLETED

FIGURATE TRIANGULAR NUMBERS — 500 BC

Triangular numbers can be found by stacking a group of objects in a triangular fashion — two objects are placed after one, three objects after two, and so on.

The fourth triangular number, for example 10, equals $1 + 2 + 3 + 4 = 10$ as shown.

What is so special about triangular numbers is that they are representing the sum of any number of consecutive integers.

You won't have any problem counting the tenth triangular number visualized below. But how long will it take you to find the hundredth triangular number?

Young Gauss found it in seconds!

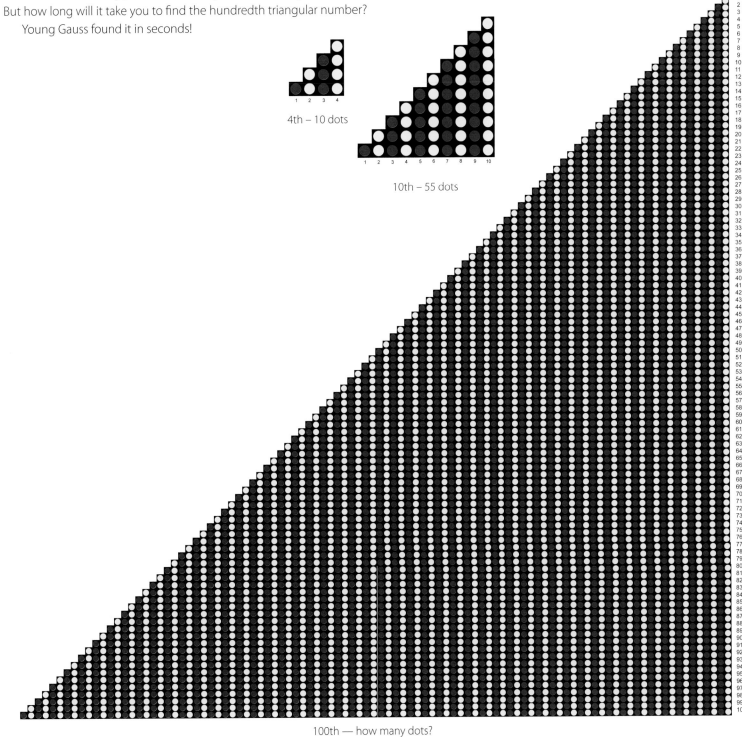

4th – 10 dots

10th – 55 dots

100th — how many dots?

THE GOLDEN RATIO — 500 BC

Where would you place a point on a line to divide the line in the most pleasing and meaningful way? Among the multitude of points on the line, there is one very special point that divides the line into a mathematical ratio called the Golden Ratio or Golden Section.

Throughout history, some of the greatest mathematical minds have been intrigued by the Golden Ration, including Pythagoras and Euclid. It has also fascinated many artists, as beauty can be achieved by the proportion, or comparative relationship, of one area to another. It is not a coincidence that Leonardo da Vinci called it the Divine Proportion. Two quantities are deemed to be in the Golden Ratio when the ratio between the sum of those quantities and the larger one is the same as the ratio between the larger one and the smaller.

$$\frac{X+1}{X} = \frac{X}{1} = \Phi$$

Ratio of X to 1 is the same as ratio X+1 to X

Mathematically, we can write this as $\quad x^2 - x - 1 = 0$

Finally we get the simple quadratic equation. The two solutions of the equation for the Golden Ratio are:

$$\Phi = \frac{1+\sqrt{5}}{2} \approx 1.618 \quad \text{and} \quad \Phi = \frac{1-\sqrt{5}}{2} \approx -0.618$$

The positive solution = 1.61803398 … gives the value of the Golden Ratio, an irrational number.

GOLDEN RATIO PUZZLE

Three sticks of the same length are broken into two parts along three selected points as shown. Which of the three sticks is divided into the Golden Ratio?

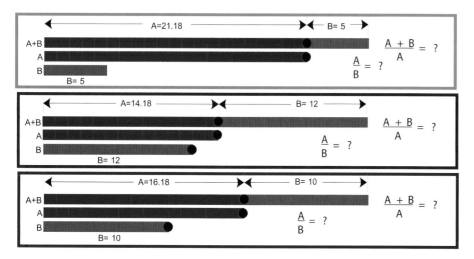

48

CHALLENGE

REQUIRES

COMPLETED

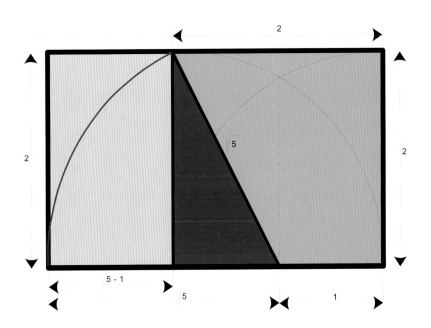

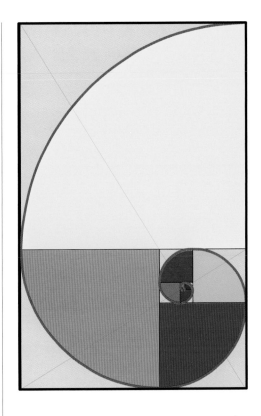

GOLDEN RECTANGLES AND TRIANGLES — 500 BC

What are the proportions of the most pleasing rectangle? Over the centuries, most people, including artists and scientists, agreed that it is a rectangle with sides in the proportion of the Golden Ratio — the Golden Rectangle. For its aesthetic appeal and elegance, it has played an important role in architecture, art and even music. The mathematical beauty of the Golden Rectangle will fully reveal itself if we build it by the method of the ancient Greeks, using a compass and a straightedge.

1. We start with a perfect square and extend the line of its base (more on perfect squares, see Chapter 7).
2. From the midpoint of the base we draw an arc from the upper left corner of the square to the base.
3. We draw a perpendicular from this point and extend the top of the square to complete the Golden Rectangle.

We can use the Pythagorean theorem to check whether the proportions of the rectangle are those of the Golden Ratio. The right triangle used for this is sometimes called the Golden Triangle. It is a triangle in which the height is twice the base. It also has an interesting property. While any triangle can be created from four smaller copies of themselves, only the Golden Triangle has the property that it can be created from five smaller copies of itself.

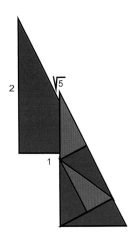

THE WHIRLING SPIRAL

If the lengths of the sides of a rectangle are in the Golden Ratio, then the rectangle is a "Golden Rectangle."

It is the only rectangle from which a square can be cut so that the remaining rectangle will always be a smaller replica of the original rectangle.

Cross section of a nautilus shell with the superimposed whirling logarithmic spiral from the Golden Rectangle

HEIGHT OF A PYRAMID — THALES OF MILETUS

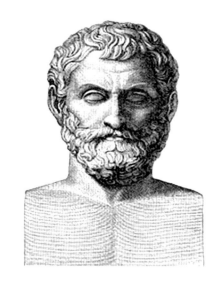

Thales of Miletus (624–547 BC) was an engineer, mathematician, scientist and philosopher. What mathematical discoveries he is responsible for remains uncertain, since none of his written works has survived; we cannot even be sure that he wrote any at all. Thales was, however, a prominent figure of his time, called by some "the father of geometry."

Thales is credited with five theorems of elementary geometry: a circle is bisected by any diameter, the base angles of an isosceles triangle are equal, the angles between two intersecting straight lines are equal, two triangles are congruent if they have two angles and one side is equal, and an angle in a semicircle is a right angle.

Thales is said to have predicted an eclipse of the sun in 585 BC. While the cycle of about 19 years for eclipses of the moon had already been established at that time, it was more difficult to calculate the cycle for eclipses of the Sun as they could be seen from different locations on the Earth's surface. It may be that Thales' prediction of the 585 BC eclipse was an educated guess based on his belief that one was due at approximately that time.

Reports of how Thales measured the height of pyramids vary. He discovered how to do so by measuring the shadow of the pyramids at the time when the height of any body was equal to the length of its shadow.

Many scientists are baffled even today by some of the discoveries of Ancient Greece. We owe a lot of what we now take for granted to the discoveries and culture of the ancient people of Greece and Rome.

When Thales gave the answer it was pure magic. Where others saw only the pyramids and their shadows in the hot sunlight, Thales saw far more than that. Thales saw abstract right triangles as well and much more. He saw patterns! Ingenious!

SIMILAR TRIANGLES

According to Plutarch, the studies of Thales led to the understanding of similar triangles. There is a close correlation between the Intercept Theorem and Similarity. It is, however, equivalent to the Similar Triangles concept. By taking two similar triangles (with the same angles but of different sizes) and placing one inside the other produces a configuration in which the intercept theorem applies; conversely, the intercept theorem configuration always contains two similar triangles.

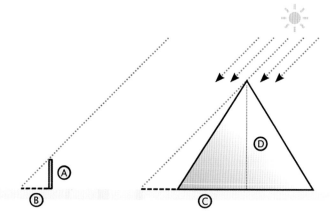

FAKE AND REAL PARADOXES

Definition of a paradox: any person, thing or situation exhibiting an apparently contradictory nature. Something can seem at first to be a paradox due to a lack of understanding, which can later be resolved with better understanding, but real paradoxes cannot be resolved easily or at all. In a more general way, the term is used for situations that are merely surprising and counterintuitive (like, for instance the birthday paradox in Chapter 7).

W. V. Quine (1962) described three general classes of paradox:

1. A veridical paradox — produces a result that appears absurd but can be demonstrated to be true.
2. A falsidical paradox — establishes a result that not only appears false but is false; there is a fallacy in the supposed demonstration.
3. A paradox that is in neither class may be an antinomy, which reaches a self-contradictory result by properly applying accepted ways of reasoning.

> **"Only two things are infinite, the universe and human stupidity, and I'm not sure about the former."**
>
> — *Albert Einstein*

Start
Achilles

Start
tortoise

Finish

ZENO'S PARADOX — 400 BC

The famous mathematician Zeno, born around 490 BC in Italy, created over 40 paradoxes to defend the teachings of the philosopher Parmenides, his teacher, who believed in monism: the concept that reality was unchanging, and that change (motion) was impossible. His puzzling paradoxes seemed impossible to resolve at the time.

The most famous of Zeno's paradoxes is the race between Achilles and the Tortoise. In the race Achilles gives the tortoise a head start and Zeno's argument is like this: when Achilles reaches the point A where the tortoise started, the tortoise has crawled to a point B. Now Achilles must run to point B to catch up with the tortoise. But in the meantime the tortoise has moved to point C, and so on. Zeno's conclusion was that it

would take Achilles an infinite amount of time to catch up with the tortoise. Achilles gets closer and closer, but he never catches up. His journey is divided into an infinite number of pieces.

Before a moving object can travel a certain distance, it must travel half of that distance. Before it can travel half of the distance it must travel 1/4 the distance, and so on, forever. The original distance cannot be traveled, and therefore motion is impossible.

We should not forget, of course, that the race was conceptualized in Zeno's mind only. It is absurd, but it is logically consistent. Try to disprove it logically. Many tried.

We obviously know that motion is possible, so what's wrong with Zeno's logic? Can you find the faulty logic in Zeno's argument?

49 CHALLENGE ● ● ● ○ ○
REQUIRES 🧠✏️
COMPLETED ○

The usefulness of Zeno's paradoxes is that they gave birth to the idea of convergent infinite series, crystallizing a number of mathematical concepts, the main one among these being the notion of limits. Interest in paradoxes was strongly revived during the Renaissance, when more than 500 collections of paradoxes are known to have been published.

Zeno's arguments may well be the first examples of a proof method called "Reduction ad absurdum," also known as proof by contradiction.

Paradoxically, many still believe that a satisfactory explanation to Zeno's paradoxes is yet to be found.

PLATONIC SOLIDS — THE REGULAR CONVEX POLYHEDRA — 400 BC

Regular solids, also called Platonic solids or polyhedral, are convex polyhedral with equivalent faces composed of convex regular polygons. The Platonic solids were the subject of a detailed study by the ancient Greeks. Pythagoras is credited with their discovery by a number of sources, although there is evidence to suggest that he might only have been familiar with the tetrahedron, cube and dodecahedron, and that Theaetetus (c. 417 BC–369 BC), a contemporary of Plato, was responsible for the discovery of the octahedron and icosahedron. We know that Theaetetus provided a mathematical description of all five and may have been responsible for the first known proof that there are no other convex regular polyhedral.

The Platonic solids are so named because of their close association with the philosophy of Plato, who wrote about them in the dialogue Timaeus (c. 360 BC), in which each of the four classical elements (earth, air, water and fire) was associated with a regular solid: earth with the cube, air with the octahedron, water with the icosahedron and fire with the tetrahedron. Plato describes the fifth Platonic solid, the dodecahedron, somewhat abstrusely as "…the god used for arranging the constellations on the whole heaven." Plato also refers to the studies of the Greek mathematician Theaetetus, who proved that there are exactly five regular convex polyhedra.

In his work *Elements,* Euclid included a comprehensive mathematical description of the Platonic solids and argued in Proposition 18 that there are no further convex regular polyhedral.

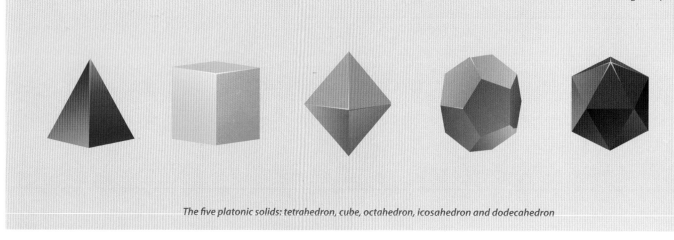

The five platonic solids: tetrahedron, cube, octahedron, icosahedron and dodecahedron

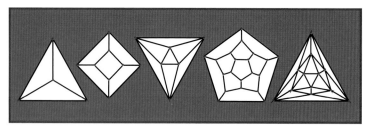

Solid	Vertices (V)	Edges (E)	Faces (F)	V - E + F
Tetrahedron				
Cube				
Octahedron				
Icosahedron				
Dodecahedron				

COLORING REGULAR POLYHEDRA

Above you can see Schlegel diagrams of the five regular polyhedra. What is the minimum number of colors needed to color the faces of Platonic Solids so that the touching faces are different colors?

50 CHALLENGE ● ● ● ○ ○
REQUIRES 🧠✏️
COMPLETED

REGULAR POLYHEDRA CHART

All classical polyhedra satisfy Euler's Theorem: F - E + V = 2
F= faces; E = edges; V = vertices
Can you fill in the chart of regular solids to check it?

51 CHALLENGE ● ● ● ● ○
REQUIRES 🧠✏️
COMPLETED

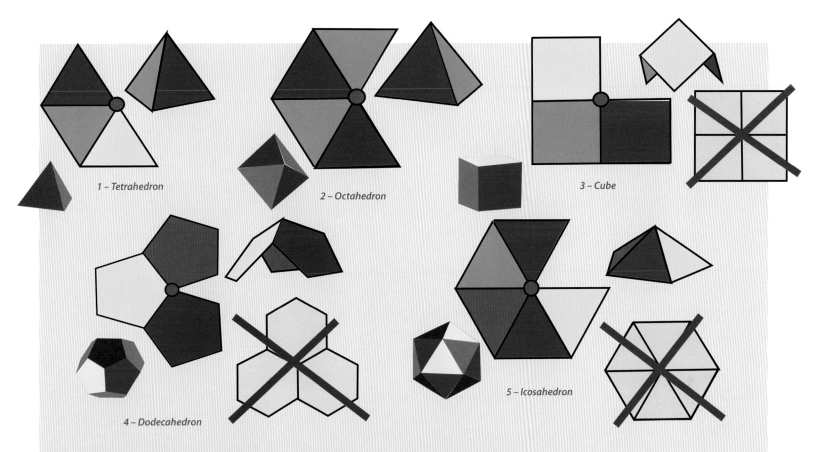

1 – Tetrahedron

2 – Octahedron

3 – Cube

4 – Dodecahedron

5 – Icosahedron

REGULAR POLYHEDRA — THERE ARE ONLY FIVE!

A minimum of three regular polygons are needed to form a solid angle in a polyhedron. Three, four and five equilateral triangles will do. Six equilateral triangles form a flat plane. Three squares form a solid angle. Four lie flat.

Three regular pentagons form a solid angle. There is no room for more.

Three regular hexagons meeting lie flat. This is the limit. No higher polygons can meet with three around a point. Thus, since only five solid angles can be formed by identical regular polygons, there are at most five possible convex regular polyhedra.

The Greeks recognized that there are only five platonic solids. The crucial factor was that the interior angles of the polygons meeting at a vertex of a polyhedron add up to less than 360 degrees.

The interior angle of an equilateral triangle is 60 degrees. Consequently only three, four or five triangles of a regular polyhedron can meet a vertex. If there were more than six triangles, their angles would add up to at least 360 degrees, which is impossible. Look at the following possibilities:

Triangles: three triangles meet at each vertex. This results in a Tetrahedron.

Four triangles meet at each vertex. This results in an Octahedron.

Five triangles meet at each vertex. This results in an Icosahedron.

Squares: as the interior angle of a square must be 90 degrees, no more than three squares can meet at a vertex. This is clearly possible and results in a hexahedron or cube.

Pentagons: the only possibility in the case of cubes is three pentagons meeting at a vertex. This results in a Dodecahedron.

Hexagons or regular polygons with more than six sides cannot form the faces of a regular polyhedron since their interior angles are at least 120 degrees. It is impossible for hexagons or regular polyhedrons with more than six sides to form the faces of a regular polyhedron as their interior angles are at least 120 degrees.

One of the most popular books ever written may be Euclid's *Elements of Geometry*, which has appealed to people of all major civilizations for more than two millennia. However, its popularity has waned in recent times, even among mathematicians, which is regrettable as with careful interpretation it may still be one of the best ways to understand the definition of a mathematical proof, and indeed of mathematics itself.

The last major assertion of *Elements,* that there are no more than five regular polyhedra, is most interesting in its subtlety.

"I say next that no other figure, besides the said five figures, can be constructed which is contained by equilateral and equiangular figures equal to one another."

EUCLID'S ELEMENTS — 300 BC

The book *Elements,* written by Euclid, is considered to be one of the most influential scientific or mathematical works ever written, principally because of its logical development of geometry and other branches of mathematics. It has had an impact on all branches of science, but its influence on mathematics and the exact sciences was particularly strong. The book was first set in type in Venice in 1482, and was one of the first mathematical works to be printed after the invention of the printing press.

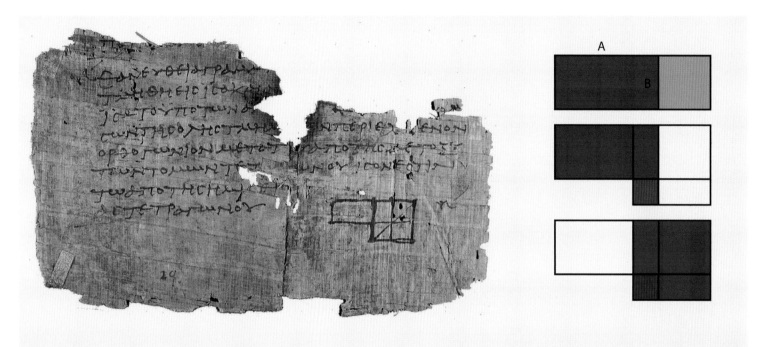

OXYRHUNCHUS PAPYRI

Above you see a fragment of Euclid's *Elements* found among the Oxyrhunchus Papyri. It is now located at the University of Pennsylvania. It is part of a collection of manuscripts discovered by archaeologists in the late 19th and early 20th centuries at an ancient rubbish tip near Oxyrhynchus in Egypt. These manuscripts date back to the 1st–6th centuries AD, and they contain thousands of documents letters and works of literature in both Latin and Greek.

The diagram is part of Proposition 5 of Book II of the *Elements*. In modern terms, it can be interpreted as a geometric formulation of an algebraic identity, specifically that of ab + (a − b)2/4 = (a + b)2/4 (although it must be said that there are doubts about the relationship between Euclid's propositions and algebra). The picture on the right might help to clarify the proposition.

If a straight line is cut into equal and unequal segments, the rectangle contained by the unequal segments of the whole together with the square on the straight line between the points of section is equal to the square of the half.

EUCLID (325–270 BC)

Euclid of Alexandria was a Greek mathematician who was often referred to as the "Father of Geometry." In his work *Elements,* Euclid deduced, from a small set of axioms, what we now know as Euclidean geometry. He also wrote works on perspective, conic sections, spherical geometry, number theory and rigor. Very few references to Euclid have been found, so we know very little of his life, and the few historical references that have been discovered were written by Proclus and Pappus of Alexandria centuries after his death.

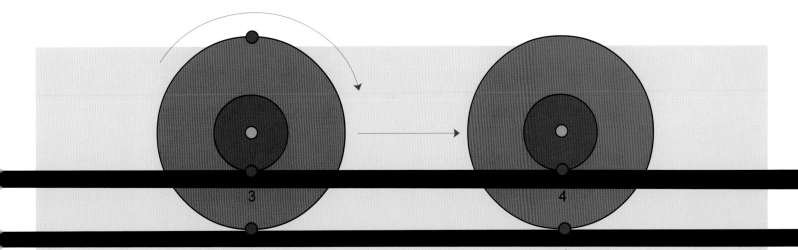

ARISTOTLE'S WHEEL PARADOX — 300 BC

Aristotle's wheel paradox is a paradox from the Greek work *Mechanica* traditionally attributed to Aristotle.

The generally accepted definition of the paradox is that there are two wheels (circles) with a different diameter, one within the other. The paths traced by the bottoms of the wheels are straight lines. At first sight these straight lines appear to equal the circumferences of the wheels. However, the two lines have the same length, so the wheels must have the same circumference, which contradicts the assumption that one wheel is smaller than the other.

The key to solving the paradox lies in the assumption that the smaller wheel actually traces out its circumference. The simple truth is that it is impossible for both wheels to move in this way. The smaller wheel did not roll from point 3 to point 4, it was pulled along the line. Physically, if two joined concentric wheels with different radii were rolled along parallel lines, at least one of them would slip, and if a system of cogs was used to prevent slippage, the wheels would jam. A modern equivalent of such an experiment is often performed inadvertently by car drivers who park too close to a curb. The car's outer tire rolls without slipping on the road surface while the inner hubcap both rolls and slips across the curb, producing a screeching noise.

In mathematical terms, the number of points in the inner circle is equal to the number of points in the outer circle, i.e. there is a bijection (a one-to-one correspondence) between them. This does not apply to physical wheels as these are made of discrete atoms, so the larger wheel has a greater number of atoms if the wheels have the same density, width and thickness (only their radii are different).

"The School of Athens" by Raphael, showing Plato (left) and Aristotle (right).

ARISTOTLE (384–322 BC)

Aristotle, student of Plato and tutor of Alexander the Great, was one of the most influential thinkers of his time. His works cover numerous subjects, including philosophy, ethics, mathematics, logic, physics, biology, poetry, theater, music, rhetoric and linguistics. Aristotle's writings, along with those of Plato and Socrates, were the first to create a comprehensive system of Western philosophy, encompassing morality, aesthetics, logic, science, politics and metaphysics. His theories had a profound impact on medieval thinking, and their influence could still be seen during the Renaissance.

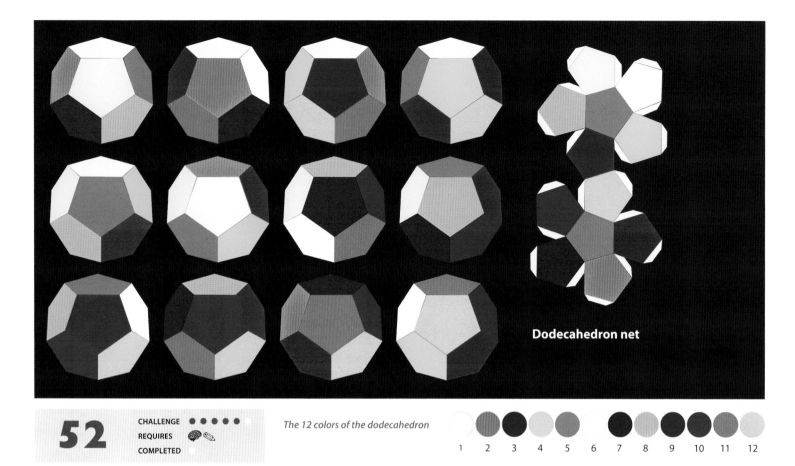

Dodecahedron net

The 12 colors of the dodecahedron

1 2 3 4 5 6 7 8 9 10 11 12

DODECAHEDRON ORIENTATION

Puzzle 1 — In how many different ways can you place a colored dodecahedron on a table to occupy the same space each time?
Puzzle 2 — A single dodecahedron is shown in various orientations.

Can you fill in the missing colors?
Puzzle 3 — Cutting a dodecahedron by a plane cut, what can be the resulting cross sections?

KURSCHAK'S THEOREM

Four equilateral triangles are drawn inwards from the outer square. The inner square is formed by connecting their points. When the midpoints of this square and the intersections are joined with the sides of the triangles, a regular dodecagon is formed. The inner square is called the Kurschak tile, and is used to prove the Kurschak theorem: showing that the area of a regular dodecagon inscribed in a circle of unit radius is three.
Just by looking at the Kurschak tile, can you prove the area of the dodecagon in terms of the area of the Kurschak tile?

J. Kurschak, a Hungarian (1864–1933), provided the elegant geometric way of finding the area of a regular dodecagon.

DODECAHEDRON UNIVERSE

According to an article of J-P Luminet in Nature, "the standard model of cosmology predicts that the universe is infinite and flat. However, cosmologists in France and the U.S. are now suggesting that space could be finite and shaped like a dodecahedron instead. They claim that a universe with the same shape as the twelve-sided polyhedron can explain measurements of the cosmic microwave background — the radiation left over from the big bang — that spaces with more mundane shapes cannot."

THE ANCIENT PROBLEM OF SQUARING THE CIRCLE

One of the three famous classical problems of Greek antiquity is to construct a square of exactly the same area as a given circle, with the restriction of using only a compass and a straightedge.

In 1882 it was proven by the Lindemann-Weierstrass theorem that π is a transcendental (rather than an algebraic irrational number; that is, it is not the root of any polynomial with rational coefficients), and consequently the problem of squaring the circle as proposed is impossible. It can have only approximate solutions, and this was already known by the Babylonians. The famous Egyptian Rhind papyrus of 1800 BC gives the area of circle as $(64/81)d^2$, where d is the diameter of the circle.

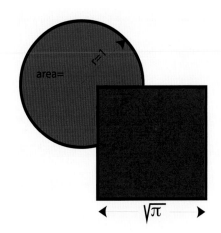

LUNES OF HIPPOCRATES

Can you work out the area of the red lune, in terms of the area of the blue isosceles triangle on the right?

Can you work out the area of the two red areas of the lunes, in terms of the total area of the right-angled blue isosceles triangle? Pythagorean theorem may help!

Can you work out the total area of the four red areas of the lunes, in terms of the area of the blue square?

Can you work out the total area of the two area red areas of the lunes, in terms of the area of the right-angled blue triangle?

54 CHALLENGE ● ● ● ● ● ○
REQUIRES 🪨 ✏️
COMPLETED ○

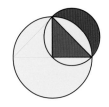   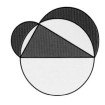

AN EARLY ATTEMPT TO SQUARE THE CIRCLE

Hippocrates of Chios (c. 470–c.410 BC), while unsuccessfully trying to solve the problem of squaring the circle, made the surprising discovery that he could solve the problem for figures bounded by circular arcs, a truly remarkable feat at the time. He was the first to be able to show that areas of lunes (areas bounded by circular arcs) can be equal to areas of rectilinear figures. Hippocrates' works are lost, but he must have tackled the problem in a similar way as follows. In the section devoted to the famous Pythagorean theorem, we have seen that in its extended generalized form, the Pythagorean relationship holds for any set of similar figures placed on the three sides of a right-angled triangle, as long as they are placed in corresponding orientation. This holds for circles as well. His discovery aroused enormous interest and false hopes for squaring of the circle solvers until this day.

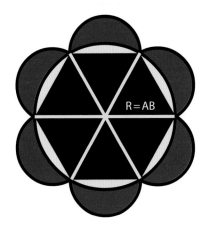

HEXAGON THEOREM OF HIPPOCRATES

55 CHALLENGE ● ● ● ● ○ ○
REQUIRES 🪨 ✏️
COMPLETED ○

Hippocrates claimed that he could square the circle. After his success at squaring the crescent he also did it with a hexagon. He started with a circle of diameter AB, then constructed a larger circle of twice the diameter AB. A regular hexagon is inscribed in the big circle with each of its sides equal to the circle's radius. Note that each side of the hexagon is also equal to the diameter AB of the initial smaller circle, and semicircles are drawn on its sides as shown. Can you work out what the total area of the six crescents (red areas) will be?

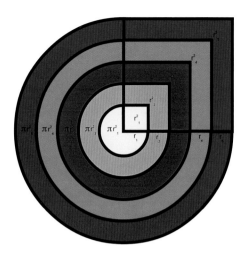

THE HAPPINESS OF SQUARING THE CIRCLE

One of the greatest challenges occupying ancient and modern mathematicians was the problem of "squaring the circle."

They could not solve the problem. Circular arcs and straight lines always had something left over when measured.

The first mathematician attempting to square the circle was Anaxagoras in the fifth century BC. While in prison, he wrote: "There is no place that can take away the happiness of a man, nor yet his virtue or wisdom ("trying to square the circle").

CIRCLE AND SQUARE AREA — 400 BC

The Demonstration Model physically visualizes the validity of the formula for the area of a circle and number π by the passage of colored liquid from one compartment to the other, when the model is turned upside down, invented by the author in 1960.

HIPPOCRATES (470 BC—410 BC)

Hippocrates, a teacher of geometry in Athens, was the first to succeed in the construction of a square with equal area to figures with circular sides.

Many mathematicians followed Hippocrates' example and tried to decipher the ancient problem of squaring the circle. When Lindemann proved in 1880 that π is a transcendental, he also proved that a circle can never be squared using a ruler and a compass.

The new challenge is now to find the best approximation to square the circle with ruler and compass. In order to solve this challenge, we need to go back to Archimedes. Archimedes took a square of side "r," the radius of a circle. The area of this square "r^2" multiplied by Pi gives the area of the circle; "π x r^2."

The validity of the above formula for a circle is visually demonstrated by the original fluidic Demonstration Model invented by the author. Hermetically sealed colored liquid fills exactly the area of the circle of radius r. When the model is turned over the liquid passes into the square's compartment, filling it: 3 1/7 x r^2.

Note: the thickness of the flat sealed compartments is uniform all over.

In 1914, Ramanujan created a ruler and compass construction that differs from π only in the ninth decimal place. For a circle with a diameter of 12,000 km, the error in the length of the side of the square is less than 25 mm.

SPIRALS

The investigation of spirals is known to date back to the ancient Greeks. The Spiral of Archimedes is the quintessential example. Descartes discovered the Logarithmic Spiral, also known as the Equiangular Spiral, in 1638 while studying dynamics. Its special feature is that the curve cuts all radii vectors at a constant angle. Any radius drawn from the center O to any point of tangency P on the spiral will form the same angle between the radius and the tangent line. This curve therefore features a property of self reproduction.

Jacob Bernoulli (1654–1705) found the Equiangular Spiral so intriguing that he wanted it to be carved on his gravestone, followed by *Eadem mutata resurgo* ("I shall arise the same, though changed").

The spiral's "wonderful" properties are further enriched by its association with Golden mean and Fibonacci numbers, which are also mysterious and intriguing.

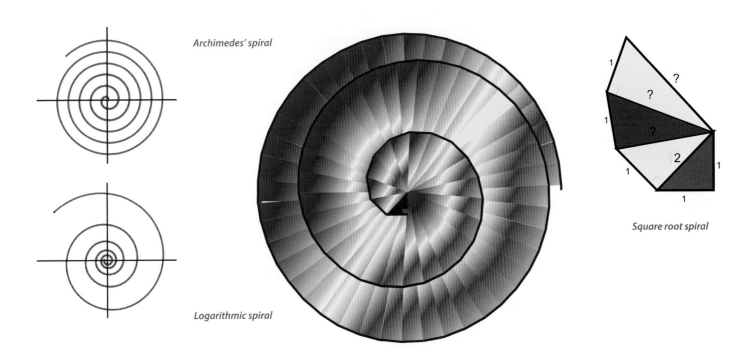

Archimedes' spiral

Logarithmic spiral

Square root spiral

SPIRAL OF THEODORUS — 400 BC

Originally created by Theodorus of Cyrene (465–398 BC), the Spiral of Theodorus, also called the square root spiral, Einstein spiral or Pythagorean spiral, is composed of contiguous right triangles,

The first step of the spiral comprises an isosceles right triangle, each leg having a length of 1. A second right triangle is then placed adjacent to the first, one leg being the hypotenuse of the first triangle (with length $\sqrt{2}$) and the other having a length of 1; the length of the second triangle's hypotenuse is $\sqrt{3}$. This process is then repeated; the i-th triangle in the sequence is a right triangle with side lengths $\sqrt{i}$ and 1, and with hypotenuse $\sqrt{(i + 1)}$.

Sadly, none of Theodorus' work has survived. Luckily Plato, one of the students of Theodorus, included his mentor in his dialogue Theaetetus, which speaks volumes about the significance of his achievements. It is assumed that Theodorus used the Spiral of Theodorus to prove that all of the square roots of non-square integers from three to 17 are irrational. Accoording to Plato, Theaetetus told Socrates the following: "It was about the nature of roots. Theodorus was describing them to us and showing that the third root and the fifth root, represented by the sides of squares, had no common measure. He took them up one by one until he reached the seventeenth, when he stopped."

Plato questioned Theodorus' reasons for stopping at $\sqrt{17}$. This is generally thought to be that the $\sqrt{17}$ hypotenuse belongs to the last triangle that does not overlap the figure.

E. Teuffel proved in 1958 that no two hypotenuses will ever coincide, irrespective of how far the spiral is continued. Additionally, if the sides of length 1 are extended into a line, they will never pass through any of the total figure's other vertices.

CONIC SECTIONS — 350 BC

Conic sections are curves formed by cutting a double cone with a plane.

They have been studied by mathematicians of ancient Greece for their aesthetic properties. Ellipses, hyperbolas and parabolas fascinated Euclid and other geometers of that era. They have been admired and considered as beautiful geometrical recreations at a time when they could not find any practical uses for them.

Mathematicians have a habit of studying utterly useless objects just for fun. But often those studies become enormously important to scientists in later centuries.

That is what happened with conic sections. The work of Johannes Kepler and Isaac Newton relied on the study of conic sections to decribe the paths traced by celestial bodies moving through space. Planets, comets and even galaxies move exclusively in ellipses, hyperbolas and parabolas.

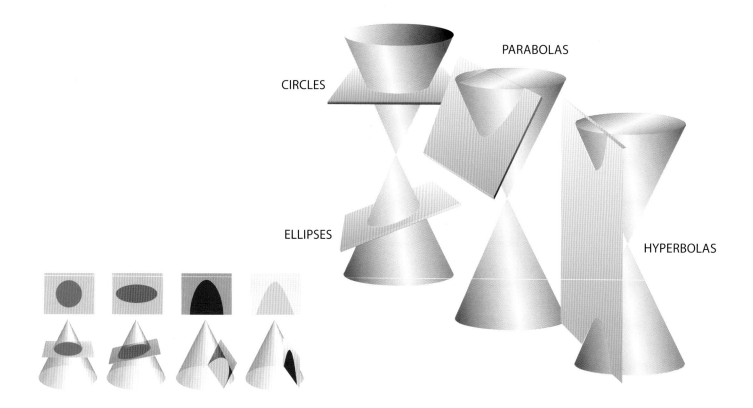

CIRCLES

PARABOLAS

ELLIPSES

HYPERBOLAS

MENAECHMUS (380–320 BC)

Menaechmus, a good friend of Plato, was an ancient Greek mathematician and geometer who is believed to have studied the Delian problem and discovered the conic sections as a result. The Delian problem, also known as the "Doubling the cube" problem, is one of the three most famous geometric problems unsolvable by compass and straightedge construction and intrigued many Egyptian, Indian and Greek scholars.

Around 200 BC, Apollonius of Perga began a systematic study of the properties of conic sections. His elaborate work — he wrote eight volumes on the subject of Conic Sections alone — greatly extended existing knowledge on the topic. The definition of a conic section in analytic geometry is "a plane algebraic curve of degree 2." It is a curve generated by the intersection of a cone (a right circular) with a plane. There are three types of conic sections: the hyperbola, the parabola and the ellipse. The circle is sometimes referred to as a fourth type of conic section, as it can be categorized as a special type of ellipse.

ABACUS — THE COUNTING FRAME — 300 BC

An ingenious ancient calculation tool, also suited for the decimal number system.

To be clear about the difference between using numbers as labels for counting individual objects and as labels to record the results of calculations was an important step toward mathematical literacy.

It is difficult to imagine counting without numbers, but there was a time when written numbers did not exist. When man got beyond the stage of relying entirely upon tally sticks, where numbers are represented by notches, he hit upon the practice of using pebbles or shells which evolved first into counting boards and later into the beautiful and ingenious mechanical device, the abacus, invented to facilitate performing calculations.

The evolution of the abacus went through ages in three distinct forms:
1. The dust tablet: the earliest form;
2. The counter abacus (a tablet with loose counters): Salamis abacus;
3. The modern abacus: a frame with counters fastened to the lines. Examples of the modern abacus are the Russian abacus (eleven wires, ten with ten beads and one with four beads) and the Chinese abacus (two beads in the upper part and five beads in the lower one). The modern abacus, besides being a calculation tool, is a mathematical model for decimal arithmetic. Long before the concept of zero was invented, the abacus used an empty column as a computing feature.

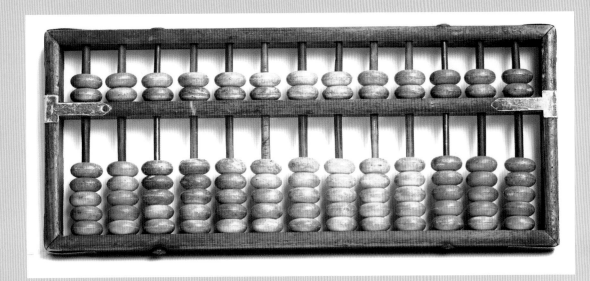

ARITHMETICA BY GREGOR REISCH

A symbolic representation of Boethius and Pythagoras in a mathematical competition, published in 1503 in the classic *Margarita philosophica,* by German philosopher Gregor Reisch. Pythagoras is using an abacus, while Severinus Boethius (480 AD–524 AD) uses numerals. Just by looking, can you tell how is an abacus used?

56

CHALLENGE ● ● ● ○ ○
REQUIRES 🧠
COMPLETED ○

CHAPTER

3

PRIMES, MAGIC SQUARES AND QUEEN DIDO'S PROBLEM

PRIME NUMBERS — 300 BC

Mathematicians have been fascinated by prime numbers for centuries. Some of them even think that behind the primes is the secret of creation.

A number can either be a prime number or a composite number. A prime number (or a prime) is a natural number greater than 1 that has no positive divisors other than 1 and itself. They are like atoms — the building blocks of integers. A natural number greater than 1 that is not a prime number is called a composite number. For example, 5 is prime, as only 1 and 5 divide it, whereas 6 is composite, since it can be divided by 2 and 3 in addition to 1 and 6. The central role of primes in number theory is demonstrated by the fundamental theorem of arithmetic: any integer greater than 1 can be expressed as a unique product of primes, ignoring the order. This theorem requires excluding the number 1 as a prime.

In about 300 BC, Euclid demonstrated that there are an infinite number of primes. There is no known formula giving all of the prime numbers and no composites. There are, however, formula's that model the distribution of primes, i.e. the "usual" statistical behavior of prime numbers. The prime number theory was the first of these. Proven at the end of the 19th century, it states that the probability that a given, randomly selected number n is prime is inversely proportional to its number of digits, or the logarithm of n.

Even today, many questions concerning prime numbers remain unresolved. Goldbach's conjecture is one example. It asserts that every even integer greater than 2 can be expressed as the sum of two primes. Another unresolved question is the twin prime conjecture, which states that there are an infinite number of pairs of twin primes whose difference is 2. Questions such as this inspired the development of different branches of number theory.

PRIMES HISTORY

The history of prime numbers can be traced back to Ancient Egypt: in the Rhind papyrus, for example, the Egyptian fraction expansions have quite different forms for primes and for composites. The true fathers of prime numbers, however are the Ancient Greeks. The Elements of Euclid (circa 300 BC) contain important theorems about primes, such as the infinitude of primes and the fundamental theorem of arithmetic. Euclid also showed how to construct a perfect number from a Mersenne prime. The Sieve of Eratosthenes, attributed to Eratosthenes, is a simple way of calculating lower prime numbers, although the large primes found today with computers are generated differently.

We then have to jump to the 17th century and Pierre de Fermat to find a new interesting study of prime numbers. In 1640, he produced his "little theorem." This theorem was proved later by Leibniz and Euler.

> **"It will be another million years, at least, before we understand the primes."**
>
> — *Paul Erdös, mathematician*

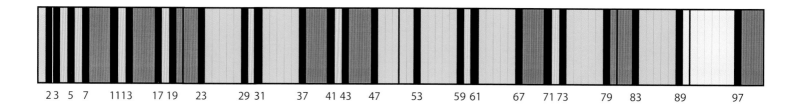

23 5 7 1113 17 19 23 29 31 37 41 43 47 53 59 61 67 71 73 79 83 89 97

PRIME NUMBERS UP TO 100

Look through a list of prime numbers, up to 100, as an example. It is impossible to predict when the next prime will appear. Their distribution seems random, chaotic, and there is no pattern as to how to determine the next prime. There are 25 primes among the 100. How will their distribution look for the next 100 numbers? It is difficult for mathematicians to admit that there might not be an explanation for the way in which nature has picked the primes. The 19th century saw a breakthrough. Bernhard Riemann looked at the problem of primes in a completely new way. From his view, he began to understand that underlying the outward randomness of the primes there is a subtle and unexpected inner harmony, which kept many of its secrets out of reach. His bold prediction of the existence of this harmony is known today as the Riemann Hypothesis, waiting to be proved and explained, as one of the greatest mysteries of mathematics.

57

CHALLENGE ● ●
REQUIRES
COMPLETED

SIEVE OF ERATOSTHENES — 250 BC

Eratosthenes of Cyrene (276–194 BC) was a man of many talents. He studied mathematics, wrote poetry and was a fit athlete. He also invented the discipline of geography as we know it today; he even introduced the word "geography" in the Greek language, together with important concepts such as the system of latitude and longitude.

As a mathematician, he is mainly remembered for his Sieve of Eratosthenes, an ancient algorithm for finding all prime numbers up to any given limit. Today it is still a very efficient method to calculate all prime numbers up to approximately 10 million, although computers of course have made it a bit redundant.

But how does the Sieve of Erastosthenes work? The list of primes is achieved by recursively marking the multiples of each prime as composite (i.e. non-prime numbers), starting with the multiples of 2. The multiples of a given prime are generated starting from that prime, as a sequence of numbers with the same difference, equal to that prime, between consecutive numbers. This constitutes the main distinction between the use of a sieve and the use of trial division to sequentially test each candidate number for divisibility by each prime.

As is the case for many Ancient Greek mathematicians, none of Erastosthenes' works have survived. We cannot be sure whether he invented the famous Sieve, but it was attributed to him by Nicomachus in his *Introductions to Arithmetic*.

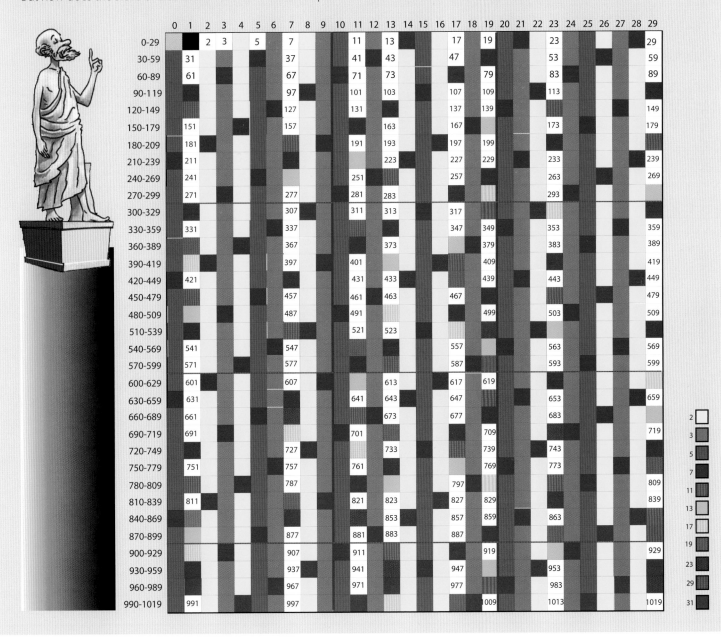

Can you work out what will be the color at the end of the table to complete the color pattern of the distribution of primes from 1 to 1,000? A prime gap is the difference between two successive prime numbers. From the pattern of primes up to 1,000, you can see that a zero gap occurs only between primes 2 and 3. They are the only consecutive pair of prime numbers.

58

CHALLENGE ● ● ●
REQUIRES
COMPLETED

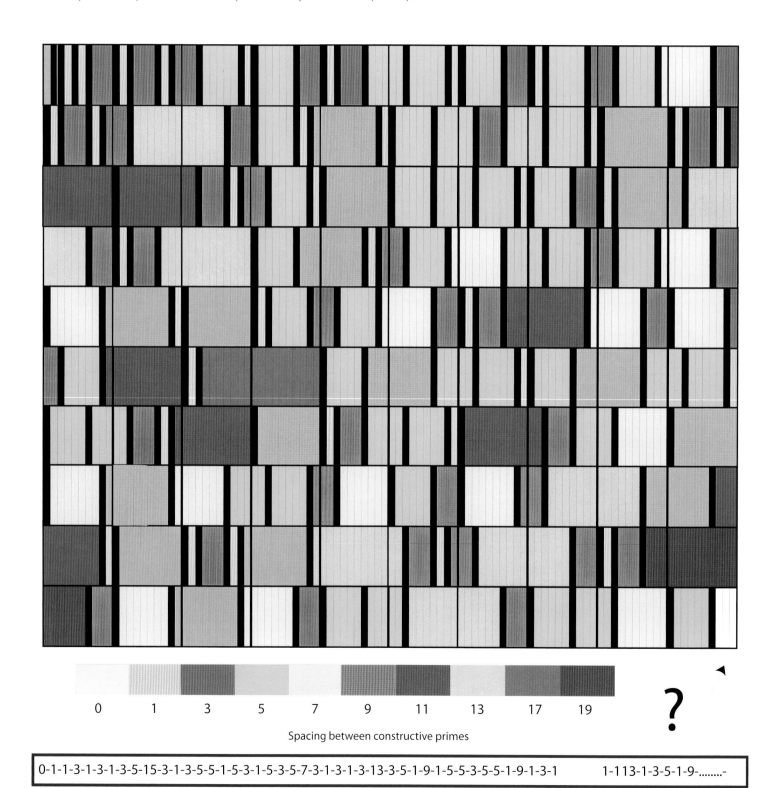

0   1   3   5   7   9   11   13   17   19   **?**

Spacing between constructive primes

0-1-1-3-1-3-1-3-5-15-3-1-3-5-5-1-5-3-1-5-3-5-7-3-1-3-1-3-13-3-5-1-9-1-5-5-3-5-5-1-9-1-3-1   1-113-1-3-5-1-9-………-

APOLLONIUS' PROBLEM — 270 BC

Apollonius of Perga (262–190 BC) was a Greek mathematician and astronomer, known by his contemporaries as "the Great Geometer." A number of later scholars, such as Ptolemy, Francesco Maurolico, Isaac Newton, René Descartes and others, were influenced by his ground-breaking methodology and terminology, especially on conic sections.

He introduced a number of terms still familiar to us today, such as the ellipse, the parabola and the hyperbola. He is also believed to have developed the hypothesis of eccentric orbits and epicycles to explain the apparent motion of the planets across the sky and the varying speed of the moon.

One of his best known problems carries his name: the "Apollonius Problem." It goes as follows: given three circles in the plane, in how many different ways can a fourth circle be added so that are all tangent (touching in a point) at the circumference of the fourth added circle? There are only eight different possibilities, as shown on the right.

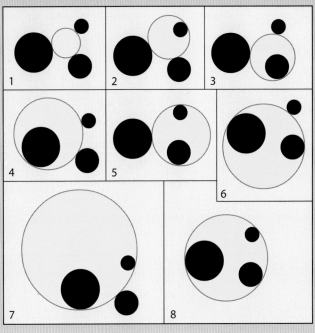

"Give me a place to stand and I will move the earth."

— a remark of Archimedes quoted by Pappus of Alexandria, 340 AD

LEVER PRINCIPLE OF ARCHIMEDES — 250 BC

The lever is the simplest example of a "simple machine." It is an energy transformer.

Does it give us something for nothing? No. But it can change mechanical energy involving a small force into mechanical energy involving a larger force.

A heavy load is lifted a small distance by an effort several times smaller: this is the law of the lever, which was proven by Archimedes using geometric reasoning.

It proves that when the distance between the fulcrum and the point of the input force is greater than that between the fulcrum and the point of the output force, the lever increases the input force. The moment of a force on a point is equal to the magnitude of the force multiplied by its perpendicular distance from the point.

A spade is an example of a lever. But can you tell how exactly you should use the spade in order to maximize the benefits of Archimedes' lever principle?

59 CHALLENGE
REQUIRES
COMPLETED

ARCHIMEDES (287–212 BC)

Little is known about the life of Archimedes of Syracuse (Greek mathematician, physicist, engineer, inventor and astronomer), but he is considered to be one of the leading scientists of his era and, indeed, one of the greatest of all time. His lists of discoveries is nearly endless. His work crucial for the development of hydrostatics, he produced an extremely accurate approximation of π, defined the spiral named after him and developed formulae for the volumes of surfaces of revolution and an ingenious system for expressing very large numbers. Last but not least, he is also famous for his explanation of the principle of the lever.

Archimedes died during the Siege of Syracuse. He was killed by a Roman soldier who ignored orders that the scientist was not to be harmed. His tomb, which has been described by Cicero, was paid tribute to his greatest mathematical achievement: on top was a sphere with an inscription that referred to Archimedes' proof that the sphere has two-thirds of the volume and surface area of the cylinder (including the cylinder's bases).

THE EUREKA MOMENT — 250 BC

Archimedes, so the story goes, discovering the principle of hydrostatics, ran naked from his bathtub shouting *Eureka* ("I've found it!").

This happened when he was working on the problem involving the suspicion that the new crown ordered by King Heiron of Syracuse was not solid gold but also contained other materials.

He solved the puzzle without melting down the crown, by discovering the principle named after him today. "A body immersed in fluid is lighter than its true weight by the weight of the fluid displaced."

The story of Archimedes running naked is strongly disputed among mathematicians studying the history of mathematics, and not because of Archimedes running naked. Nudity was not a big thing for the ancient Greeks. However, the thought of Archimedes, so famous and dignified, running and yelling is unacceptable to them.

There were many other recorded instances of creative lightning-flashes in science. James Watt was struck with the idea of the steam engine while watching his kettle; Leo Szilard with the sudden illumination of a neutron chain reaction (or how to make an atomic bomb) while waiting at traffic lights, etc.

ARCHIMEDES' PRINCIPLE

Looking at the outcome of Archimedes' experiment below, what was Archimedes' conclusion?

60 CHALLENGE ● ● ● ● ● ○
REQUIRES 🧠 ✏️
COMPLETED ○

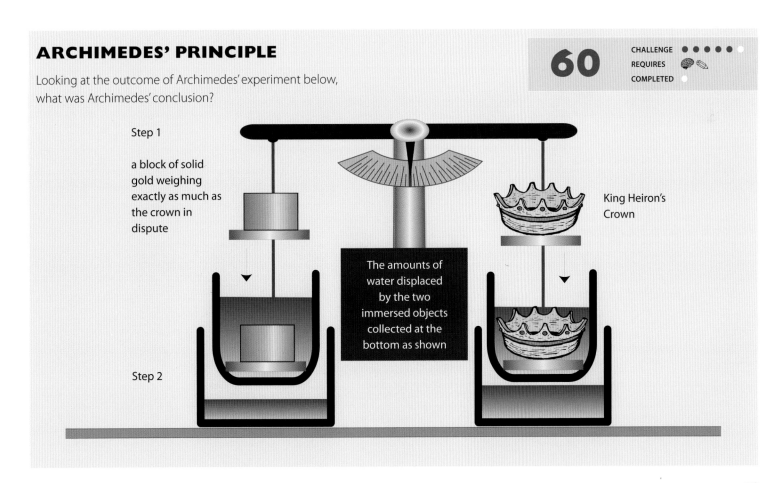

Step 1

a block of solid gold weighing exactly as much as the crown in dispute

King Heiron's Crown

The amounts of water displaced by the two immersed objects collected at the bottom as shown

Step 2

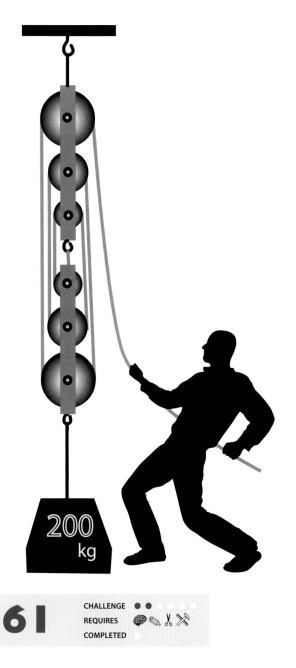

6| CHALLENGE ●● | REQUIRES 🧠✏️✂️🔨 | COMPLETED

PULLEY SYSTEMS — 250 BC

Also called block and tackle systems, rope and pulley systems use a rope transmitting a linear motive force to a load through one or more pulleys for the purpose of lifting the load. Pulley systems are the only simple machines in which the possible values of mechanical advantage are limited to whole numbers. Block and tackle systems usually lift objects with a mechanical advantage greater than two. Our example shown on the left is a compound pulley system of three fixed and three movable pulleys with a 200 kilogram weight attached to it. Can the man lift the heavy weight of 200 kg pulling the rope?

SIMPLE MACHINES

A simple machine is a mechanical device that changes the direction or magnitude of a force. They are generally defined as the simplest mechanisms providing mechanical advantage (also known as leverage).

The Greek philosopher and mathematician Archimedes was the founding father of the concept of the "simple machine." Some 300 years BC, he discovered the principle of mechanical advantage in the lever, and also studied the pulley and screw. Later, Greek philosophers defined the classic five simple machines, together with an approximate calculation of the mechanical advantage they provided. The five classic "simple machines" that can set a load in motion as listed by Heron of Alexandria (ca. 10–75 AD) in his Mechanics, are:

- the lever;
- the windlass;
- the pulley;
- the wedge;
- and the screw.

Heron also described how these "simple machines" were made and used. However, the Greeks still had a limited understanding of how these machines functioned: they focused on the statics of simple machines (the balance of forces) and did not include dynamics (the trade-off between force and distance) or the concept of work.

MECHANICAL POWER

During the Renaissance, the study of simple machines, then known as Mechanical Powers, continued. Their dynamics began to be studied in terms of how much useful work they could perform, finally resulting in the new concept of mechanical work.

Three famous Renaissance scientists are well-known for their work on Mechanical Powers: the Flemish engineer Simon Stevin, Galileo Galilei and Leonardo Da Vinci. The first added a sixth simple machine, the inclined plane, to the classical five in 1586. Galileo Galilei produced the complete dynamic theory of simple machines in 1600 in his work *Le Meccaniche* ("On Mechanics"). He was also the first scientist to understand that simple machines merely transform energy, rather than creating it. Leonardo Da Vinci (1452–1519) discovered the classic rules of sliding friction in machines. They were rediscovered by Guillaume Amontons (1699) and further developed by Charles-Augustin de Coulomb (1785).

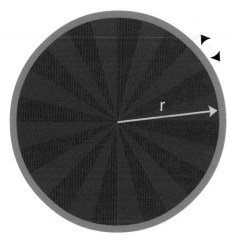

$$A = r\pi \times r = r^2\pi$$

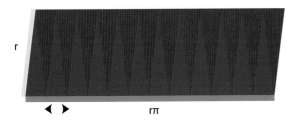

$r\pi$

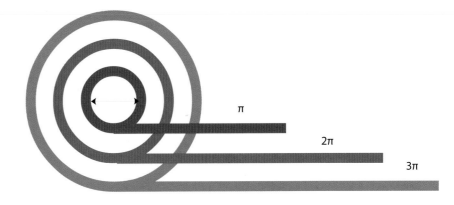

π
2π
3π

THE NUMBER π AND A CIRCLE'S CIRCUMFERENCE

Prehistoric men must have made the observation that the bigger the diameter of the roller, the farther the load advanced. Early civilizations knew that the ratio of the circumference to the diameter of a circle is the same for all circles no matter what their size, and that this constant ratio is slightly more than three. It is denoted today as π (for the letter p of the Greek alphabet).

The problem of determining the area of a circle was once a great mathematical challenge.

Archimedes tried to solve it using the "squaring the circle" method, i.e. trying to find the square (polygon) that has the same area as a circle of a given radius. His method leading to the exact formula is demonstrated.

The circle of radius "r" is divided into a great number of near isosceles triangles with sides r and bases "a" (small circular arcs approximating straight lines), which can be arranged to form a parallelogram as shown.

The more sectors the circle is divided into, the more the sectors resemble triangles. These triangles become smaller and smaller and the figure approaches the form of a rectangle.

The height of each triangle is roughly the same as the radius of the circle. The circumference of the circle = 2 x r x π. Each color of triangle covers half of the circle's circumference, so the length of the rectangle is half this, that is π x r. Hence: area of circle = area of the rectangle (as shown) = height x width = r x (π x r) = π x r², which is the famous formula we know today.

Note that this is only an approximation. The method actually works only if the base of each triangle ("a" in the diagram) is infinitely small.

CONWAY'S GROUPING OF π DECIMALS

Is π really random? Separate the decimals of π into groups of 10.

Will there ever be a group containing all 10 digits from 0 to 9: (0, 1, 2, 3, 4, 5, 6, 7, 8, 9)?

π = 3.1415926535
8979323846
2643383279
5028841971
6939937510
5820974944
5923078164…

62

CHALLENGE ●● ○ ○ ○ ○
REQUIRES 🎨 ✏️ ✂️ 🔧
COMPLETED ○

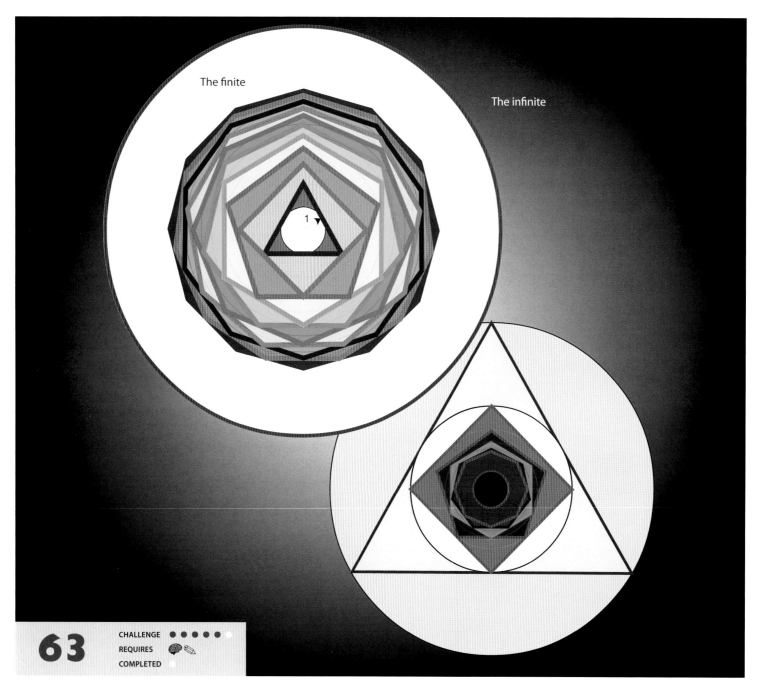

The finite

The infinite

CIRCUMSCRIBED, INSCRIBED AND INFINITY — 250 BC

The radius of the innermost circle is one. It is circumscribed (and inscribed) by a series of circles and regular polygons, starting from an equilateral triangle, which is circumscribed by a circle, and so on, square, regular pentagon, hexagon, heptagon, octagon, enneagon, decagon, hendecagon, dodecagon, etc.

The pattern above visualizes the first ten exploding circles up to a regular circumscribed polygon of 12 sides. If you go on with this process an infinite number of times, surrounding the polygons with circles and further polygons, adding a side to each consecutive polygon, the radius of the circles grows larger and larger.

What do you think? Would the circumscribed circles become infinitely large as we continue the process?

The same process is repeated by inscribing regular polygons in the initial unit circle. Will the diminishing process end with the innermost circle becoming infinitely small? How many sides will the smallest inscribed polygon have?

STOMACHION OF ARCHIMEDES — 250 BC

Stomachion, also known as loculus Archimedius ("Archimedes' box"), is a mathematical treatise attributed to Archimedes (287–212 BC).

The origin of the puzzle goes back to two ancient texts that are attributed to Archimedes, in which he mentions a game similar to tangram. One manuscript is a Greek palimpsest (a manuscript that has been written on, scraped off and used again), discovered in Constantinople in 1899 and dating back to the 10th century. The other one is an Arabic translation.

It is not known whether Archimedes invented the game or simply explored its geometrical aspects. In the Greek manuscript the areas of the Stomachion pieces are determined. According to this text, Archimedes wrote a book about the stomachion, which was lost for 2,000 years. Parts of this book were recently rediscovered by the palimpsest, arousing great interest in and research into the puzzle.

The game consists of 14 flat ivory pieces of various polygonal shapes originally forming a 12-by-12 square. Like in the tangram, the object of the game was to rearrange the pieces to form interesting things (people, animals, objects, etc.). We don't know whether in Archimedes' version the pieces were allowed to be turned over. In ancient Greece, this was not allowed.

Archimedes was probably also interested in the difficult problem: in how many different ways the 14 pieces can form a square. The solution to this problem had to wait for more than 2,200 years. It was solved in 2003, by Bill Cutler using a computer. He found a total of 536 distinct solutions, rotations and reflections not considered to be different.

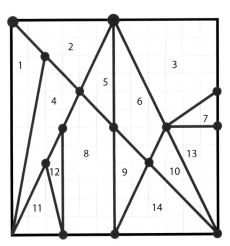

STOMACHION

Above is the structure of a stomachion puzzle with its lattice points on a 12-by-12 square grid, and consisting of 14 parts.

Can you work out the areas of the 14 pieces?

64

CHALLENGE ●●○○○○
REQUIRES 🧠✏️⚒️
COMPLETED ○

1	
2	
3	
4	
5	
6	
7	
8	
9	
10	
11	
12	
13	
14	

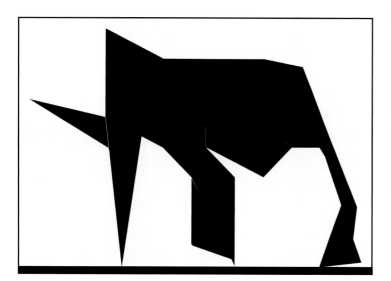

THE STOMACHION ELEPHANT

In his *Cento Nuptialis*, Magnus Ausonius (310–395 AD) described his famous Stomachion Elephant puzzle. It is the oldest published puzzle known today. The goal is to form the elephant on the left by rearranging the parts of the stomachion. Can you solve it?

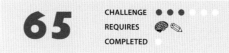

65 CHALLENGE ● ● ● ○ ○ ○
REQUIRES 🧠✏️
COMPLETED

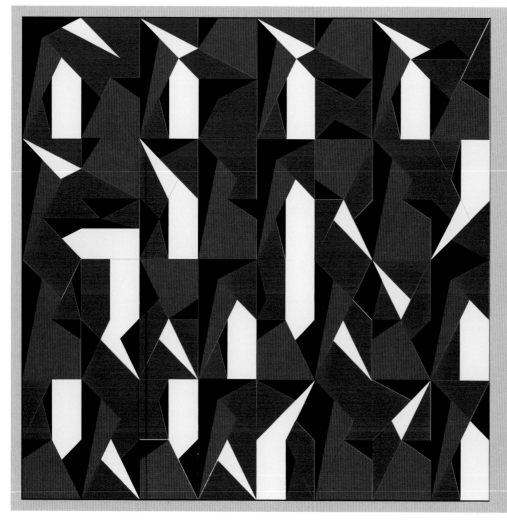

MATHEMATICS CREATES ART

On the left you see a composition of 16 randomly selected distinct solutions of the Stomachion square problem.

The palimpsest discovered in a prayer book in Constantinople (see "Stomachion of Archimedes" on the previous page), was studied by Dr. Reviel Netz. He came to the conclusion that Archimedes was trying to solve the combinatorial problem of finding how many ways there are to form a square from the 14 pieces of the puzzle. A total of 17,152 solutions were found for this. Therefore, the Stomachion was not only one of the oldest puzzles, but certainly one of the first combinatorial puzzles in the history of mathematics.

THE SHOEMAKER'S KNIFE

Archimedes was the first to study the Arbelos, or the Shoemaker's Knife in Greek, in his *Book of Lemmas*.

The arbelos is formed by three semicircles, two along the diameter of the third, which can be of any size (green areas).

The figure has amazing properties and coincidences that are quite counter-intuitive. To mention just a few striking ones: Archimedes discovered that the area between the semicircles (green) is equal to the area of the circle, which has a diameter equal to the perpendicular segment from the point of intersection of the two smaller semicircles along the diagonal of the large semicircle, as shown in the three examples (green circle). This will be very obvious if we look at the special case when the two smaller semicircles are identical (bottom).

The sum of the two smaller arc lengths is equal to the larger arc length.

The two smaller circles, twin circles (yellow), touching the perpendicular and the two arcs are identical regardless of the size of the two smaller semicircles

For 500 years Arbelos was forgotten, until Pappus continued work on Arbelos, further discovering its astonishing properties.

Recently, Leon Bankoff and Victor Thebault published a comprehensive manuscript on the properties of Arbelos, still discovering many new properties that were previously unknown.

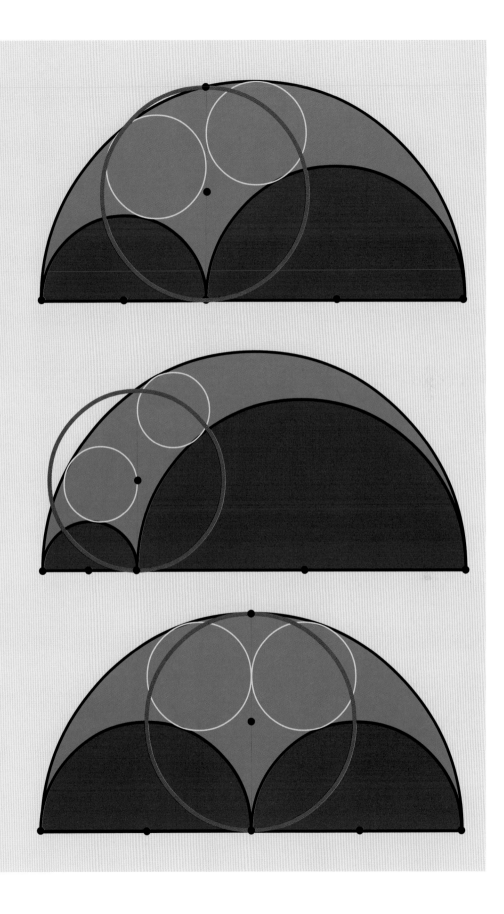

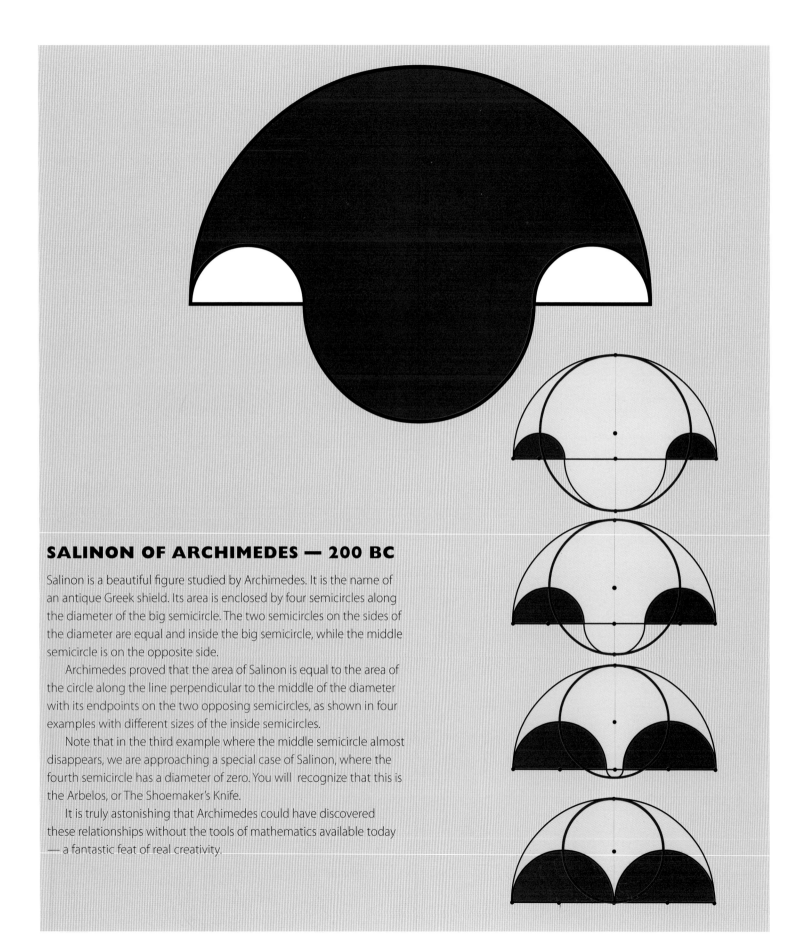

SALINON OF ARCHIMEDES — 200 BC

Salinon is a beautiful figure studied by Archimedes. It is the name of an antique Greek shield. Its area is enclosed by four semicircles along the diameter of the big semicircle. The two semicircles on the sides of the diameter are equal and inside the big semicircle, while the middle semicircle is on the opposite side.

Archimedes proved that the area of Salinon is equal to the area of the circle along the line perpendicular to the middle of the diameter with its endpoints on the two opposing semicircles, as shown in four examples with different sizes of the inside semicircles.

Note that in the third example where the middle semicircle almost disappears, we are approaching a special case of Salinon, where the fourth semicircle has a diameter of zero. You will recognize that this is the Arbelos, or The Shoemaker's Knife.

It is truly astonishing that Archimedes could have discovered these relationships without the tools of mathematics available today — a fantastic feat of real creativity.

ARCHIMEDES' BURNING MIRRORS — 214 BC

Mirrors can achieve seemingly impossible feats in science, magic, puzzles and everyday life. The great Greek scientist Archimedes imaginatively applied mirrors to many of his practical inventions.

According to ancient writings, his most remarkable feat was related to warfare, using mirrors that focused the sun's rays to set fire to Roman ships besieging the town of Syracuse in 214 BC.

Many scientists and historians, fascinated with this story, dismissed it as an impossible feat, but a few tried to prove that Archimedes could indeed have caused the Roman vessels to burst into flames.

The assumption of these experiments was that Archimedes could not have used giant mirrors, but instead had to create the effect of large mirrors using a great number of small reflectors, probably highly polished metals. Did he use the shields of Syracuse's soldiers?

Could Archimedes just line up his men and have them focus the sun's rays on the Roman ships and light them?

In 1747, Georges-Louis Leclerc conducted an experiment. Using 168 ordinary rectangular flat mirrors, he succeeded in igniting wood from about 100 meters. It seems that Archimedes could certainly have done the same, since in the port of Syracuse the Roman ships must have been no farther than about 20 meters from land.

A similar experiment was repeated in 1973 by a Greek engineer using 70 mirrors, focusing the sun's rays on a rowboat about 80 meters offshore. A few seconds after the mirrors were properly aimed, the boat started burning. In order for this to work, the mirrors had to be slightly concave, which Archimedes could have done as well.

Most recent experiments on the other hand, showed that the best result achieved was to ignite a small piece of wood about 50 meters away. Such a result would hardly be sufficient, and the story of destruction of the entire Roman fleet seems to be a bit far-fetched, but is it? We'll probably never know.

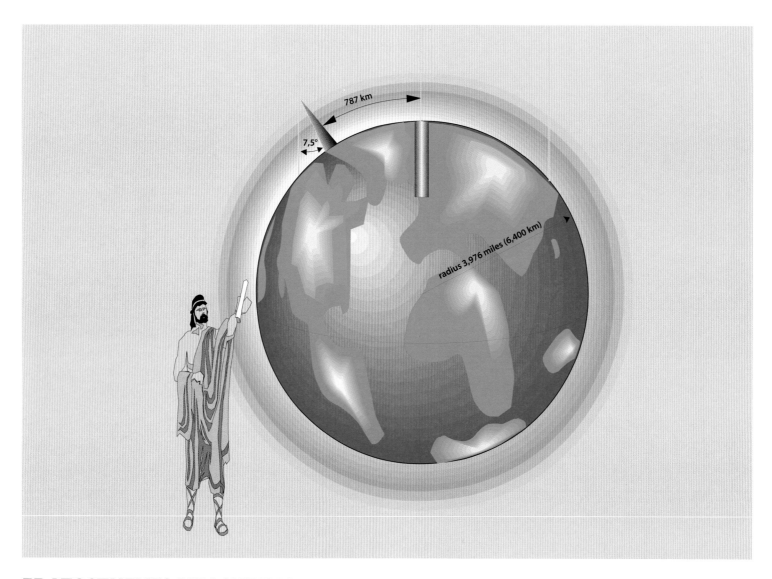

787 km

7,5°

radius 3,976 miles (6,400 km)

ERATOSTHENES MEASURING EARTH — 200 BC

Although early Greek geometers made huge theoretical advances, the mathematician Eratosthenes (276–194 BC) from Alexandria accomplished perhaps the greatest achievement in terms of insight. He learned that on a day in midsummer in the town of Syene, the reflection of the noonday sun was visible on the water of a deep well. For that to occur, the sun had to be directly overhead, with its rays pointing directly toward the center of the earth. On the same day, the noonday sun cast a shadow in Alexandria that measured 7.5 degrees, or about 1/50th of a full circle.

Eratosthenes also knew the north to south distance between Alexandria and Syene, which is about 787 kilometers. These details were sufficient for him to calculate the circumference of the earth with astonishing accuracy. Can you calculate the circumference of the earth using these numbers?

Eratosthenes was born in Cyrene, now in Libya. He studied in Athens and was the librarian of the famous Mouseion, a massive collection of hundreds of thousands of papyrus and vellum scrolls, housed in the temple of Muses. He was nicknamed Beta, falling short of the highest ranks in many categories of his work, though recognized as a great scholar. Today, his accomplishments are considered remarkable for their creativity, providing early examples of modern scientific method.

> **"My conviction is that the Earth is a round body in the corner of the heavens."**
>
> — *Plato in Phaedo (360 BC, attributed to Socrates), the earliest known written record of the idea that the world is round*

66 CHALLENGE ● ● ●
REQUIRES 🧠 ✏️
COMPLETED

AREA AND PERIMETER

One set of the four shapes (circle, square, triangle and hexagon) all have the same areas. The other four all have the same perimeters. Can you sort out the two sets, those with equal areas and those with equal perimeters?

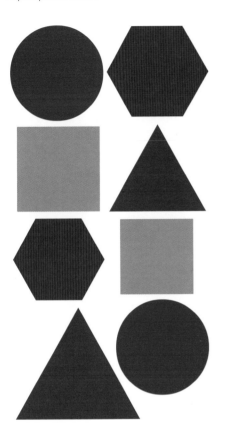

67 CHALLENGE ● ● ● ○ ○ ○
     REQUIRES 🧠 ✏️
     COMPLETED

ISOPERIMETRIC PROBLEM

The solution to the isoperimetric problem must be convex. (i.e. curving out, extending outward).

The area of the non-convex shape can be increased keeping its perimeter unchanged. The elongated shape can be more round, keeping its perimeter fixed but increasing its area.

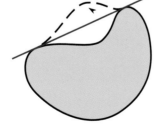

QUEEN DIDO'S PROBLEM — 200 BC

Areas and volumes are easy to estimate for rectangular shapes; they are more difficult to estimate for other shapes, especially with curved sides.

The ancient Greeks knew the significance of perimeter in terms of area enclosed — indeed, the word meter is derived from the Greek word for "measure around." Since many Greeks lived on islands, they had good reason to be aware of the pitfalls of measurement. After all, it is easy to see that the area of an island cannot be assessed using the time it takes to walk around it; a long coastline might simply mean that the shape of the island is irregular rather than meaning that the island is large. Nevertheless, the custom was for landowners to base real estate on the perimeter of their holdings, not the area.

One ancient story tells of Dido, the princess of Tyre, who fled to a spot on the North African coast. There she was given a grant of land that was terribly small — equal to what could be covered by the hide of an ox. Undaunted, Dido had the hide cut into strips and sewn together to make one ribbon about a mile in length. Then, using shoreline as a boundary, she had her supporters stretch the ribbon of hide in as big a semicircle as was possible. In this way one ordinary ox hide encompassed about 25 acres of land. On that spot Dido founded the famous and powerful city of Carthage.

Today, the problem is called the "isoperimetric" theorem, which states that of all the plane figures with the same perimeter, the circle has the largest area.

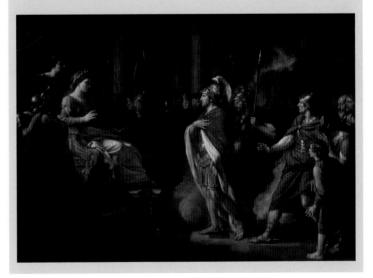

CHINESE RINGS — 200 AD

The Chinese Rings, also known as Baguenadier ("time-waster" in French) and by many other names, is one of the oldest-known mechanical disentanglement puzzles.

The object is to remove all the rings from a stiff frame. The removal of rings is complicated, because rings must be returned to remove other rings, etc.

Martin Gardner wrote: "Twenty-five rings require 22,369,621 steps, for a very quick puzzle solver taking two years to accomplish." The puzzle's popularity is due to the fact that it ingeniously combines logic and mathematics, but its solution is not easy. A Chinese Ring puzzle consisting of nine rings is still challenging enough. Its solution requires 341 moves. It was invented by the Chinese general Chu-ko Liang (181–234 AD), allegedly to keep his wife busy while he was away at the wars, a very clever invention indeed. Mathematically, there is a logical connection between the Chinese Rings puzzle and the binary numbers, and its solution is related to what is known today as the Gray Binary Code, invented by Louis Gros in 1872. Édouard Lucas, who invented the Tower of Hanoi puzzle, is also known for his solution to the Chinese Rings puzzle using binary and Gray codes.

A harem girl holding a set of rings over her head (painting by Giacomo Mantegazzain, 1876)

The Chinese Linking Rings is a related puzzle and also one of the classics of illusionmagic. The magician links and unlinks seemingly solid rings, which appear to pass through each other forming chains and other patterns and configurations.

HERO'S DOOR-OPENING MECHANISM — 50 AD

Magic tricks often use basic scientific principles to wow the crowd.

The most ingenious mechanisms of the ancient world were the inventions of Hero of Alexandria. He can certainly be considered the earliest and possibly the greatest toy and gadget inventor ever.

The mechanism on the right for opening a temple door was typical of the many toys and automata Hero invented that were intended for "magical" purposes. Can you read Hero's blueprint and explain how it was meant to work?

68 CHALLENGE ● ● ● ○ ○ ○
REQUIRES
COMPLETED ○

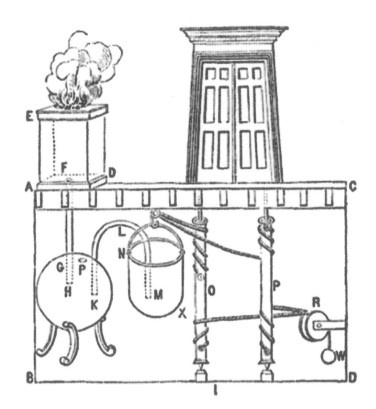

HERO (HERON) OF ALEXANDRIA (20–60 AD)

Hero (also known as Heron) of Alexandria was a Greek mathematician, scientist and inventor born in Egypt, mostly renowned for his works in mechanics, mathematics and physics.

Hero of Alexandria was a Greek mathematician, physicist and inventor. He was born in Egypt and worked mostly in Alexandria. Among his many inventions are several mechanical machines with practical applications, such as a fire engine, a coin-operated device and a water organ. He also invented the aeolipile, the earliest known steam-powered engine, which was a rotary steam engine consisting of a sphere mounted on a boiler with two canted nozzles to produce a rotary motion from the escaping steam. The siphon, known as Hero's fountain, was an instrument that produced a vertical jet of water by air pressure.

Heron was also well-known for his studies in geometry (a branch of mathematics that studies the relationships, measurements and properties of points, lines, angles and solids) and in geodesy (a branch of mathematics that seeks to determine the size and shape of the Earth, and the location of objects or areas on the Earth).

One of his most important works was the collection of his books of *Metrica*, which was found in the year 1896. In Book I he portrays a derivation of Heron's formula that expresses the area of a triangle in terms of its sides. This formula was the result of his attempt to show that the angle of incidence in optics is equal to the angle of reflection. Book II of the Metrica describes ways to calculate the volumes of bodies such as cylinders, cones, prisms, etc. In Book III, Hero studies the division of volumes and areas into parts of given ratios.

Hero's "aeolipile." The first-recorded reaction steam turbine also known as Hero's "Engine" spins when heated as shown above.

JOSEPHUS' PROBLEM — 100 AD

Flavius Josephus (37–100 AD), a famous historian, soldier and scholar, decided to solve a puzzle to save his life, so the legend says.

He was defending the city of Jotaphat, which fell to the Roman general Vespasian. Josephus and his fighters hid in a cave and entered into a suicide pact, instead of giving up.

This moment of history is the topic of the Josephus' puzzle.

The group of 41 zealots including Josephus agreed to form a circle, and starting from a fixed position every third man counted would be killed, until the last man, who would then kill himself. Was it pure luck or divine intervention that Josephus was the last to remain alive?

Or, did Josephus want to stay alive and was he able to figure out how to place himself in the last spot at the end of the counting?

The original problem appeared for the first time in Ambrose of Milan's book (ca. 370 AD).

Many different variations of Josephus' puzzle can be found all over the world. The problem was studied by many great mathematicians, including Leonhard Euler, but the mathematical formula for solving such a problem was never found. General solutions are still achieved by trial and error. The puzzle is a minimalistic model for the combinatorial study of systematic arrangements — a branch of mathematics today called Systems Analysis.

JOSEPHUS' PUZZLE

In a circle of 41 men, every third man is killed. Where should Josephus stand in the circle to survive? And suppose Josephus also wanted to save the life of his best friend. Where should he place him?

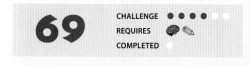

69   CHALLENGE ● ● ● ● ○ ○
REQUIRES 🪨 ✏️
COMPLETED ○

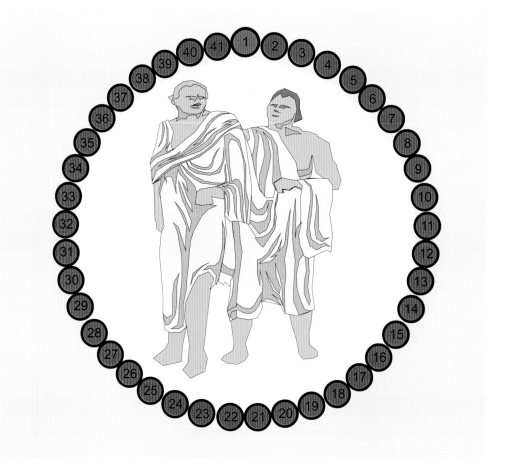

BORROMEAN RINGS — 200 AD

In mathematics, the Borromean rings consist of three topological circles that are linked and form a Brunnian link, i.e. removing any ring results in two unlinked rings. In other words, no two of the three rings are linked with each other, but nonetheless all three are linked.

The name "Borromean rings" is borrowed from the aristocratic Borromeo family in Italy, in whose coat of arms they appear. The concept itself is much older. We, for example, find drawings similar to Borromean rings in Gandhara (Afghan) Buddhist art (approximately 200 AD) and even on Norse image stones dating back to the seventh century.

70

CHALLENGE ● ● ● ○ ○ ○
REQUIRES
COMPLETED ○

BORROMEAN GOLDEN RING NECKLACE

The necklace is formed from 11 interlinked golden rings as shown.

You are allowed to cut only one ring to separate the necklace into the largest possible number of parts.

Which of the rings will you cut?

DIMENSIONS

It was Pappus, one of the great mathematicians of Alexandria, in the fourth century AD, who recognised that space could be filled by a moving point. A point moving in one dimension produces a straight line. This line, moving in a direction at right angles to the first, defines a rectangle. The rectangle moving in a direction at right angles to the previous two, defines a rectangular prism — a cube, etc. as shown below.

Pappus of Alexandria (ca. 290–ca. 350 AD) was one of the last great Greek mathematicians of Antiquity, known for his *Synagoge* or *Mathematical Collection,* his best-known work, obviously written with the object of reviving the classical Greek geometry. It's a compendium of mathematics in eight volumes, the bulk of which has

survived. He observed the eclipse of the Sun in Alexandria which took place on October 18, 320, thus clarifying the date for Pappus's commentary on Ptolemy's *Almagest.* This accurate date is virtually all we know of Pappus. He was born in Alexandria and appears to have lived there throughout his life.

The style of his work suggests that he was primarily a teacher of mathematics. Pappus rarely claimed to present original discoveries, but he was adept at spotting interesting material in his predecessors' writings, many of which have not survived outside of his work. His *Synagoge of Mathematical Collection* proved that man is one of the best sources of information concerning the history of Greek mathematics.

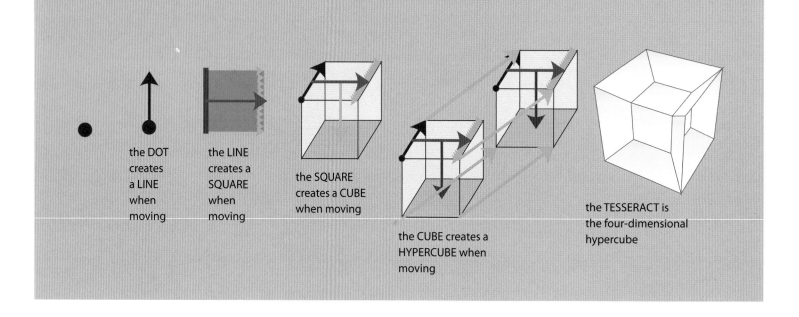

the DOT creates a LINE when moving

the LINE creates a SQUARE when moving

the SQUARE creates a CUBE when moving

the CUBE creates a HYPERCUBE when moving

the TESSERACT is the four-dimensional hypercube

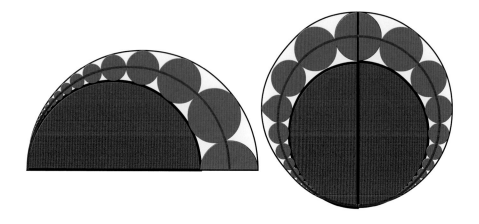

PAPPUS AND THE CHAIN OF ARBELOS

For 500 years, the Arbelos of Archimedes (see earlier in this chapter) was nearly forgotten, until Pappus continued discovering its then-unknown astonishing properties. In Pappus' chain of Arbelos circles, the centers of the inscribed circles lie on an ellipse (red line).

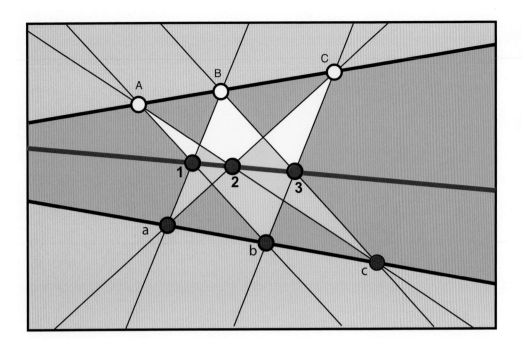

PAPPUS' THEOREM

Let three points A, B, C be incident to a single straight line and another three points a, b, c incident to another straight line. Then three pairwise intersections: Ab – Ba, Ac – Ca, Bc – Cb are incident to a third straight line. Try it out for yourself.

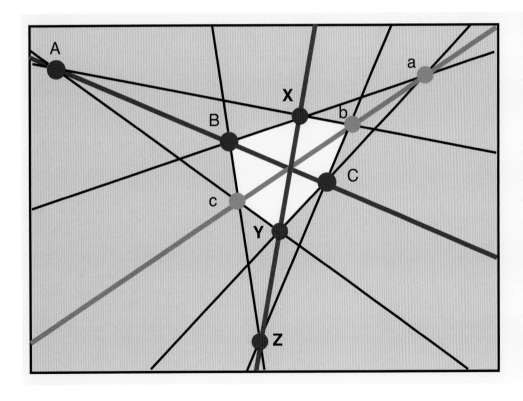

PAPPUS' HEXAGON THEOREM

Pappus' hexagon theorem states that given one set of collinear points A, B, C (i.e. the points are incident on a line), and another set of collinear points a, b, c, the intersection points X, Y, Z of line pairs Ab and aB, Ac and aC, Bc and bC are collinear. The points A, b, C, a, B and c together form Pappus' hexagon.

DIOPHANTUS OF ALEXANDRIA — 250 AD

Diophantus, sometimes referred to as "the father of algebra," wrote an influential series of books called *Arithmetica*, which deal with solving algebraic equations. Unfortunately, many of them have not survived. The ones that did influenced some of the greatest mathematicians of all times. One of them was Pierre de Fermat. While studying the Arithmetica, he concluded that a certain equation considered by Diophantus had no solutions, and noted without elaboration that he had found "a truly marvelous proof of this proposition." The famous "Fermat's Last Theorem" was born, which resulted in huge advances in number theory. Diophantus was the first Greek mathematician to recognize fractions as numbers; thus allowing positive rational numbers for the coefficients and solutions. In modern use, Diophantine equations are usually algebraic equations with integer coefficients, for which integer solutions are sought.

DIOPHANTUS' RIDDLE

Diophantus's riddle is a poem that encodes a mathematical problem. In verse, it reads as follows:

"Here lies Diophantus," the wonder behold. Through art algebraic, the stone tells how old: "God gave him his boyhood one-sixth of his life, One-twelfth more as youth while whiskers grew rife; And then yet one-seventh ere marriage begun; In five years there came a bouncing new son. Alas, the dear child of master and sage, after attaining half the measure of his father's life chill fate took him. After consoling his fate by the science of numbers for four years, he ended his life."

In more prosaic terms, the poem states that Diophantus's youth lasted one-sixth of his life. After another one-twelfth of his life he grew a beard. Diophantus married after one-seventh more of his life. Five years later, he had a son. The son lived exactly half as long as his father, and Diophantus died just four years after his son's death. The sum of these parts gives the number of years Diophantus lived. Can you figure out how old Diophantus was when he died?

71 CHALLENGE ● ●
REQUIRES 🧠
COMPLETED

MATH SYMBOLS QUIZ

Mathematics is a universal language and its symbols also have a wider application outside the discipline of mathematics. The father of algebra was the Greek Diophantus (200 AD). He was the first to use symbols to represent unknown quantities. On the next page you see a series of symbols that are still used today. How many of these symbols do you know? You can choose between:

72 CHALLENGE ● ● ● ● ○
REQUIRES 🧠✏️
COMPLETED

parallelogram	congruent	radius
secant to a circle	right angle	tangent
smaller than	diameter	parallelopiped
sphere	approaches (in value)	acute angle
natural numbers	and so on	similar to (proportional)
infinity	scalene triangle	circle area
line A B	there exists	corresponds to
approximately equal	tetrahedon	isosceles triangle
obtuse angle	cube	integers
trapezoid	equal to or less than	intersecting circles
parallel	circular segment	right-angled triangle
diamond	therefore	identically equal to
plus or minus	pyramid	not equal to
equilateral triangle	inscribed circle	a is not an element of b
perpendicular	rectangular prism	central angle
rhombus	summation	inscribed angle
perimeter	octahedron	circumsized circle
square root	factorial n	regular octagon
cylinder	arc	regular pentagon
percent	because	regular hexagon
intersection	sector	regular septagon
line segment AB	regular nonagon	vector AB
cone	symbol for Pi	equal to
parallels	end of proof	
semi-circle	equal to or greater than	

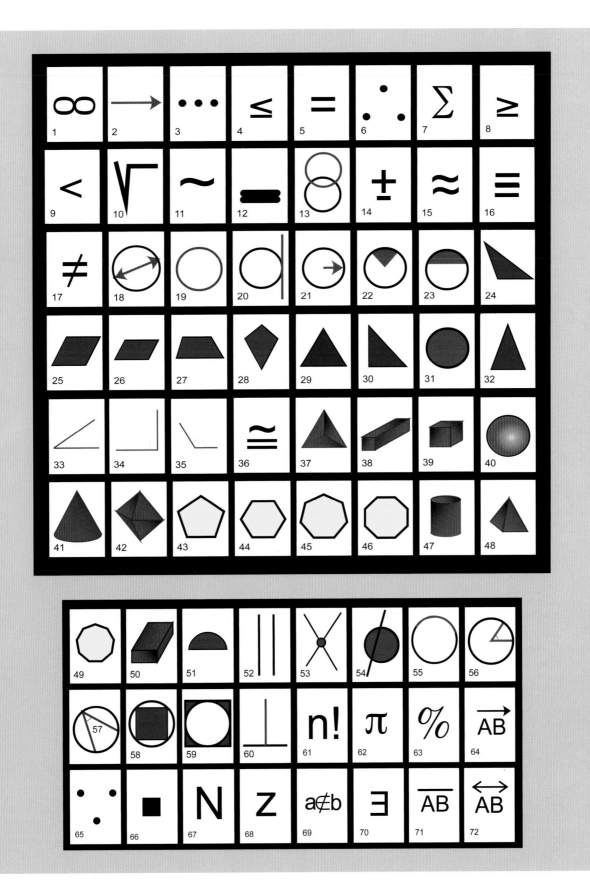

RIVER-CROSSING PROBLEMS — 790 AD

The classic eighth-century river-crossing problem is well known today. In eighth century England, Alcuin of York, a mathematician and theologian, published a book that included the following riddle.

A man has to take a wolf, a goat, and some cabbage across a river. His boat has enough room for the man plus either the wolf or the goat or the cabbage. If he takes the cabbage with him, the wolf will eat the goat. If he takes the wolf, the goat will eat the cabbage. Only when the man is present are the goat and the cabbage safe.

How will the man carry the wolf, goat and cabbage across the river? River-crossing problems in different variations were very popular in medieval Europe and still are today.

73 CHALLENGE ● ● ● ● ● ●
REQUIRES 🧠 ✏️
COMPLETED ○

INTERPLANETARY COURIER

In my dream, I looked at the passengers and groaned. As an Interplanetary Courier at the Alpha Centauri spaceport, I am responsible for transporting passengers from the spaceport up to the space liner orbiting the planet thousands of zerks above us. Weird passengers indeed! Before me I see a Rigellian, a Denebian and a strange-looking four-limbed creature called a Terrestrial.

First, the Rigellians and the Denebians are at war; left together in the airlock one of them will undoubtedly become the victim of an unfortunate "accident." The Denebian, unlike the vegetarian Rigellian, is a voracious meat-eater; left alone with the Denebian, the feeble Terrestrial would soon become no more than a light snack. Somehow they must all be ferried up to the spaceship's airlock, where all three will become the responsibility of the delightful reptilian hostesses. I will need a few trips,

1. Denebian   3. Terrestrial   2. Me   4. Rigellian

and one passenger will have to accompany me more than once, but I know I can do this. There will be no accidents and no unfortunate meals. All three will emerge from the liner's airlock in one piece.

How can I organize my trips?

74 CHALLENGE ● ● ● ● ● ●
REQUIRES 🧠 ✏️
COMPLETED ○

JEALOUS HUSBANDS

In the 16th century, more elaborate versions of the river-crossing problem appeared by the Venitian mathematician Niccolo Tartaglia, featuring three beautiful brides and their young very jealous husbands, who come to a river. The small boat that is to take them across holds no more than two people. To avoid potential problems, you must organize the crossings so that no woman is left with a man unless her husband is also present. How many trips does it take to ferry them across the river?

75 CHALLENGE ●●●●● ○
REQUIRES 🧠 ✏️ ✂️
COMPLETED ○

76 CHALLENGE ●●●● ○ ○
REQUIRES 🧠 ✏️
COMPLETED ○

RIVER CROSSING SOLDIERS

Three soldiers have to cross the river. Two boys in a small boat passing by agree to help them, but their boat can support only one soldier or two boys. None of the soldiers can swim. Under these circumstances, how can they cross the river, end up on other side and return the boat to the boys?

DISSECTIONS — POLYGON TRANSFORMATIONS

Dissection problems must have confronted man thousands of years ago, but only in the 10th century, we find the first systematic study of dissections, in a book of the the Persian astronomer Abul Wefa. Only fragments of his book survive, including the beautiful dissection problem below.

Wefa's puzzle was the forerunner of the most interesting type of geometrical dissections, the problem of dissecting a geometrical figure into another specified figure in the minimum possible number of pieces. Ever since, dissection records have been constantly improving.

There are many ways to divide an area into parts, and some of the ways of making these divisions are particularly interesting.

Putting small shapes together to make larger shapes is also fun — like making a pattern of tiles on a floor. In mathematics this is called mosaics or "tessellation."

Mathematicians have only recently begun to take dissection problems seriously. The branch of mathematics called dissection theory provides valuable insights into solutions of many practical problems in plane and solid geometry.

In dissection problems, the pieces may be given and the object is to create a pattern with them. The ancient recreation of Tangram is a good example.

On the other hand, two polygons are given and the problem is to transform one into another, dissecting them into the smallest number of pieces.

A third, apparently paradoxical variant, is to dissect a shape into pieces, remove one piece, and reassemble the remainder to form the original shape. Although this is impossible, many puzzles appear to achieve it (geometrical paradoxes or geometrical vanishes).

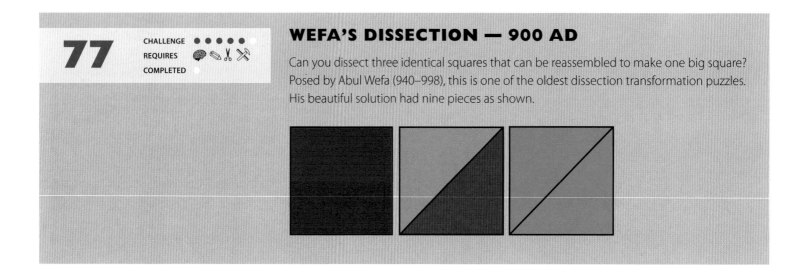

77

CHALLENGE ● ● ● ● ● ○
REQUIRES 🧠 ✏️ ✂️ ⚒️
COMPLETED

WEFA'S DISSECTION — 900 AD

Can you dissect three identical squares that can be reassembled to make one big square? Posed by Abul Wefa (940–998), this is one of the oldest dissection transformation puzzles. His beautiful solution had nine pieces as shown.

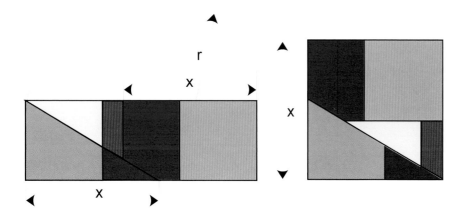

PERIGAL'S RECORD

Henry Perigal (1801–1898), an amateur mathematician, improved Wefa's dissection using only six pieces, as you can see on your left, a record that still stands. He is also famous for his elegant proof of the Pythagorean Theorem.

DOMINO GAMES — 1200 AD

Dominoes are thought to have first been used in the 12th century in China, although there are some theories that they originated in Egypt. They made their way to Italy in the early 1700s, then spread across Europe during the rest of the 18th century, becoming one of the most popular games in pubs and sitting rooms.

Nowadays, dominoes is played throughout the world. The game is especially popular in Latin America. Dominoes is even considered the national game of many Caribbean countries. Domino tournaments are held annually in many countries, and there are numerous local domino clubs in cities all over the world.

The word "domino" is probably inspired by the black dots on a white background, a pattern that is reminiscent of a "domino" (a kind of hood) worn by Christian priests.

Dominoes is a generic game, like playing cards and dice. The pieces are building blocks that can be combined in numerous different ways to create a wide variety of games and puzzles. Dominoes evolved from dice: the numbers in a standard double-six domino set represent all the rolls of two six-sided dice.

DOMINO TOPPLING WORLD RECORD

The setup used 4.8 million dominoes. 4,345,028 dominoes needed to be toppled in order to break the existing world record. In the end, 4,491,863 toppled dominoes were counted.

DOMINO PATTERN PUZZLES

Both patterns are composed of a complete set of 28 dominoes.

By carefully observing the patterns, determine how the domino pieces were assembled, to be able to add the outlines of the individual domino pieces.

This will be not as easy as it seems.

78

CHALLENGE ● ● ● ○ ○

REQUIRES 🧠 ✏️ ✂️

COMPLETED ○

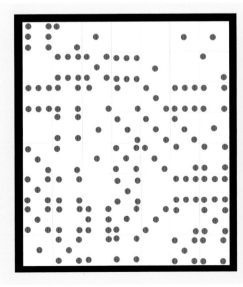

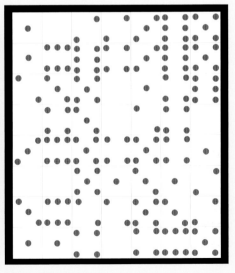

FIBONACCI
(1170–1250)

Fibonacci, the nickname of Leonardo da Pisa, was born in Pisa (Italy) but was educated in North Africa. Apart from some autobiographical notes he included in his own books, we have very little information about his life.

Fibonacci was one of the most skilled mathematicians of the Middle Ages. We have to thank Fibonacci for our current decimal system. As a student of mathematics he found the Roman numeral system, which had no zero and lacked place value, insufficient, and decided to use the Hindu-Arabic symbols (0-9) instead.

Fibonacci is also famous for his famous sequence of numbers, now known as the Fibonacci numbers (1, 1, 2, 3, 5, 8, 13, 21, 34 and 55). They are closely associated with Lucas numbers as they are a complementary pair of Lucas sequences (see next page). They are also very closely related to the Golden Ratio: the closest rational approximations to the ratio are 2/1, 3/2, 5/3 and 8/5. Fibonacci numbers can also be found in elements of nature, as in the branching of trees, Phyllotaxis (the arrangement of leaves on a stem), the flowering of an artichoke, an uncurling fern and the arrangement of a pine cone, etc.

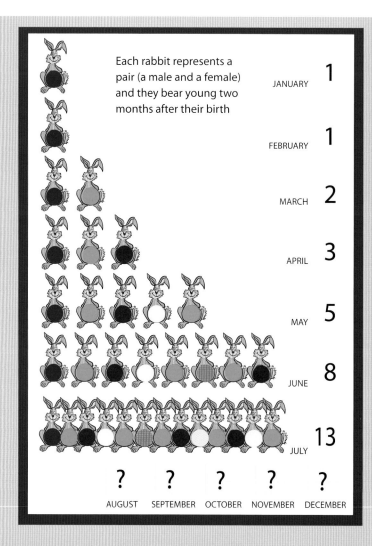

Each rabbit represents a pair (a male and a female) and they bear young two months after their birth

JANUARY	1
FEBRUARY	1
MARCH	2
APRIL	3
MAY	5
JUNE	8
JULY	13

? ? ? ? ?

AUGUST   SEPTEMBER   OCTOBER   NOVEMBER   DECEMBER

FIBONACCI'S RABBIT PROBLEM — 1202 AD

The most famous recreational mathematics problem concerning number sequences is this classic from 1202. How many pairs of rabbits can be produced from a single pair in one year, if every month each pair produces one new pair, and new pairs begin to bear young two months after their own birth?

This imaginary rabbit breeding puzzle was found in a book called *Liber Abaci*, from 1202, written by Leonardo da Pisa, then a 27-year-old Italian mathematician, who is known by the nickname Fibonacci. Stating the puzzle, Fibonacci hypothetically assumed that every pair referred to is composed of a male and a female rabbit and that they bear young two months after their birth, when in reality they only reach maturity after four months. This innocent mathematical puzzle with its man-invented number sequence is known today as the famous Fibonacci number sequence, and the sequence was later found everywhere in nature. What a coincidence! Can you find the number of rabbits for the rest of the year?

79

CHALLENGE ● ● ● ● ○

REQUIRES

COMPLETED

FIBONACCI NUMBER SEQUENCE AND THE GOLDEN RATIO

In mathematics, the Fibonacci numbers are the below sequence of numbers, growing endlessly. The first 13 numbers of the sequence are shown. Can you deduce the logic of the growth of the sequence and continue it as long as you wish?

Can you work out the fascinating relationship of the Fibonacci number sequence to the Golden Ratio described as Phi=1.618?

> 1, 1, 2, 3, 5, 8, 13, 21, 34, 55, 89, 144, 233, …

80
CHALLENGE ● ● ● ● ● ○
REQUIRES 🧠 ✏️
COMPLETED ○

LUCAS SEQUENCES AND LUCAS NUMBERS

Lucas numbers should not be confused with Lucas sequences, which are a generic category of sequences to which the Lucas numbers belong. Eduard Anatole Lucas studied both that sequence and the closely related Fibonacci numbers.

Lucas numbers and Fibonacci numbers form complementary instances of Lucas sequences. In math, a recurrence relation is an equation that defines a sequence. One or more initial terms given define each further term in the sequence.

Lucas sequences are integer sequences that satisfy recurrence relations. Famous examples of Lucas sequences include the Fibonacci numbers, the Mersenne numbers, the Pell numbers, the Lucas numbers, the Jacobsthal numbers, etc.

Each Lucas number is defined as the sum of its two immediate previous terms, just as the Fibonacci integer sequence. The ratio between two consecutive Lucas numbers therefore also converges to the Golden Ratio. However, whereas Fibonacci starts with 0 and 1, Lucas starts with 2 and 1. Consequently, the properties of Lucas numbers are consequently rather different from those of Fibonacci numbers.

Thus the sequence of Lucas numbers is:

> 2, 1, 3, 4, 7, 11, 18, 29, 47, 76, 123, …

In fact, for every series formed by adding the latest two values to get the next, and no matter what two values we start with, we will always end up having terms whose ratio is Phi=1·6180339… eventually!

FIBONACCI NUMBER STAIRCASE

The red numbers show how Fibonacci numbers grow relative to each other by looking at the quotients of two consecutive numbers. Look at how these fractions — one Fibonacci number divided by the previous one — change as the sequence grows. What will be the amazing outcome of the long list of decimals?

Can every natural number be expressed using only Fibonacci numbers? Can you, for example try to form the natural number 232, using the first 13 Fibonacci numbers?

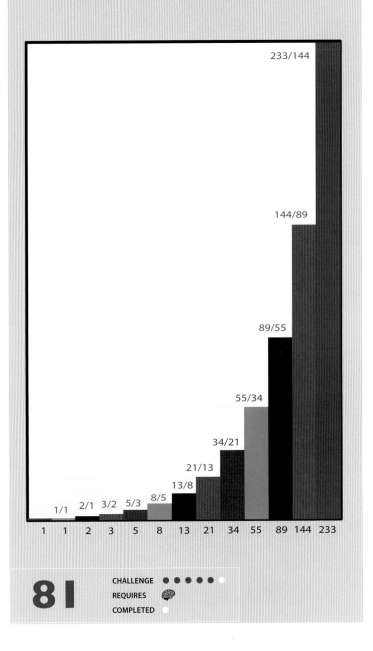

81
CHALLENGE ● ● ● ● ● ○
REQUIRES 🧠
COMPLETED ○

FIBONACCI SQUARES RECTANGLE

In this puzzle we present you the infinite squares tessellation problem. For a long time, the Fibonacci squares rectangle was associated with a beautiful old unsolved problem: is it possible to tessellate the infinite plane with squares, no two of which are of the same size? Until 1938 the problem was considered impossible. The Fibonacci rectangle was the closest solution to this problem. It involves the sequence of Fibonacci squares. By setting up a Fibonacci rectangle of consecutive squares with sides of consecutive Fibonacci numbers we can cover as large an area as we wish.

In our counterclockwise spiral of Fibonacci numbers, part of the last square is already out of our page (purple area). There is just one problem: the Fibonacci rectangle starts with two identical unit squares (see below), which contradicts the condition that no two squares be the same size. If we could solve another problem, we would then have the solution to the challenging infinite tiling problem: can a square be subdivided into smaller squares no two of which are alike? This problem was only solved in 1938 (see perfect squares in Chapter 7). Substituting such a Perfect Square into the first unit square (black) of the Fibonacci rectangle solves the infinite squares tiling problem.

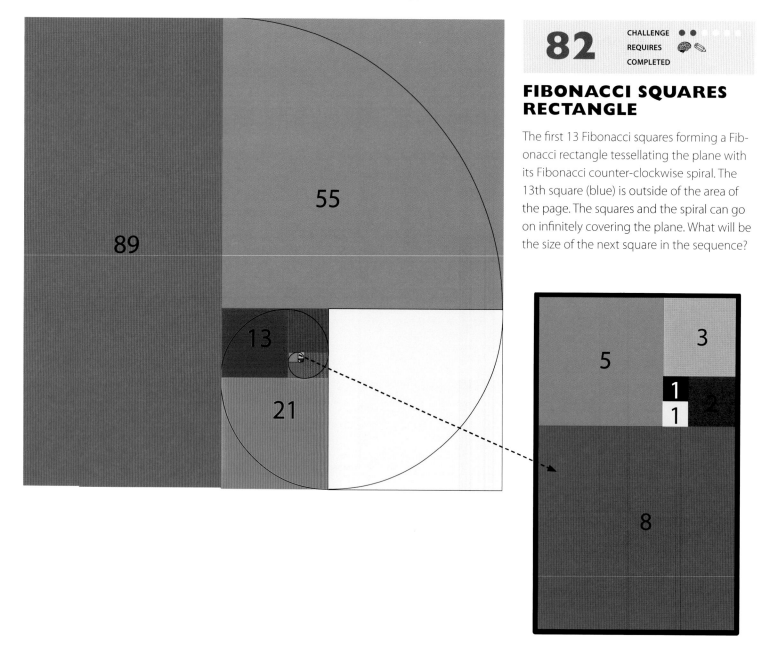

82	CHALLENGE ● ●
	REQUIRES 🧠✏️
	COMPLETED

FIBONACCI SQUARES RECTANGLE

The first 13 Fibonacci squares forming a Fibonacci rectangle tessellating the plane with its Fibonacci counter-clockwise spiral. The 13th square (blue) is outside of the area of the page. The squares and the spiral can go on infinitely covering the plane. What will be the size of the next square in the sequence?

ALHAMBRA

The Alhambra in Granada, constructed in the 13th century by the Moorish kings of Spain, is one of the most splendid architectural structures in Western Europe. After the Catholic Reconquista of Southern Spain in 1492, the palace was severely damaged, but has now been restored to its former glory. It is a wonderful example of the once vibrant and refined Moorish culture and architecture. The interiors of the buildings are decorated with beautifully intricate examples of detailed geometric ornamentation.

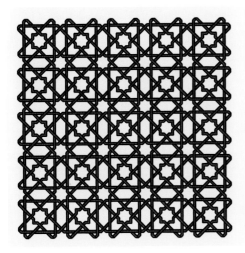

ALHAMBRA MOSAIC PATTERN — 1230 AD

The former palace of the Moorish kings of Granada is a treasure house of mathematical beauty. The intricate pattern shown on the left is an example of a remarkable wealth of its many complex geometrical designs and tessellations. Can you tell whether it is one loop or whether it is composed of separate parts? In case of the latter, how many?

83 CHALLENGE ● ● ● ● ○ ○
REQUIRES 🧠 ✎
COMPLETED

PERPETUAL MOTION MACHINES

Our machines are hungry servants. We have to feed them in order to keep them working.

Perpetual motion can be described as "motion that continues indefinitely without any external source of energy; impossible in practice because of friction." It can also be described as "the motion of a hypothetical machine which, once activated, would run forever unless subject to an external force or to wear." Scientists generally agree that perpetual motion in an isolated system would violate the first and/or second law of thermodynamics. Many early inventors dreamed about making a perfect machine, that would need only to be started and would work all by itself forever, until the parts wore out.

LEONARDO'S PERPETUAL MOTION MACHINE

Leonardo da Vinci was one of the first inventors to tackle the issue of perpetuum mobiles. His design was based on the concept of Villard de Honnecourt, a French architect from 1240. It uses the power of gravitation to create a never ending energy. Just by looking, can you explain what Da Vinci had in mind? And why was his invention not very succesful?

84	CHALLENGE ●●● ○○
	REQUIRES 🧠
	COMPLETED ○

 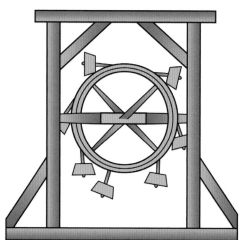

GEORGE GAMOW'S MACHINE

George Gamov, a Ukrainian-American nuclear physicist, formulated one of the first theories of the Big Bang together with Ralph Alpher in 1948. He also studied quantum mechanics, stellar evolution and genetic theory. In 1954 he, for example, proposed the first coding system for DNA, shortly after the discovery of DNA by James Watson and Francis Crick. Gamov also invented the perpetuum mobile on your right. Just by looking at the design shown on your right, can you explain the theoretical principles behind its operation?

85	CHALLENGE ●●● ○○
	REQUIRES 🧠
	COMPLETED

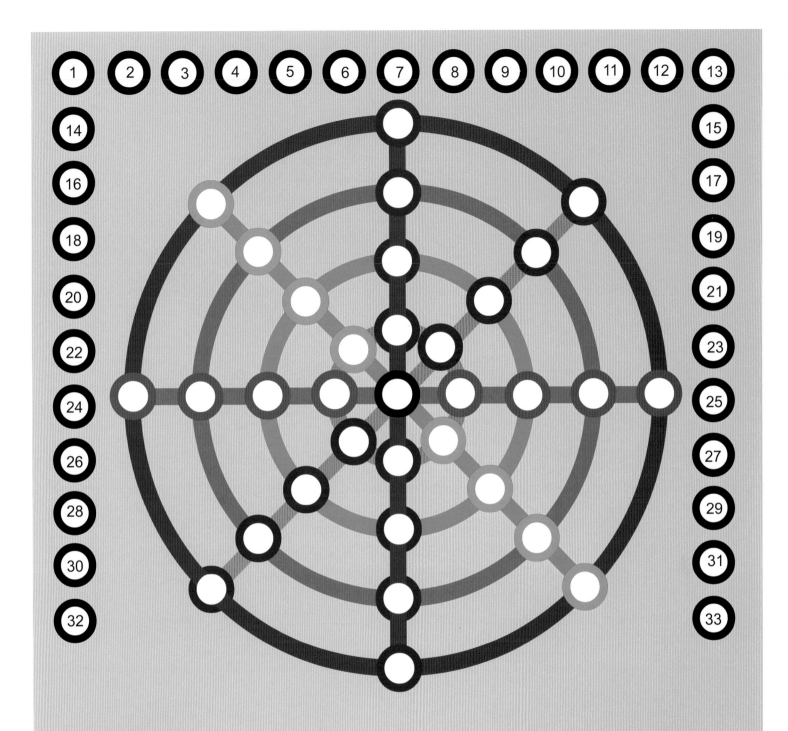

MAGIC CIRCLE — 1250 AD

The Chinese mathematician Yang Hui (c. 1238–c. 1298) wrote two books, one including the early Magic Circle puzzle, shown above. He is best remembered as the first to describe the famous Pascal's triangle, which was later studied by Blaise Pascal, and became one of the cornerstones of modern mathematics.

86

CHALLENGE ● ● ● ● ● ○
REQUIRES
COMPLETED ○

Can you arrange the numbers from 1 to 33 in the small circles so that every circle including the one at the center and every diagonal has the same sum?

WHEAT AND CHESSBOARD PROBLEM — 1260 AD

In about 1260 AD, Ibn Khallikan (1211–1282), a Kurdish historian living in the Abbasid Empire (modern Iraq) wrote an encyclopedia with biographies, one of which contained a story involving the game of chess and the concept of "exponential growth."

In the story, Ibn Khallikan describes the despotic Indian king Shihram. A wise man named Sissa ibn Dahir invented the game of chess to prove to the King that all his subjects, regardless of their rank, were important and should be treated well.

King Shihram was impressed and asked Sissa ben Dahir how he could repay him for his services. Sissa first declined a reward, but when the king insisted, he said the king should put one grain of wheat on the first square of a chessboard, two grains of wheat on the second square, four grains on the third square, eight grains on the fourth square, and so on, doubling the number of grains of wheat with each square (today called an exponential rate of growth). The King agreed. Can you calculate how many grains Sissa received? The final number may surprise you!

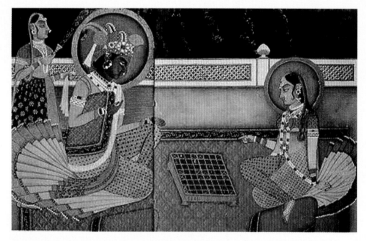

The Hindu Gods Krishna and Radha playing a game on a chessboard

87

CHALLENGE ● ● ○ ○ ○
REQUIRES
COMPLETED ○

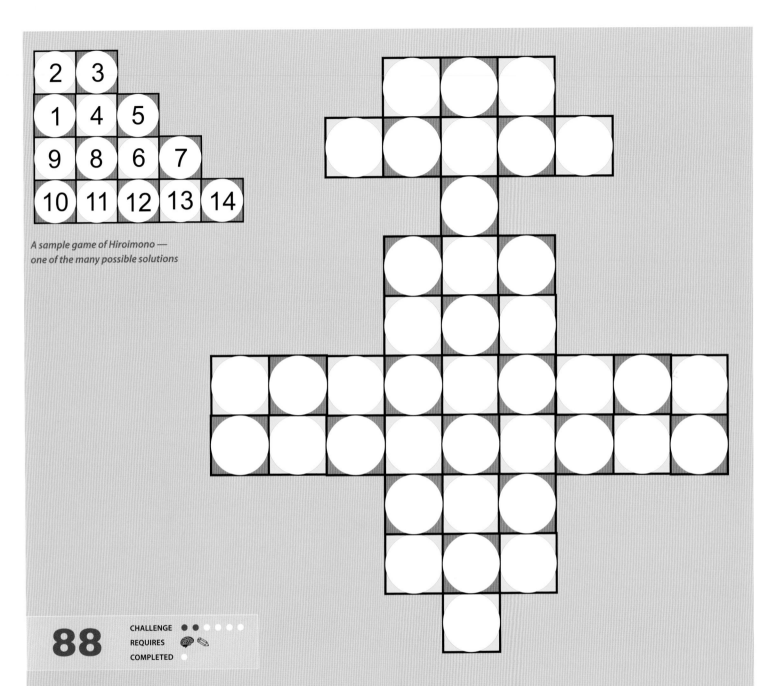

A sample game of Hiroimono —
one of the many possible solutions

88

CHALLENGE ● ● ○ ○ ○ ○

REQUIRES 🧠 ✎

COMPLETED ○

HIROIMONO — A PAPER-AND-PENCIL GAME — 1300 AD

This game, originating from 14th-century Japan, is played on a rectangular board by arranging stones in an orthogonal pattern and removing them one by one, according to the rules of the game. It can also be played as a paper-and-pencil game by filling the patterns with consecutive numbers starting from 1, as shown in the sample game above. The object is to fill all circles of the given pattern with consecutive numbers, starting at a chosen circle.

The rules are:

1) Once you have written a number in a circle, you may move horizontally or vertically to the next empty circle.

2) You may not jump over an empty circle, but you may jump over a filled circle to reach an empty circle behind it.

3) You may not land on a circle you have previously visited.

Can you solve the puzzle?

CYCLOID — 1450 AD

The curve traced by a point on a rotating circle is called a cycloid. It is an example of a roulette, which is the curve traced by a fixed point on one curve rolling on another. The cycloid was given its name by Galileo in 1599, but was first studied by Nicholas of Cusa (1401–1464). G.P. de Roberval demonstrated in 1634 that the area under one arch of a cycloid is three times the area of its generating circle. In 1658 Christopher Wren showed that the length of one arch of a cycloid is four times the diameter of its generating circle.

The cycloidal curve appears in many places in modern society: mechanical gears have teeth whose sides possess a cycloidal curve; a machine engraves an elaborate cycloid on the plates used for printing banknotes; a popular science toy known as the Spirograph produces an endless variety of cycloidal shapes with just a very few moving parts. The cycloid has many fascinating properties. It is the solution to the brachistochrone problem (i.e. it is the curve of fastest descent under gravity) and the related tautochrone problem (i.e. the period of an object in descent without friction inside this curve does not depend on the object's starting position).

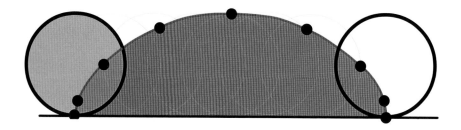

POLYGONAL CYCLOID AREA

Polygonal cycloids are good analogues to obtain solutions about the length and area of a cycloid. Can you work out the area under the polygonal arch generated by the revolving regular decagon along a straight line?

This beautiful visual proof was created by Philip R. Mallinson and first published in *Proofs without Words II*, (The Mathematical Association of America, 2000).

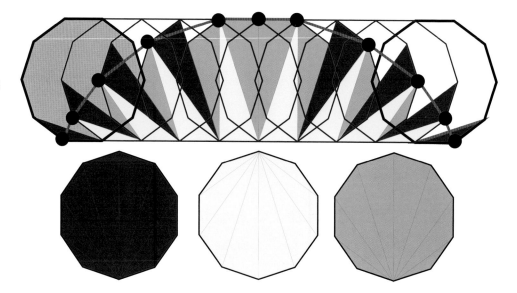

CYCLOIDS AND SQUARING THE CIRCLE

The problem of squaring the circle, namely constructing a square with the same area as a given circle using ruler and compass alone, had been one of the classical problems of Greek mathematics (see Chapter 2).

In 1982, Ferdinand von Lindemann (1852–1939) proved that it is not possible to square the circle if only compass and straightedge are available. On the other hand, it is possible to square the circle if a "mechanism" is used. The mechanism consists of a circle rolling along a straight line, producing a curve called a cycloid.

As the circle completes one revolution, the point A on its circumference, moving from A to Z describes a cycloid. The length of the straight line from A to Z is equal to the circumference of the circle — that is, $2\pi r$.

Thus if B is the midpoint of AZ, then BZ=πr. Hence, if CZ= r, the area of the rectangle BZDC is $\pi r \times r = \pi r^2$, which is also the area of the rolling circle.

Squaring this rectangle yields the square with the side ZF, as shown.

Thus the rectangle and the circle too are squared.

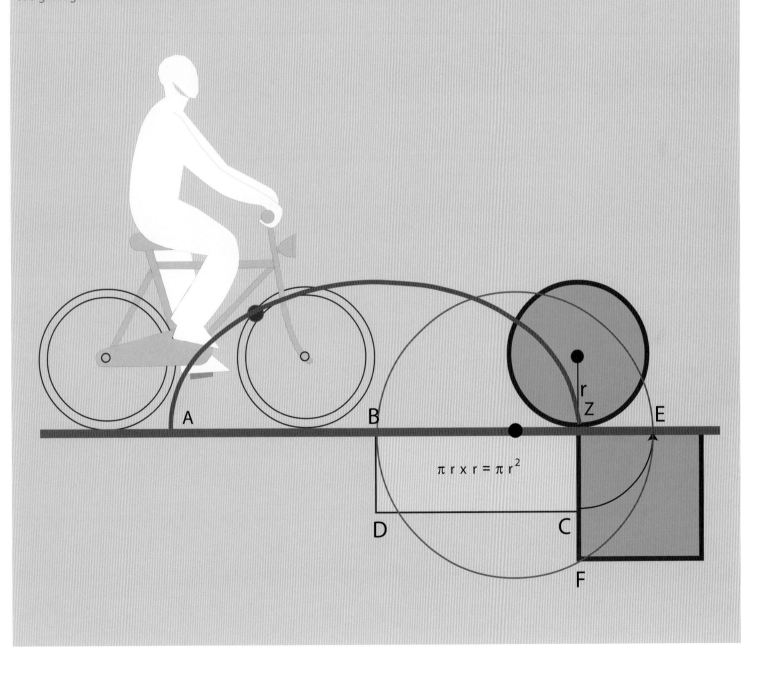

$\pi \ r \times r = \pi \ r^2$

EGG OF COLUMBUS — 1492 AD

After critics mocked him and claimed that the discovery of the Americas was no great achievement, Christopher Columbus is said to have challenged them to make an egg stand on its top. When they failed to solve the puzzle, Columbus tapped the egg on the table to flatten its tip. Ever since, the expression "an Egg of Columbus" or "Columbus's Egg" is used to describe an idea or discovery that — with hindsight — appears to be quite simple, but is brilliant nonetheless.

BALANCING AN EGG

An ingenious equilibrium toy could be found in toy shops several years ago. It was a plastic egg which, of course, like Columbus did, you had to balance on its pointed end. As much as you tried, this feat would evade all your efforts. You can try every way, shake the egg in order to discover its inner secret, but you can't hear anything moving inside. But there is a way to solve the secret of the puzzle, if you follow the procedure:

1. Hold the egg with its thin pointed end up vertically, and inconspicuously, for at least 30 seconds.
2. Turn the egg over and wait for 10 seconds more and then place it on its pointed end. It will beautifully balance and stand on its pointed end. Don't display it in this position for more than 15 seconds.
3. Just before the lapse of that period pick up the egg, holding it in the same position for at least 10 more seconds, before handling it over (still pointed end down) to somebody, challenging him or her to repeat your feat.
4. The egg won't be balanced on its pointed end anymore, no matter how one tries.

Can you explain the mysterious inner structure of the egg and its operation, from the above description?

89 CHALLENGE ● ○ ○ ○ ○ ○
REQUIRES
COMPLETED

LEONARDO DA VINCI (1452–1519)

Leonardo da Vinci, often described as the greatest creative genius of all time, was a man of many talents. Millions of visitors flock to his *Mona Lisa* in the Louvre (Paris) each year, but few of those admirers truly realize the impact Leonardo da Vinci had on science and mathematics today.

He started off his carreer in Florence, where after an apprenticeship of five years, he was accepted in the painter's guild in 1472. Between 1482 and 1499, he worked for the Duke of Milan, and during his time in that Northern Italian city, his interest in mathematics grew. He studied and illustrated Pacioli's Divina Proportione and become so intrigued by his own geometry research that he neglected his painting. Around 1498, his book on the elementary theory of mechanics was published in Milan, and he also invented several methods of squaring the circle.

King Francis I was a great admirer of his work and invited Leonardo to join his court. In 1516, Leonardo was given the title of first painter, architect and mechanic of the King. In France, he finished some works he had with him, such as *St. John the Baptist, Mona Lisa* and the *Virgin and Child with St. Anne,* but apart from that he spent most of his time arranging and editing his scientific studies and inventions.

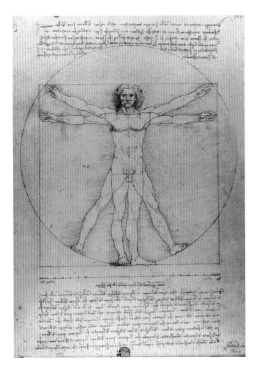

VITRUVIAN MAN OF LEONARDO

The pen and ink drawing by Leonardo from 1492 depicting a man fitting his body to a circle and a square is probably the most famous drawing in the world. The drawing is known as the *Vitruvian Man,* named for the man who created him, the Roman architect Vitruvius (c. 80 BC–c.15 BC). Many artists tried to depict Vitruvius's perfect man, but Leonardo's version is considered the most accurate and beautiful of all attempts, a perfect blend of art and science.

> **"No human inquiry can be called science unless it pursues its path through mathematical education."**
>
> — *Leonardo da Vinci*

ANAMORPHIC DISTORTIONS

The two red shapes are drawn in an anamorphic distorted grid. Just by looking, can you tell what these shapes would look like without distortion?

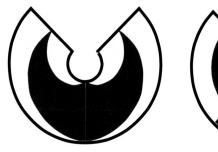

 90 CHALLENGE ●● ○ ○ ○ ○
REQUIRES 🧠 ✏️
COMPLETED ○

ANAMORPHOSIS — 1485

Anamorphic art is the use of perspective by distorting an image so that it can only be seen from a particular angle or reflected in a mirrored surface. A cylindrical mirror is frequently used for this purpose, although reflective cones and pyramids have also been used. When the viewer finally perceives the undistorted image, it will invariably be with surprise and delight.

Leonardo Da Vinci was the first to experiment with anamorphic perspective, and an eye he created in 1485 is the first known example of an anamorphic drawing. During the Renaissance, artists experimenting with perspective made great advances and perfected the techniques of using geometric perspective to stretch and distort images in different ways.

Anamorphic images became extremely popular In the 16th, 17th and 18th centuries, providing the perfect way to disguise controversial political statements, heretical ideas and even erotic images.

THE AMBASSADORS — 1533

Hans Holbein (1497–1543), the great court painter to Henry VIII, created perhaps the most famous anamorphic painting, *The Ambassadors*. Can you tell what the strange image at the bottom is? The reason why Holbein included the image is still unclear.

91 CHALLENGE ●●● ○ ○ ○
REQUIRES 🧠 ✏️
COMPLETED ○

ANAMORPHIC DISTORTED GRID IMAGE

What is the strange looking, distorted creature? When you place a cylindrical mirror on the black circle the real image will be revealed. What is it? Can you draw the undistorted image from the circular grid into the regular square grid? (from "The Magic Cylinder" of the author).

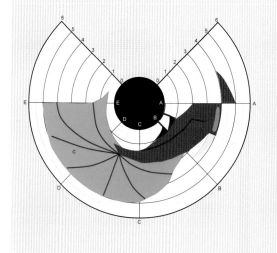 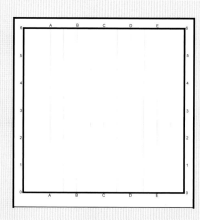

92 CHALLENGE ●●● ○ ○ ○
REQUIRES 🧠 ✏️
COMPLETED ○

ANAMORPHIC PYRAMID

Copy, cut and construct the square pyramid with the strange pattern. When you're finished, what will you see when you look at the pyramid from above?

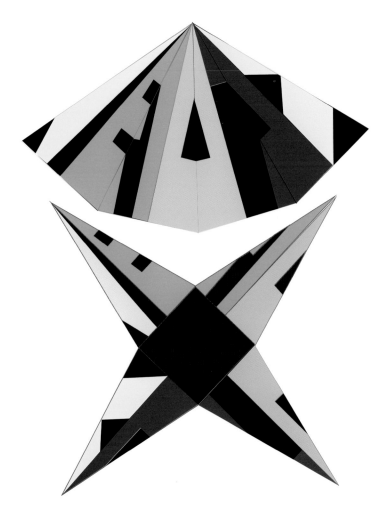

INTERRUPTIONS

Can you read the word?

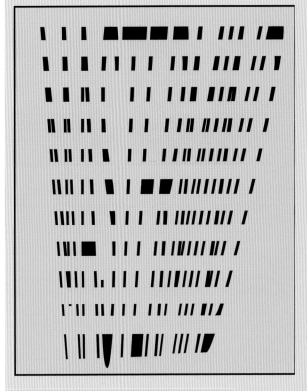

VOLTAIRE'S MESSAGE

The great French satirist Voltaire (1694–1778) loved puzzles and created many brainteasing riddles. Can you decipher his hidden message on the right?

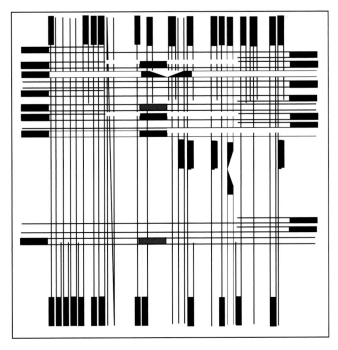

MAGIC SQUARES

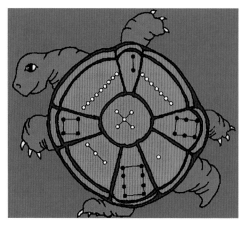

2	9	4
7	5	3
6	1	8

Lo-Shu, the oldest magic square, dates back to 2200 BC.

Arrangements consisting of numbers in square pattern so that the sum of the numbers in each row, column, and sometimes diagonal is the same are called magic squares.

The oldest known example of a magic square is the Chinese Lo-Shu from around 2200 BC. It is a magic square of order-3, and its pattern is unique. No matter how you try, you will not be able to make a different three-by-three magic square using the first nine integers.

Chinese myth tells that Emperor Yii saw a divine tortoise swimming in the Yellow River with a curious pattern on its shell, of circular dots of figurate numbers, set out in a three-by-three nine grid pattern so that the sum of the numbers in each row, column and diagonal was always 15.

The sum of digits in Lo-Shu adds up to 45, which divided by 3 gives the "magic constant" of 15. In general, this constant for any "order-n" magic square can easily be found using the formula: $n(n^2 + 1)/2$.

There are eight possible triads in Lo-Shu that add up to 15:
9+5+1 9+4+2 8+6+1 8+5+2 8+4+3 7+6+2
7+5+3 6+5+4

The center digit belongs to four lines. Five is the only digit to appear in four triads and therefore it must be the center digit. Digit 9 is in only two triads. Therefore it must go in the middle side cell which gives us the complete middle column: 9+5+1. Three and 7 are also in only two triads. The remaining four numbers can fit in in only one way — proving elegantly the uniqueness, not counting rotations and reflections, of the Lo-Shu pattern.

LATIN MAGIC SQUARES — SUDOKU

Sudoku is one of the most popular puzzles today and has revived an interest in magic squares and Latin squares. A Latin square is a square grid containing symbols arranged in such a way that each symbol occurs only once in each row or column. A completed three-by-three Latin square is shown on the left. On the right you find a partially filled magic square. The sum of its numbers in any row, column or diagonal should be the same. Can you complete the magic square?

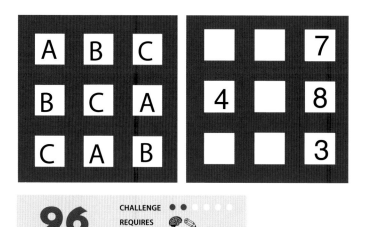

A	B	C
B	C	A
C	A	B

		7
4		8
		3

96 CHALLENGE ●● ○○○○
REQUIRES
COMPLETED

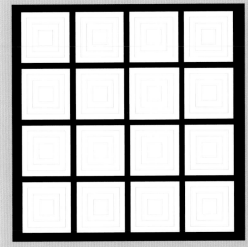

The game board

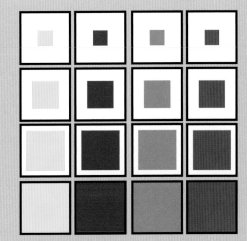

The 16 playing pieces: Four different sizes, each set in four different colors

GRAECO-LATIN MAGIC SQUARES

In the last years of his life, Euler extended the basic Latin magic squares concept into Orthogonal Latin square, also called Graeco-Latin or Euler Magic Squares.

Such squares consist of two or more superimposed Latin squares, so that each cell contains one element of each square, and each row and column contains one element of both squares only once. No two cells contain the same ordered pair of symbols. Euler knew that there was no Graeco-Latin square of order 2. After numerous experiments, he conjectured that there were no Graeco-Latin squares for orders of the form 4k + 2, k = 0, 1, 2, …French mathematician Gaston Tarry strengthened Euler's conjecture in 1901 by demonstrating that no Graeco-Latin square of order 6 exists. In 1959 however, Parker,

Bose and Shrikhande discovered a way to construct an order 10 Graeco-Latin square, and provided a construction an order 10 Graeco-Latin square, and provided a construction for the remaining even values of n that are not divisible by four. Latin squares have been known to predate Euler. A four-by-four Graeco-Latin square was published by Jacques Ozanam in 1725 as a puzzle involving playing cards, proving that Euler's square exists for all n, except for n=2 and n=6.

Can you arrange the 16 color squares in the four-by-four square game board, so that they form more than just a "magic" color square, containing 16 perfect four-color, four-size configurations as illustrated in the six different black-and-white patterns shown on the right.

1. 4 vertical columns;
2. 4 horizontal rows;
3. 2 main diagonals;
4. 4 corner squares;
5. 4 center squares;
6. 4 squares in each quarter.

The puzzle has a total of 1152 solutions. Find one!

97 CHALLENGE ● ● ● ● ● ●
REQUIRES 🧠 ✏️
COMPLETED ○

THREE LATIN SQUARES OF ORDER 4

Three Latin magic squares (small, medium and large) superimposed form a Graeco-Latin magic square, or Orthogonal Latin Square, of order 4. The big Latin square is given solved. Can you complete the pattern to form the Graeco-Latin magic square?

98 CHALLENGE ● ● ● ● ● ●
REQUIRES 🧠 ✏️
COMPLETED ○

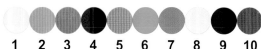

1	**2**	**3**	**4**	**5**	**6**	**7**	**8**	**9**	**10**

GRAECO-LATIN MAGIC SQUARE OF ORDER 10

In 1959, a computer was programmed to search for order-10 Graeco-Latin magic squares. It searched for 100 hours and found none.

This was not a complete surprise, since a complete search would have taken more than 100 years. The computer's inability to find a solution appeared to confirm that no order-10 square could exist. However, in 1960 a new approach was invented to try to find such squares which, surprisingly, produced a wealth of squares of order-10, 14, 18, and more.

Here is one of those newly found squares in which numbers from 1 to 10 were substituted by ten colors. The beautiful pattern is a classic image of recreational mathematics.

DURER'S DIABOLIC MAGIC SQUARE — 1514

There are 880 different possible magic squares of order 4. The most famous of them is Dürer's Diabolic Magic Square. Albrecht Dürer (1471–1528) included this magic square in his famous engraving Melancholia (1514). It is called diabolic, because it is far more "magic" than the definition of a magic square requires. It contains an amazing number of different ways in which the magic constant 34 can be achieved.

There is a staggering number of 86 different ways to achieve the sum of 34, using the set of the first 16 integers. Can you complete the chart of the quartets (the 10 basic quartets of the rows, columns and two main diagonals are shown)? All of them are figured in Dürer's Diabolic Magic Square, forming different patterns, many of them symmetrical. Can you discover the 86 patterns associated with the quartets in Dürer's magic square?

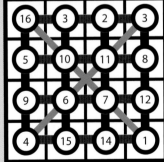

#	Quartet				#	Quartet			
1					44				
2					45				
3					46	3	6	10	15
4	1	4	14	15	47				
5					48				
6					49				
7					50				
8					51				
9					52				
10	1	7	10	16	53				
11					54				
12					55				
13					56				
14					57	4	5	9	16
15					58				
16	1	8	12	13	59				
17					60				
18					61				
19					62				
20	2	3	13	16	63				
21					64	4	6	11	13
22					65				
23					66				
24					67				
25					68				
26					69				
27					70				
28					71				
29					72				
30					73				
31					74				
32	2	7	11	14	75				
33					76				
34					77				
35					78				
36					79				
37					80				
38					81	5	8	10	11
39					82				
40					83	6	7	9	12
41					84				
42					85				
43					86				

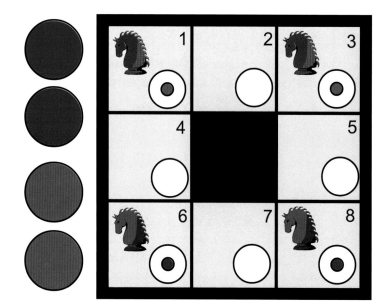

CHESS KNIGHTS EXCHANGE — 1512

In 1512 Guarini, an Italian mathematician, presented a chessboard knights puzzle in which the object is to exchange the positions of two sets of knights in the smallest possible number of moves. The knights move as in chess. What is the smallest number of moves to switch the places of the two sets of two knights?

100 CHALLENGE ● ● ● ● ● ○
REQUIRES 🧠 ✏️
COMPLETED ○

CHESS KNIGHTS EXCHANGE (2)

On the right you can find a variation of Guerini's problem involving two sets of three chessboard knights. What is the smallest number of moves to switch the places of the two sets of two knights?

101 CHALLENGE ● ● ● ● ● ○
REQUIRES 🧠 ✏️
COMPLETED ○

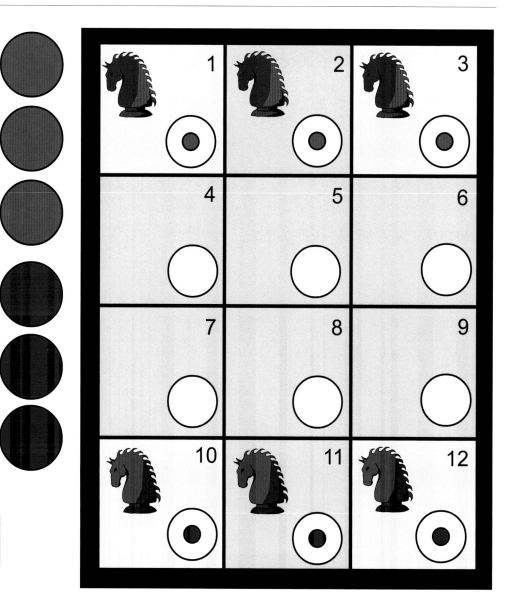

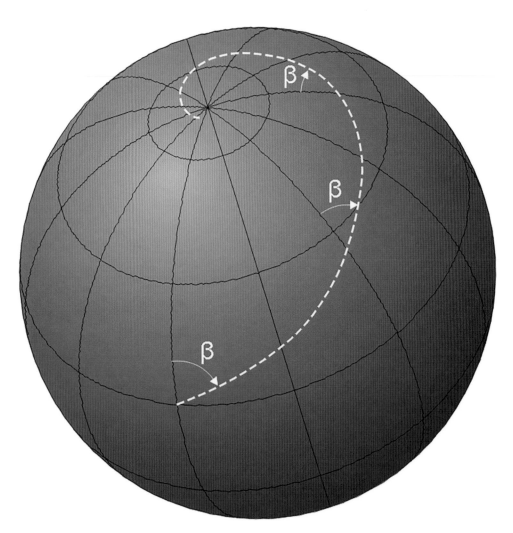

LOXODROME — 1537

A loxodrome (or rhumb line) is a term used in navigation to describe a curve cutting all meridians of longitude at a constant angle, i.e. a path taken when a compass is kept pointing in the same direction. This means taking an initial bearing relative to true (or magnetic) north and following it without deviation.

In his work "Treatise in Defense of the Marine Chart," completed in 1537, Portuguese mathematician Pedro Nunes was the first to address the possibilities of following a rhumb line course on the surface of a globe. Thomas Harriot further developed this idea in the 1590s.

In contrast to the rhumb line is the great circle, which, while it represents the shortest path between two points on the surface of a sphere, also requires frequent changes of bearing.

NORTH POLE TRIP

You start at the North Pole and are navigating your ship keeping your compass pointing to a constant direction. What will be your path on the spherical surface of earth?

102

CHALLENGE ● ● ● ● ○
REQUIRES 🧠 ✏️
COMPLETED ○

NICCOLÒ FONTANA TARTAGLIA (1499–1557)

Italian mathematician Niccolò Fontana Tartaglia worked in Venice as a bookkeeper and engineer, but is best known for his formula for solving cubic equations. His work *Quesiti et Inventioni Diverse* (1546) includes a number of recreational mathematics problems. You find one of his puzzles, "How to divide 17 horses," described below.

Among the many books Tartaglia published are the first Italian translations of Archimedes and Euclid, as well as a highly praised compilation of mathematics. He is also credited for inventing the science of ballistics, as he was the first to apply mathematics to a study of the paths of cannonballs. His work later validated by Galileo's studies on falling bodies.

103

CHALLENGE ● ● ● ● ● ●
REQUIRES 🧠 ✏️
COMPLETED ○

DIVIDING 17 HORSES — 1546

A dying man leaves 17 horses to be divided among his three sons, in proportions: 1/2 : 1/3 : 1/9. Can the brothers carry out their father's will? How?

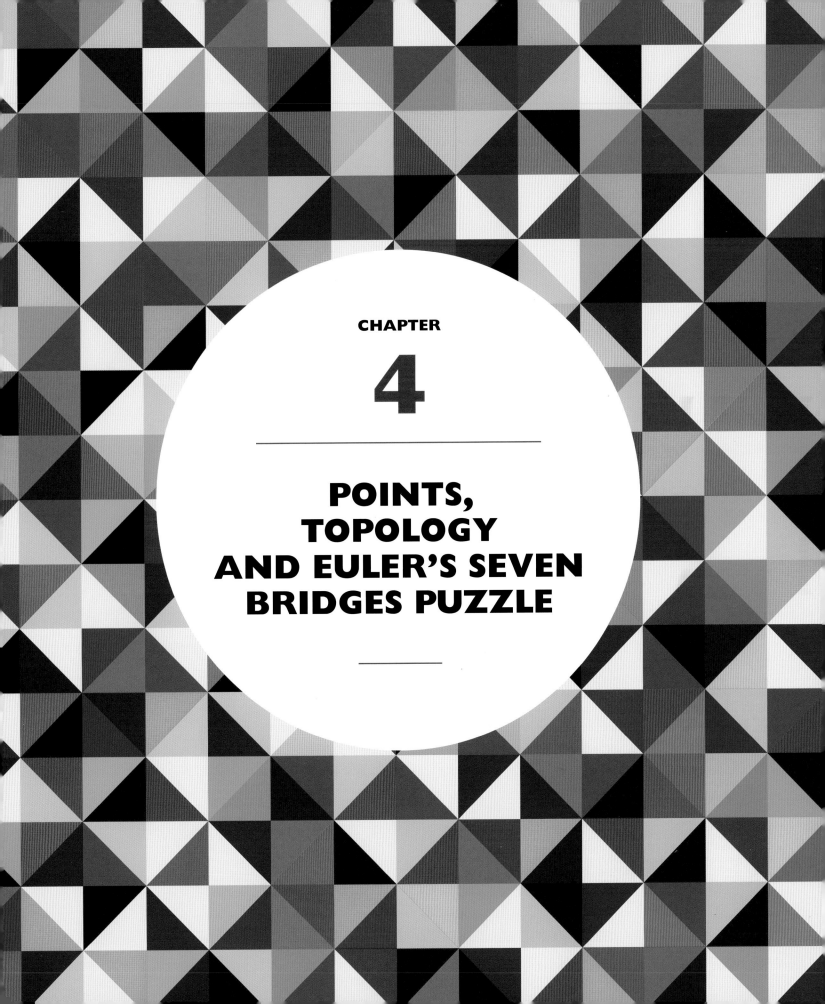

CHAPTER

4

POINTS, TOPOLOGY AND EULER'S SEVEN BRIDGES PUZZLE

A BEAUTIFUL METHOD OF CALCULATING FORCES — 1580

Starting in the 1580s with the work of Simon Stevin (1548–1620) and Galileo Galilei, many engineers transformed the principles of mechanics by translating them into mathematical form. That translation often involved devising abstract mathematical models of the physical mechanisms that embodied the maxim, like the mechanism of the parallelogram of forces.

In mathematics and physics, a parallelogram of forces is the clever method of calculating the combined effort (resultant) of two or more forces acting together on an object.

A force is a vector quantity. Since a force has both a magnitude and direction, it can be conveniently represented by a directed straight line. In physics, a tilted surface is known as an inclined plane. The forces acting on an object on an inclined plane must be analyzed. In our diagram, the two forces acting on the sides of the inclined plane are vectors acting from a point. Their size is the sum of the weights of the small metal spheres acting on the inclined plane. These forces (black) represent the force of gravity (also known as the weight) acting in a downward direction. However, any object on an inclined plane always has at least two forces acting upon it. The other force is the normal force (blue), always acting in a direction at right angles to the surface.

Using the parallelogram of force, the force of gravity is resolved into two components of force: one directed parallel to the inclined surface and the other directed perpendicular to the inclined surface.

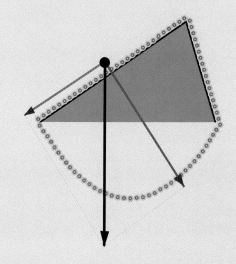

SIMON STEVIN'S WREATH OF SPHERES

Simon Stevin, a Flemish mathematician and engineer (1548–1620), and a true Renaissance man like Descartes and Galileo, is best known for his contribution to statics (the science of forces under equilibrium) and hydrostatics.

His most famous discovery was the Law of Inclined Planes, which he proved by

drawing his "Wreath of Spheres" illustration, featured as the title page of his *The Elements of the Art of Weighing* from 1615.

Stevin's law of forces on inclined planes, and more generally his vector law for the decomposition of forces (parallelogram of forces) is noteworthy as a thought experiment because it is one of the earliest examples of a law of mechanics deduced from a general physical principle, the conservation of energy.

Stevin's problem was to determine the force F needed to hold a frictionless object (with a known weight W) in place on a frictionless inclined plane.

The fundamental supposition of his law is that less weight on a steep slope can balance more weight on a gentler slope. He approached the problem by the thought experiment of his "Wreath of Spheres," which was a double inclined plane around which a loop of small connected spheres was placed as shown. His reasoning was that when the loop of the chain below the slopes is removed,

nothing changes, everything stays in equilibrium. Otherwise he would have had "something that moves," which would be a perpetual motion machine. So by leaving out the "free" spheres hanging in the air, the system remains in equilibrium. As a result of this, he realized that when weights are in equilibrium on the inclined planes, the weights of the bodies involved are proportional to the lengths of the planes.

He just had to count the number of small spaces on both sides of the inclined planes, and he had the proof of his law!

Stevin was so delighted and proud of his beautiful geometrical argument that he wrote on his frontispiece the sentence that later became his motto: "Wonder en is gheen wonder" ("What appears a wonder is not a wonder"). Equilibrium on the slopes was due to the relationship between the downward forces on either side, due to the differing angles of their support. In modern terminology, this decomposition of forces is called the parallelogram of forces.

> **"The universe cannot be understood unless one first learns to comprehend the language and interpret the characters in which it is written. It is written in the language of mathematics, and its characters are geometric figures without which it is humanly impossible to understand a single word of it: without these, one is wandering in a dark labyrinth."**
>
> — *Galileo Galilei*

GALILEO GALILEI (1564–1642)

Galileo Galilei was an Italian physicist, mathematician and astronomer, closely associated with the Scientific Revolution, a period of scientific history starting roughly in the mid-16th century. Among his many great achievements is the first systematic study of uniformly accelerated motion. Galileo's experiment-based research was a meaningful break from the abstract approach of Aristotle, symbolizing the beginning of experimental science. His enormous achievements in science were accomplished with relatively simple and crude instruments.

INCLINED PLANE AND GALILEO — 1600

How to determine the rate of acceleration due to Earth's gravity?

At the time, it was very difficult to answer that question. The problem facing Galileo when conducting his experiments was that objects in free fall are moving too quickly to be measured exactly using the available time measuring instruments. His ingenious insight was that an inclined plane would allow him to slow down the effects of gravity while still maintaining the rate of acceleration, a fact that allowed him to determine the actual rate of acceleration due to Earth's gravity.

He released a small ball at the same time as a pendulum swung. At each swing the ball would strike a small bell on its descent. Repeating his experiments, we release a ball on an inclined plane and mark its position after exactly one second of descent.

We then divide the whole length of the slope into units of this length as shown. Can you mark the positions of the ball after 2, 3, 4, 5, 6, 7, 8 and 9 seconds?

Will these marks change if the slope is steeper, or higher?

Did the ball's speed change at the end of the inclined plane by the change in the inclination of the plane?

104 CHALLENGE ● ● ● ● ● ○
REQUIRES 🧠 ✏️ ⚒️
COMPLETED

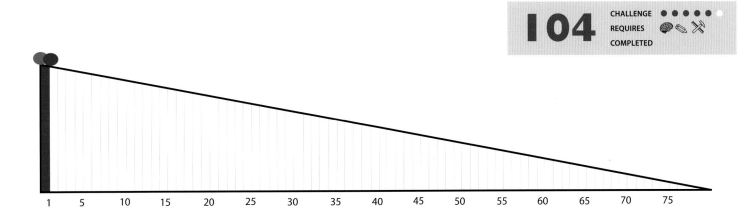

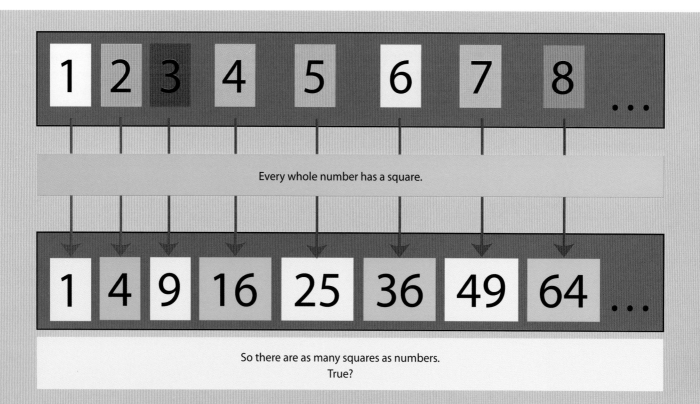

Every whole number has a square.

| 1 | 4 | 9 | 16 | 25 | 36 | 49 | 64 | ... |

So there are as many squares as numbers.
True?

GALILEO'S PARADOX — 1600

Look at the problem above. What do you think? Are there as many squares as there are numbers?

Galileo's paradox demonstrates one of the surprising characteristics of infinite sets. In *Two New Sciences*, his last scientific work, Galileo Galilei (1564–1642) made statements about the positive integers that at first sight appeared to be contradictory. He claimed that some numbers are squares, while others are not; therefore, when squares and non-squares are counted together, they must be more numerous than the squares alone. However, for every square there is just one positive number that is its square root, and for every number there is just one square; consequently, there cannot be more of one than of the other. This is an early application, though not the first, of the concept of one-to-one correspondence in the context of infinite sets.

Galileo's conclusion was that the concepts of less, equal and greater apply only to finite sets, not to infinite sets. In the nineteenth century, the German mathematician Georg Cantor applied the same methods to demonstrate that this restriction is unnecessary. Comparisons of infinite sets can be given a meaningful definition (according to which the two sets he considers, integers and squares, are of "the same size"), with some infinite sets being technically larger than others.

105

CHALLENGE ● ● ● ● ○
REQUIRES 🧠 ✏️
COMPLETED ○

GALILEO GALILEI

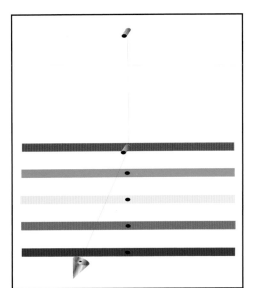

GALILEO'S PENDULUM — 1600

106   CHALLENGE ●●●●○
 REQUIRES 🧠✍️⚒️
 COMPLETED ○

Pendulums have fascinated scientists for a very long time. Galileo was the first to examine the unique characteristics of pendulums. From simple observations he came to the conclusion that pendulums can keep time, measure the force of gravity and sense relative motion.

His simple experimental setup is self-explanatory. While the pendulum is swinging, a peg is inserted into one of the holes. The use of pegs will shorten the effective length of the pendulum. How will these experiments influence the operation of the pendulum? What happens if the string of the pendulum is shortened? Does the pendulum swing faster or slower?

Does the frequency of the pendulum change as the swings become shorter?

From such simple experiments and observation, Galileo came to revolutionary conclusions and invented the pendulum clock in 1642.

Dome of the Cathedral of Pisa, with the lamp of Galileo in front, leading Galileo to invent the law of the isochrony of the pendulum

UPHILL ROLLERS — A MECHANICAL ANTI-GRAVITY PARADOX

The double cone uphill roller paradox was invented by William Leybourn (1626–1719). A land surveyor and prolific author, Leybourn published the recreational volume *Pleasure with Profit,* containing the ingenious mechanical puzzle known as the "Uphill Roller," with the double cone on two inclined rails.

Its operation is counterintuitive since the double cones, when placed on the lowest points of the double track inclined plane, roll up the incline, seemingly defying gravity. Can you explain the strange behavior of the double cones?

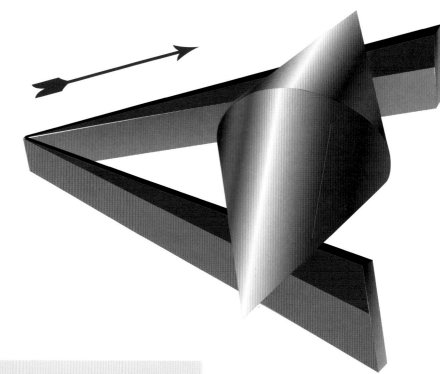

 107 CHALLENGE
REQUIRES
COMPLETED

ANTI-GRAVITY: FROM NEWTON TO EINSTEIN

The concept of anti-gravity is associated with the biggest scientific issue of all: the origin of the universe. When Albert Einstein developed his theory of general relativity, he came across a difficult question. Why hadn't gravity caused the matter in the universe to collapse inward on itself?

Isaac Newton (1642–1727) was confronted with the same problem with his own theory of gravity, explaining that God was responsible for keeping things apart. Einstein was reluctant to invoke God, and his solution was to add an anti-gravity force alongside gravity. In the 1920s, all that changed. Cosmology adopted a new view in which the universe had been created at

a finite moment, exploding and expanding from a small primeval superatom, a view that has since developed into the Big Bang theory. It did not require a belief in anti-gravity. It seemed to be correct and Einstein eventually endorsed it. But there was a twist to the tale. To their astonishment, astronomers later found that the universe is in fact accelerating and galaxies are moving apart faster and faster. Due to gravity, the expansion of the Big Bang should be slowing, and so there was a problem again. The best explanation was the existence of an anti-gravity force. It seems that even when Einstein was ready to admit he was wrong, he ended up being right after all.

ANTI-GRAVITY RAILWAY — 1829

The fascinating behavior of the anti-gravity cones of Galileo inspired a Victorian inventor in 1829 to conceive the concept of the anti-gravity railway based on its principle of motion.

Fig. 5.—SELF-MOVING RAILWAY CARRIAGE ON TRUNDLING CONES. 1829.

SANGAKU — 1603

Sangaku or San Gaku (literal translation: calculation tablet) are beautiful Japanese wooden tablets from the Edo period (1603–1867), on which we find geometrical problems or theorems. They were placed by members of all social classes at Shinto shrines or Buddhist temples as offerings or as challenges for the congregation.

During the Edo period, Japan's trade and foreign relations with Western countries were tightly regulated. That explains why the tablets used a form of Japanese mathematics, which developed in isolation from Western mathematics. For example, the relationship between differentiation and integration (the fundamental theorem of calculus) was unknown, so Sangaku problems on areas and volumes were solved by expansions in infinite series and term by term calculation. The first collection of Sangaku problems was published in 1790 by Fujita Kagen (1765–1821), a leading Japanese mathematician, in his book *Shimpeki Sampo* (Mathematical problems suspended from the Temple), followed in 1806 by the *Zoku Shimpeki Sampo*.

JAPANESE SANGAKU THEOREM

On the left you see a convex irregular octagon that is randomly drawn and triangulated. The incircles to each triangle have been constructed as shown.

The same octagon is triangulated again on the right with its incircles but in a different way. Can you work out the size relationship of the sizes of the two sets of incircles in the two triangulations?

108

CHALLENGE ● ● ● ● ○ ○
REQUIRES ✏️
COMPLETED ○

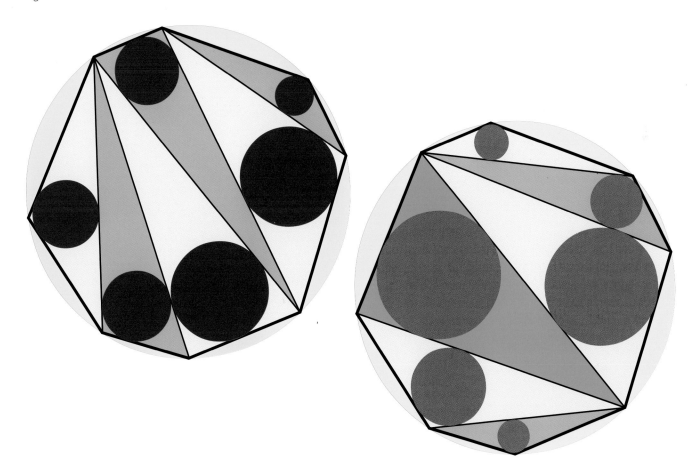

KEPLER'S CONJECTURE — 1600

Johannes Kepler (1571–1630), a German astronomer, found that there are two basic ways to arrange circles or spheres in a plane: square packing and hexagonal packing.

The Kepler conjecture, named after Johannes Kepler, is a mathematical conjecture about sphere packing in three-dimensional Euclidean space. It says that no arrangement of equally sized spheres filling space has a greater average density than that of the hexagonal close packing arrangements. These arrangements have a density of just over 74 percent.

Thomas Hales announced in 1998 that, using an approach proposed by Fejes Toth (1953), he had discovered a proof of the Kepler conjecture. Hales achieved this proof after exhaustively checking numerous individual cases using complex computer calculations. Referees say they are "99 percent certain" of the accuracy of Hales' proof, so the Kepler conjecture is now extremely close to being accepted as a theorem.

Square layers can be stacked so that the spheres are vertically above each other, or the spheres in each layer nestle into the gaps between four spheres in the layer below.

With hexagonal layers there are also two possibilities: aligned or staggered, but staggered layers of hexagonal packing lead to the same arrangement as staggered layers of square packing. If the spheres in these arrangements are allowed to expand, they form three-dimensional shapes. The cubic lattice forms cubes; the hexagonal lattice hexagonal prisms; and for the face-centered cubic lattice, Kepler's rhombic dodecahedron — the tightest possible packing.

The efficiency of packing is measured by its density (i.e., the proportion of space that is filled with spheres):

The result in a plane:
1- Square lattice	0.7854
2- Hexagonal lattice	0.9069

In three-dimensional space:
3- Cubic lattice	0.5236
4- Hexagonal close packing	0.7404
5- Dense random packing	0.64

The problem of close packing of spheres is closely related to geometric solids that can be fitted together to completely fill space. Kepler tried to obtain such solids by imagining that each packing sphere expands to fill the intermediate space.

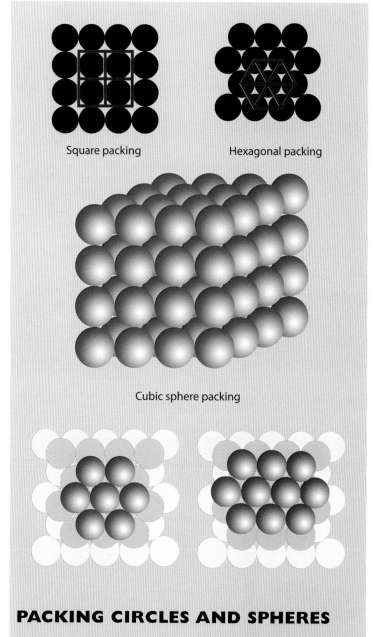

Square packing   Hexagonal packing

Cubic sphere packing

PACKING CIRCLES AND SPHERES

There are two ways of filling the plane with circles or spheres as shown above.

Cubic sphere packing: in square layers, corresponding spheres are stacked vertically above each other.

Hexagonal sphere packing: two ways to add a hexagonally packed layer. They differ in how the sheets are stacked upon one another. In hexagonal cubic packing every third layer is the same, directly above those in the first. In face centered cubic packing, every other layer is the same.

SPHERE PACKING BOX — 1600

"Once upon a time the king had all his money made into identical gold spheres. He packed the money into a big chest tightly. He knew that the chest was full, because it didn't rattle. Soon the queen took out some money and still the chest didn't rattle. Then the treasurer took out some more money and still the chest didn't rattle. Then the prime minister took out some more money and still it didn't rattle…"

Could the story of the king's treasure be true? There are 23 golden spheres in the rectangular chest, in a tight fitting arrangement. How many spheres can be removed from the chest so that you can still have the remainder to make a tight fit? Tight in the sense that each sphere is securely held by adjacent touching spheres so that it can't be moved from its place.

109

CHALLENGE ● ● ● ● ● ○
REQUIRES 🧠 ✏️ ⚒️
COMPLETED ○

PACKING 105 SPHERES IN A SQUARE

If the diameter of the packing spheres is 1 unit, you can easily pack 100 circles in a square with 10-unit sides. You can do better by packing the circles in a hexagonal array in which you can pack 105 circles as shown.

But can you do even better?

110

CHALLENGE ● ● ● ● ○
REQUIRES 🧠 ✏️ ⚒️
COMPLETED ○

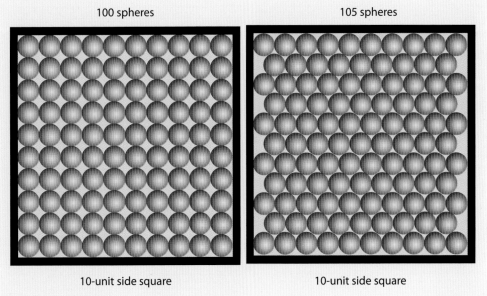

100 spheres

105 spheres

10-unit side square

10-unit side square

BACHET'S WEIGHT PROBLEM — 1612

In 1612, French scholar Claude-Gaspar Bachet de Mézeriac (1581–1638) published an early collection of puzzles titled *Problèmes plaisans et delectables qui se font par les nombres* ("Amusing and delightful number problems"), which became the forerunner for all later books on recreational mathematics, and has so far been published in five editions. The emphasis was on arithmetical rather than geometrical puzzles, including classic problems like: think-of-a-number, River crossing, magic squares, Josephus, weighing, liquid pouring and other puzzles.

Bachet's book included the classic weights problem. W. Rouse Ball attributes the first recording of this problem to Bachet in the early 17th century, calling it "Bachet's Weights Problem." However, Bachet's problem can be dated as far back as Fibonacci in 1202, making it a plausible candidate to be the original integer partition problem.

Here's the famous problem: suppose you need to have the ability to weigh any integral weight from 1 to 40 kilogram on a scale. Can you work out the minimum number of weights you must have if the weights are allowed to be placed on one side of the scale, and also when the weights are allowed to be placed on both sides of the scale?

PUZZLE 1
weights are allowed only on one side of the scale balance

1 =	21 =
2 =	22 =
3 =	23 =
4 =	24 =
5 =	25 =
6 =	26 =
7 =	27 =
8 =	28 =
9 =	29 =
10 =	30 =
11 =	31 =
12 =	32 =
13 =	33 =
14 =	34 =
15 =	35 =
16 =	36 =
17 =	37 =
18 =	38 =
19 =	39 =
20 =	40 =

PUZZLE 2
weights are allowed on both sides of the scale balance

1 =	21 =
2 =	22 =
3 =	23 =
4 =	24 =
5 =	25 =
6 =	26 =
7 =	27 =
8 =	28 =
9 =	29 =
10 =	30 =
11 =	31 =
12 =	32 =
13 =	33 =
14 =	34 =
15 =	35 =
16 =	36 =
17 =	37 =
18 =	38 =
19 =	39 =
20 =	40 =

WEIGHING THREE WEIGHTS

You have three identical boxes of different weights. With one set of scales, how many weighings will you need to arrange the three boxes from the lightest to the heaviest?

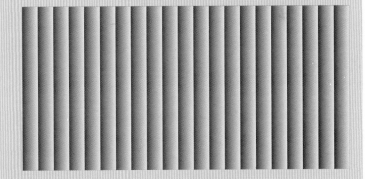

WEIGHING 21 WEIGHTS

You have 21 identical rods, one of which is slightly heavier than the rest. With one balance scale, how many weighings will you need to identify the heavier box?

112 CHALLENGE ● ● ● ● ○
REQUIRES 🧠 ✏️ ⚒️
COMPLETED

113 CHALLENGE ● ● ● ● ○
REQUIRES 🧠 ✏️ ⚒️
COMPLETED

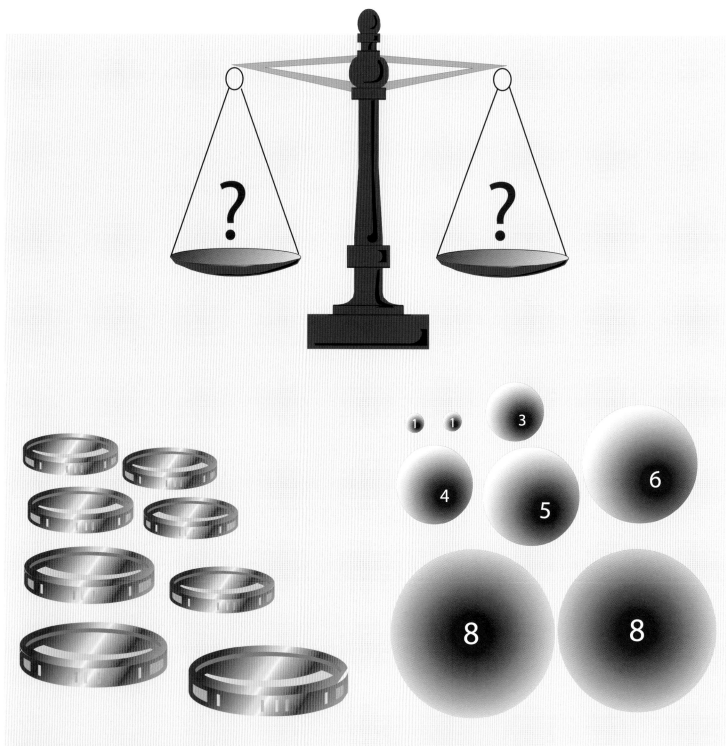

WEIGHING EIGHT COINS

Suppose you have eight gold coins including one false and assumed lighter than the rest, which are all of equal weight. What is the minimum number of weighings on the scale without weights to find the false coin?

WEIGHT SORTING

You have a set of steel balls with unit radii as shown above. Can you divide the set in two groups of the same weight?

114 CHALLENGE ●●●●●
REQUIRES
COMPLETED

115 CHALLENGE ●●●●○
REQUIRES
COMPLETED

116

CHANCE BALANCE

How many ways can you find to arrange the five weights so that the scale is in equilibrium when you remove the two cylindrical supports?

Remember that the further a weight is from the fulcrum, the more force it exerts. So a weight over the number 2 on the scale would exert twice as much force as the same weight over the number 1 distance.

If you place the weights on the scale at random, what is the probability that they will be in equilibrium?

And how many ways can you find to arrange the six weights so that the scale is in equilibrium when you remove the two cylindrical supports?

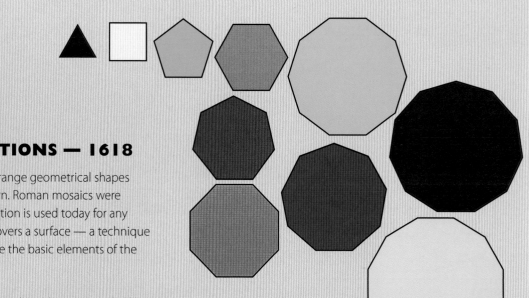

REGULAR TESSELLATIONS — 1618

To "tessellate" in general means to arrange geometrical shapes to cover the plane in a mosaic pattern. Roman mosaics were called tesserae, and the term tessellation is used today for any pattern of shapes that completely covers a surface — a technique of space-filling. Plane tessellations are the basic elements of the three-dimensional polyhedra.

A regular tessellation is made up of regular polygons exactly alike and completely filling the plane.

There is an infinite number of regular polygons, starting from the equilateral triangle, square, pentagon, hexagon, heptagon, octagon, etc. up to the circle, which can be considered a regular polygon having an infinite number of sides.

One of the most astonishing counter-intuitive facts of geometry is the small number of regular tessellations.

Surprisingly enough, the only edge-to-edge tessellations by regular polygons are the three regular tessellations. They have as their elements equilateral triangles, squares and regular hexagons.

There is a beautiful geometrical logic behind the rarity of regular tessellations. Since their basic elements are regular polygons, one condition must be fulfilled and that is: at every meeting point of such polygons (vertices), the sum of all their angles must be 360 degrees. In a regular (equilateral) triangle, the angles are 60 degrees, therefore exactly six such triangles can meet in a vertex. In a square, the angles are 90 degrees; therefore exactly four squares can meet in a vertex. In a regular hexagon, the angles are 120 degrees; therefore exactly three such hexagons can meet in a vertex.

No other regular polygon, no matter how many sides it has, can tessellate the plane in a regular fashion — there are only three regular tessellations.

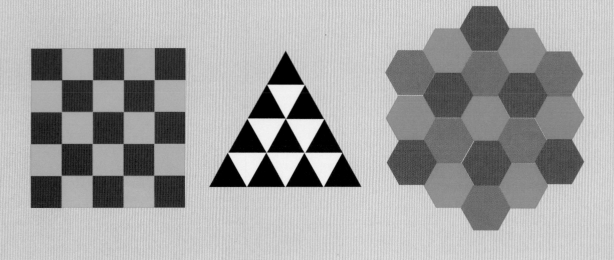

3.4.3.3.4

3.3.3.4.4

3.3.3.3.6

3.6.3.6

3.4.6.4

3.8.8

4.6.12

3.12.12

SEMIREGULAR TESSELLATIONS — 1618

There are only eight semiregular tessellations. They are made up of only five different regular polygons as shown: triangles, squares, hexagons, octagons and dodecagons. As with regular tessellations, a surprisingly small number. Johannes Kepler (1571–1630) and his successors were pioneering research on mosaics and tessellations, which are useful tools not only in recreational mathematics, but also in crystallography, coding theory, cellular structure, etc.

Semiregular tessellations are those tessellations in which two or more kinds of regular polygons are fitted together to cover the plane in such a way that the same polygons, in the same relative positions, surround every vertex (corner point) — or, to put it in mathematical terms, so that every vertex is congruent to every other vertex.

This information can be conveniently expressed by the Schläfli symbols (for example, {3,12,12} means that at every vertex

there are one triangle and two dodecagons in the same clockwise cyclic order).

We have to find which combinations of regular polygons can fill the 360 degrees around a single vertex. Combinations of angles that can do this are called "vertex pictures." This is the basic condition to create any kind of tessellation.

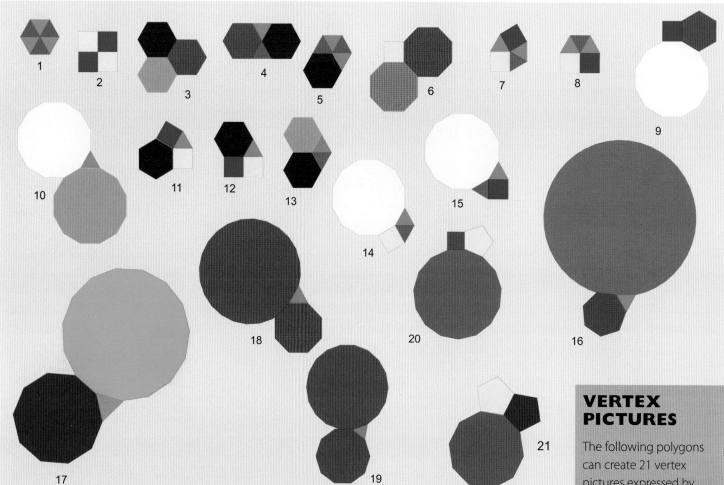

TESSELLATIONS AND SCHLÄFLI SYMBOLS

Johannes Kepler (1580–1630), widely renowned for his work in astronomy, was also deeply interested in geometric tessellations and polyhedra. He included a variety of tilings of regular and star polygons in his work *Harmonices Mundi* (1619).

"The chief aim of all investigations of the external world should be to discover the rational order and harmony which has been imposed on it by God and which He revealed to us in the language of mathematics," he stated.

If we omit the uniformity restriction that every vertex must be equal to every other vertex (regular tessellations), we can create additional groups of tessellations. For all these, there is a basic requirement: the regular polygons around their vertices must form complete vertex pictures, and the sum of their angles must be 360 degrees.

How many complete vertex pictures can we find?

A systematic procedure will produce only 21 different complete vertex pictures or vertex figures as shown, represented by Schläfli symbols. This is a surprisingly small number taking into consideration the infinite number of regular polygons.

These are the basic requirements for possible tessellations, but they are not sufficient.

Only some combinations of complete vertex pictures can be extended to form groups of tessellations: vertex pictures 1, 2 and 3 form the three regular tessellations. Vertex pictures from 4 to 11 form the eight semiregular tessellations. Different combinations of two or three vertex pictures will form at least 14 demiregular tessellations. Tessellations of regular polygons with more than three different vertex pictures are infinite in number.

VERTEX PICTURES

The following polygons can create 21 vertex pictures expressed by Schläfli symbols:

1- 3.3.3.3.3.3
2- 4.4.4.4
3- 6.6.6
4- 3.6.3.6
5- 3.3.3.3.6
6- 4.8.8
7- 3.4.3.3.4
8- 3.3.3.4.4
9- 4.6.12
10- 3.12.12
11- 3.4.6.4
12- 3.4.4.6
13- 3.3.6.6
14- 3.3.4.12
15- 3.4.3.12
16- 3.7.42
17- 3.9.18
18- 3.8.24
19- 3.10.15
20- 4.5.20
21- 5.5.10

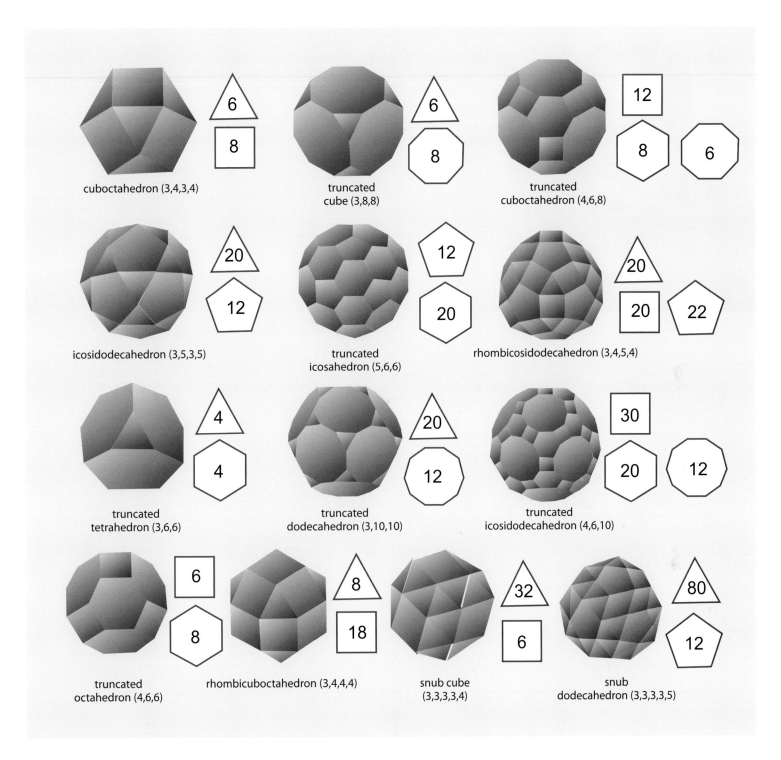

cuboctahedron (3,4,3,4)
6
8

truncated cube (3,8,8)
6
8

truncated cuboctahedron (4,6,8)
12
8
6

icosidodecahedron (3,5,3,5)
20
12

truncated icosahedron (5,6,6)
12
20

rhombicosidodecahedron (3,4,5,4)
20
20
22

truncated tetrahedron (3,6,6)
4
4

truncated dodecahedron (3,10,10)
20
12

truncated icosidodecahedron (4,6,10)
30
20
12

truncated octahedron (4,6,6)
6
8

rhombicuboctahedron (3,4,4,4)
8
18

snub cube (3,3,3,3,4)
32
6

snub dodecahedron (3,3,3,3,5)
80
12

ARCHIMEDEAN SOLIDS — SEMIREGULAR POLYHEDRA

Convex semiregular polyhedra, or Archimedean solids, are beautiful solids composed of regular polygons that can be different, while their vertices are identical.

There are exactly 13 different semiregular polyhedra. These solids were first described by Archimedes. They were rediscovered during the Renaissance and Kepler reconstructed the complete set in 1619, most of which still has a lot of unexplored possibilities in the game and puzzle fields. The notations for the truncated tetrahedron (3,6,6), for example, mean that each vertex contains a triangle, hexagon, hexagon, in that cyclic order. Two of the solids, the snub cube and snub dodecahedron, come in two handednesses (mirror-images), or two enantiomorphs.

CAVALIERI'S PRINCIPLE — 1630

Italian mathematician Bonaventura Francesco Cavalieri is renowned for his work on the problems of optics and motion and the precursors of infinitesimal calculus. He also introduced the logarithms to Italy. Cavalieri's principle in geometry partially anticipated integral calculus.

The principle establishes the fact that the volume of a pyramid, irrespective of the shape of its base, is (1/3) × base × height.

117

VOLUME OF CONES AND PYRAMIDS

The three cones and three pyramids have the same base and height as the cylinder and are filled with water. The water is poured from the cones into the cylinder and from the pyramids to the prism.

How much of the cylinder and prism will be filled with water?

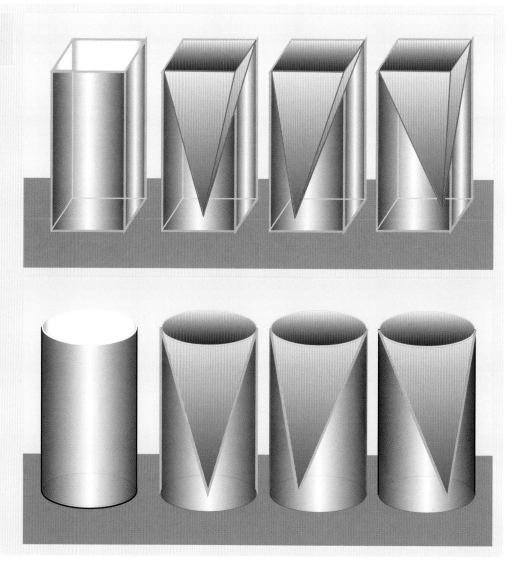

DESARGUES'S THEOREM — 1641

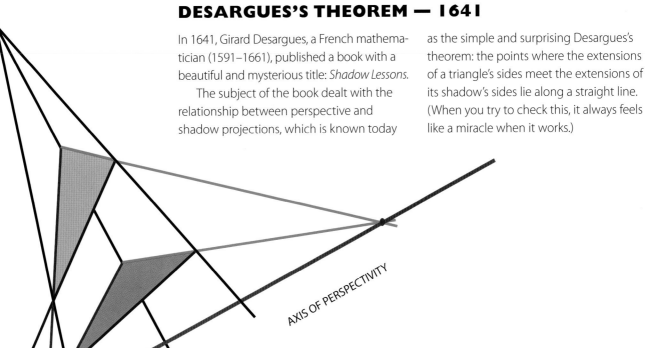

In 1641, Girard Desargues, a French mathematician (1591–1661), published a book with a beautiful and mysterious title: *Shadow Lessons*.

The subject of the book dealt with the relationship between perspective and shadow projections, which is known today as the simple and surprising Desargues's theorem: the points where the extensions of a triangle's sides meet the extensions of its shadow's sides lie along a straight line. (When you try to check this, it always feels like a miracle when it works.)

AXIS OF PERSPECTIVITY

118

CHALLENGE ● ● ● ○ ○ ○
REQUIRES 🪨 ✏️
COMPLETED ○

(IM)POSSIBLE SHADOWS

The lower sides of the four triangles sit on the ground. Are all of the shadows equally possible?

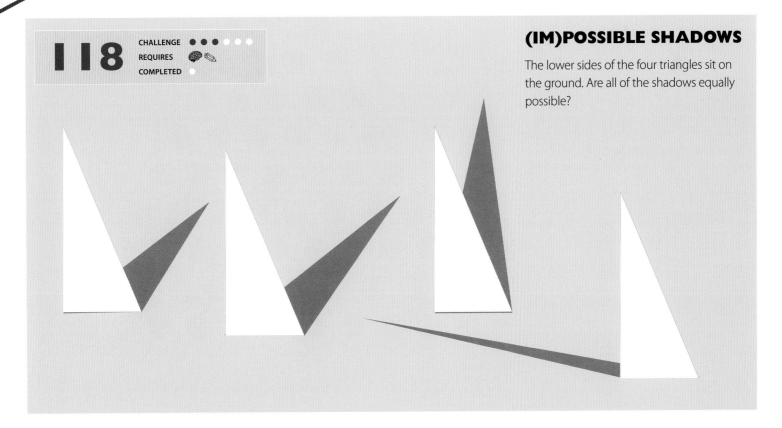

BLAISE PASCAL (1623–1662)

Blaise Pascal was a French mathematician, physicist, inventor, writer and Catholic philosopher.

Pascal's father, a tax collector in Rouen, was greatly involved in the education of his son. Pascal first focused on the natural and applied sciences, in which he made a significant contribution to the study of fluids. He also studied the work of Evangelista Torricelli and further defined the concepts of pressure and vacuum. In 1642, while still a teenager, he began some ground-breaking work on calculating machines to assist his father's work, and eventually invented the mechanical calculator, called the Pascaline, building 20 in total. Pascal suffered from poor health, particularly after reaching adulthood, and died just two months after his 39th birthday.

PASCAL'S TRIANGLE

One of the most beautiful and useful number patterns in mathematics is the famous Pascal's triangle. It is shown here with its first 10 rows. Can you discover the pattern of its construction and add further rows?

The number pattern was originally invented by the ancient Chinese, but it was Blaise Pascal who discovered its numerous patterns and uses, which made Pascal's triangle one of the most important tools in many fields of mathematics. The numbers in Pascal's triangle indicate the number of possible routes to get to that point from row zero.

119

CHALLENGE

REQUIRES

COMPLETED

Fibonacci numbers are formed by adding the digits along these lines.

Natural numbers

Triangular numbers

Tetrahedral numbers

Pyramidal numbers

$(a + b)^0$ row 0

$(a + b)^1$ row 1

$(a + b)^2$ row 2

$(a + b)^3$ row 3

$(a + b)^4$ row 4

$(a + b)^5$ row 5

$(a + b)^6$ row 6

$(a + b)^7$ row 7

$(a + b)^8$ row 8

$(a + b)^9$ row 9

row 10

1
1   1
1   2   1
1   3   3   1
1   4   6   4   1
1   5   10   10   5   1
1   6   15   20   15   6   1
1   7   21   35   35   21   7   1
1   8   28   56   70   56   28   8   1
1   9   36   84   126   126   84   36   9   1
1   10   45   120   210   252   210   120   45   10   1

Fibonacci: 1, 1, 2, 3, 5, 8, 13, 21

$$(a + b)^{10} = 1a^{10} + 10a^9 b + 45a^8 b^2 + 120a^7 b^3 + 210a^6 b^4 + 252a^5 b^5 + 210a^4 b^6 + 120a^3 b^7 + 45a^2 b^8 + 10a\, b^9 + 1b^{10}$$

PRINCE RUPERT'S PROBLEM — 1650

As a founding member of the Royal Society, Prince Rupert of the Rhine (1619–1682), devised a question that remained unanswered for 100 years and became known as Prince Rupert's problem — could a hole be cut in a cube so that another cube the same size or larger could pass through?

The mathematics of cubes passing through cubes was considered by John Wallis. Later, in 1816, a solution was published posthumously by the Dutch mathematician Pieter Nieuwland (1764–1794) to the question of what is the largest cube that can be passed through a cube of unit side.

Nieuwland answered this by finding the largest square that fits inside a unit cube. When viewed from directly above one apex, a unit cube has the outline of a regular hexagon of side $\sqrt{3}/\sqrt{2}$.

The largest square that will go into a cube has a face that can be inscribed within this hexagon; the length of its edge is $3\sqrt{2}/4 = 1.0606601\ldots$

Thus, curiously enough the cube is slightly larger than the original cube.

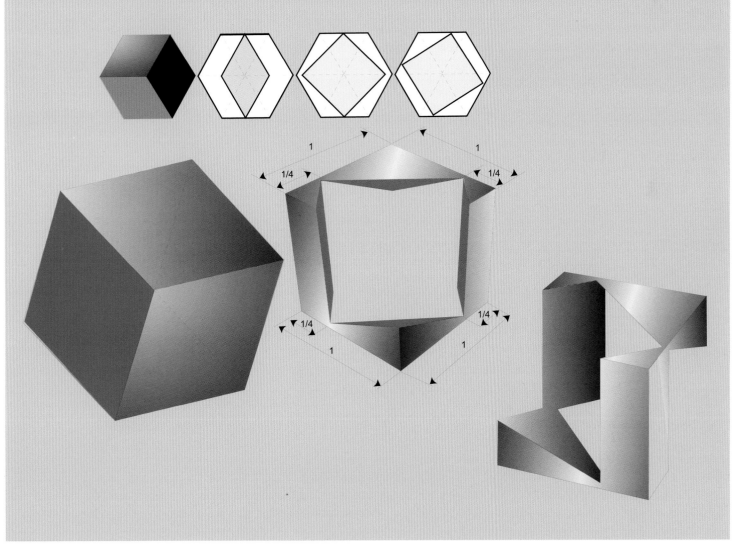

COMBINATORICS — COUNTING MADE EASY

Combinatorial problems have attracted mathematicians since antiquity. Magic squares were mentioned in the *I Ching,* a Chinese book that dates back to the 12th century BC. Pascal's triangle (not using that name, obviously) was taught in 13th century Persia.

In the West, combinatorics began in the 17th century with Blaise Pascal and Pierre de Fermat in connection with their development of the theory of probability, and later by the work of Gottfried Wilhelm Leibniz. Leonhard Euler was responsible for the de-velopment of school combinatorial mathe-matics in the 18th century. It was solving the Konigsberg bridge problem that made him the father of graph theory. Many combina-torial problems were presented as recreatio-nal problems in the 19th century (such as the problem of eight queens or the Kirkman school girl problems).

Percy Alexander MacMahon's 1915 book *Combinatory Analysis* is one of the earliest books on combinatorics.

Combinatorics is a branch of modern mathematics whose name is derived from the fact that it studies the ways in which numbers and objects can be combined.

Probability, computer theory and many everyday situations depend on the princi-ples of combinatorics, specifically on com-binations and permutations. The number of possible arrangements in a system may seem small at first, but possibilities rise quickly with the number of elements and soon become impossibly large.

The basic instance is simplicity itself.

One object by itself can be arranged in just one way:

Two objects, (a and b) can be arranged as ab or ba for a total of two permutations:

Three objects — a, b and c — can be arranged in six ways: abc, acb, bac, bca, cab, cba.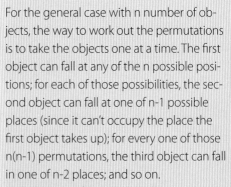

For the general case with n number of ob-jects, the way to work out the permutations is to take the objects one at a time. The first object can fall at any of the n possible posi-tions; for each of those possibilities, the sec-ond object can fall at one of n-1 possible places (since it can't occupy the place the first object takes up); for every one of those n(n-1) permutations, the third object can fall in one of n-2 places; and so on.

In general, for n objects there are n times as many more permutations as there are in systems with only n-1 objects. For example, there are four times as many pos-sible permutations in a system with four ob-jects than there are in a system with three — in other words, 24 permutations. There are 5 x 24, or 120, different ways to arrange five things and 6 x 120, or 720, ways to ar-range six things.

These numbers are called factorials and are designated with a!, as in 6!, or six facto-rial, which equals 720.

Therefore, the general formula for the total number of possible orders of n things, or permutations, is

$$P = n! = n \times (n-1) \times (n-2) \times (n-3) \times ... \times 3 \times 2 \times 1$$

This number becomes very large very rapidly. What about cases that do not deal with or-dering one group but with finding the per-mutations of n things taken k at a time? The mathematics here is only a bit trickier. Say you wanted to know how many ordered groups of three can be made from five dif-ferent elements (such as color, or letters or something else). You would calculate:

$$P_k = n! / (n-k)! = 5! / (5-3)! = 120/2 = 60$$

Sometimes we are not concerned about the order of the things (permutations), but are only interested in the constitution (the num-ber of choices) of the sample in question (combinations). A combination is a set of things chosen from a given group, when no significance is attached to the order of the things within the set. Therefore, the general formula for the total number of combina-tions is:

$$C = n!/k!(n-k)! = 5!/3!(5-3)! = 10$$

which in our specific case would produce 10 groups of the elements in each (regardless of the order of the element in each group).

> ## "Research is what I'm doing when I don't know what I'm doing."
>
> — *Werner von Braun*

Up to this point we have been dealing with objects that are all different. Some-times it may happen that there is a number of identical things with "a" of one sort, "b" of another, etc. In this case the number of per-mutations is:

$$P_{a,b,c} = n!/a!b!c!$$

Most probabilities relating to games and puzzles can be determined by counting the total number of possibilities and the num-ber of outcomes having some desired prop-erty. The ratio of these two numbers gives the probability. The formulas for permuta-tions and combinations facilitate and short-en the counting.

The values for the number of combina-tions of n elements taken k at a time can be obtained from the well-known Pascal's triangle.

VIVIANI'S THEOREM — 1660

Viviani's theorem is named after Vincenzo Viviani. It states that the sum of the distances from any interior point to the sides of an equilateral triangle equals the length of the triangle's altitude.

Vincenzo Viviani (1622–1703) was an Italian mathematician and scientist. He was a pupil of Torricelli. In 1639, at the age of 17, he was an assistant of Galileo Galilei.

In 1660, Viviani conducted an experiment in collaboration with Giovanni Alfonso Borelli to determine the speed of sound, which involved calculating the time between seeing the flash and hearing the sound of a cannon shot at a distance. They calculated a value of 350 meters per second (m/s).

In 1661 he experimented with the rotation of pendulums, almost two centuries before Foucault's famous demonstration.

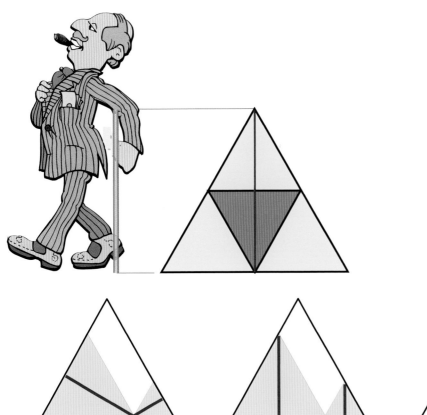

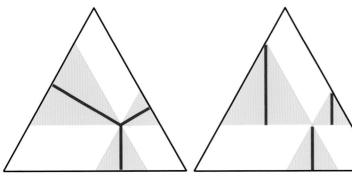

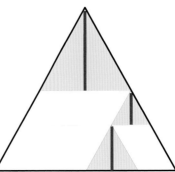

Viviani's theorem proof

BROKEN STICK

If a walking stick is broken at random into three pieces, what is the probability that the pieces can be put together in a triangle?

A remarkable property of the equilateral triangle is behind this problem as expressed by the theorem attributed to Viviani. The perpendiculars to the sides from a point (Fermat's point) within an equilateral triangle add up to the height of the triangle. In 2005, Kawasake used rotation to prove the theorem as shown. The equilateral triangle will be helpful in solving this classic probability problem. Its height is equal to the length of the stick.

120

CHALLENGE ● ● ● ● ○ ○
REQUIRES
COMPLETED ○

HOLE IN EARTH — THE NAPKIN RING PROBLEM

Marilyn vos Savant, famous for her Monty Hall problem published in her "Ask Marilyn" column in *Parade magazine,* reported another challenging problem. The object is to drill a 6-inch hole through a given sphere. Can you achieve that with a six-inch diameter sphere?

An early study of this problem was researched by the 17th-century Japanese mathematician Seki Kawa. The hole is a circular cylinder of empty space which has a height of six inches.

We can already conclude that it is impossible to drill a six-inch hole through a sphere with a diameter of six inches.

To drill a six-inch long hole in a much bigger sphere you must have a very thick drill that will remove two caps of the sphere and a big part of its volume, leaving only a curved cylindrical ring, like a napkin-ring as shown, of six-inch height.

What will be the volume of this ring — the remaining portion of the spheres, including that of a giant sphere like the Earth? The accompanying drawings may give you visual clues as to the answer.

The interesting result is that the volume of the band depends on "h" but not on "R." As the radius R of the sphere shrinks, the diameter of the cylinder must also shrink in order for h to remain the same. The band becomes thicker, increasing the volume. However, its circumference shortens, and this reduces the volume. These two effects cancel each other out. The most extreme case, involving the smallest possible sphere, is that in which the sphere's diameter is identical to the height h. In that case, the band's volume is the volume of the entire sphere.

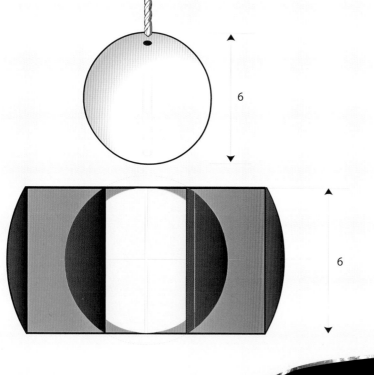

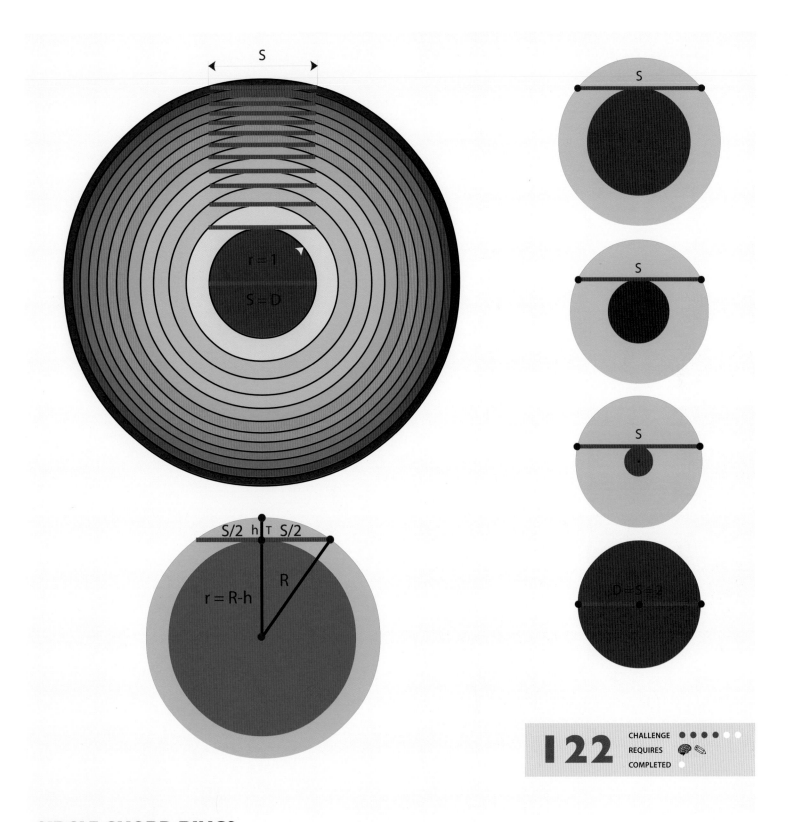

CIRCLE CHORD RINGS

Chords of length S of the bigger circles are tangent to the smaller circles, touching them in one point T. The problem is to find the area of the 12 rings around the middle circle of radius 1 above, and also of the three light blue rings on the right.

Do you think there is enough information to calculate the areas of the rings? A hint: the Pythagorean theorem can be of help!

122 CHALLENGE ●●●●○○
REQUIRES 🧠 ✎
COMPLETED ○

THE BINARY NUMBER SYSTEM AND THE LANGUAGE OF THE COMPUTER

The simplest possible number system is the binary system based on successive powers of 2.

Some primitive tribes use binary counting systems, and ancient Chinese mathematicians were aware of the binary system, but it was fully developed by the renowned German mathematician Gottfried von Leibniz (1646–1716) who described the binary system in his article *Explication de l'Arithmetique Binaire*.

Leibniz was fascinated by the binary system, which to him symbolized a metaphysical truth. The power of just 0s and 1s is sufficient to describe any number and according to him the entire world became possible by a binary split between being and nothingness. Until Leibniz, no one ever realized that two digits — 0 and 1 — are all that is really needed to create a fully-working positional number system.

In 1666, Leibniz believed that a purely mathematical approach to logic could be achieved by his binary system (0-false; 1-true), an idea which was ignored by his contemporaries, so he abandoned it. Ten years later he was introduced to the Chinese *I Ching*, and he enthusiastically revived the idea.

Gottfried von Leibniz (1646–1716)

The universe is composed of and structured by the interdependent opposites of the physical (1) and non-physical (0). It is this binary pairing that provides the foundation of everything. The world around us has many aspects that work in the same ways as a computer, based on opposites (if it is not one, it must be the other).

But the binary system of Leibniz was just a curiosity, until hundreds of years later, when the computer changed the world.

> **"Binary... So simple, even a computer could do it."**
>
> — *Kerry Redshaw*

BINARY ABACUS — 1680

The binary abacus operates on the same principle as the classical abacus. When 0s and 1s are written in a row to represent a number, each place in the row has a different value. The first sixteen numbers in the binary system are demonstrated. Each time a 1 is added to an occupied place it is emptied and the 1 is placed in the first empty place on the left, etc.

Four additional numbers are given in the decimal system. Can you translate them into binary?

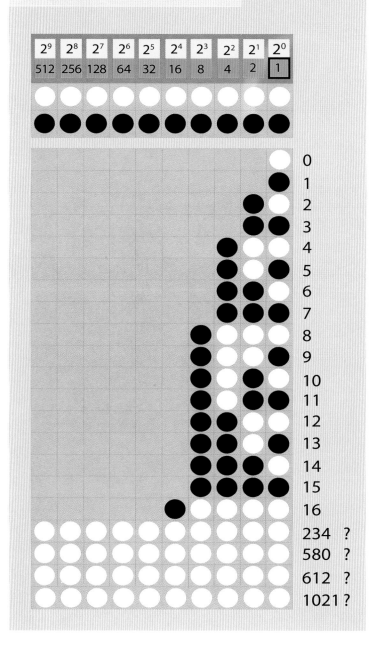

	CHALLENGE	
123	REQUIRES	
	COMPLETED	

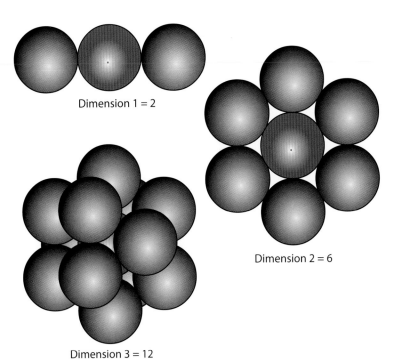

Dimension 1 = 2

Dimension 2 = 6

Dimension 3 = 12

KISSING SPHERES — 1694

The problem of "kissing spheres" arose as the result of a famous conversation between David Gregory and Isaac Newton in 1694.

How many unit spheres can simultaneously touch a given sphere of the same size?

In one dimension, the kissing number is 2. In two dimensions, the kissing number is 6.

Newton thought the answer was 12 spheres, while Gregory believed it was 13.

For more than 250 years this problem remained unsolved.

Finally in 1953 Kurt Schütte and Bartel L. van der Waerden found a definitive solution. It is interesting to note that today the problem is solved in a very high dimension, like 24, giving the answer as 196,560. Can you tell who was right? Gregory or Newton?

124 CHALLENGE ● ● ○ ○ ○ ○
REQUIRES 🧠 ✏️ ✂️
COMPLETED ○

CLOSEST PACKING AND CUBOCTAHEDRONS

How many identical spheres can you pack surrounding and touching a single sphere? This number is called the "Kissing Number." If we add a second and further layer of spheres following this principle we can pack spheres in the closest possible packing as shown by the first three layers. The spheres in the closest possible packings form cuboctahedrons, one of the Archimedean solids. Can you work out the number of spheres in the first three layers?

125 CHALLENGE ● ● ● ● ● ○
REQUIRES 🧠 ✏️ ✂️
COMPLETED ○

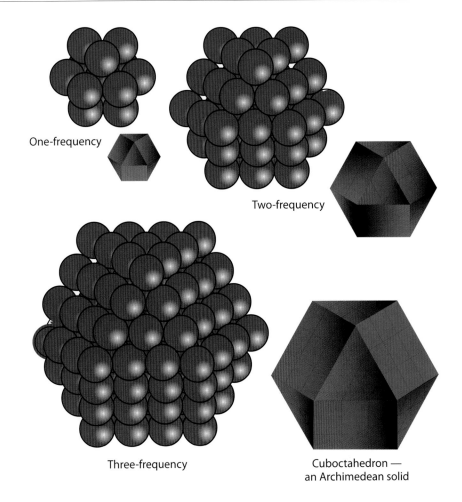

One-frequency

Two-frequency

Three-frequency

Cuboctahedron — an Archimedean solid

BRACHISTOCHRONE AND TAUTOCHRONE CURVE PROBLEMS

In 1696 Johann Bernoulli (1667–1748) challenged mathematicians worldwide to find the curve that solved the brachistochrone problem. This is the curve of fastest descent between two points, on which a ball will start rolling with zero speed and move to the lower point under the action of constant gravity, assuming no friction.

Bernoulli was not the first to tackle the brachistochrone problem. Galileo in 1638, in his experiments with inclined planes, came to the wrong conclusion that the brachistochrone would be an inverted arc of a circle.

After much hard work, Leibniz, Newton, Johann Bernoulli and his brother Jakob found the solution. Can you?

In 1659, Christian Huygens solved another problem: the tautochrone curve problem. A tautochrone or isochrone curve is the curve for which the time needed for a ball to roll down without friction by the action of gravity to its lowest point is independent from its starting point. He proved that the cycloid is also the tautochrone. His discovery was crucial to the design of an isochronous pendulum clock.

Four balls released at different points along the tautochrone at the same time will arrive at the bottom at the same time.

QUICKEST DESCENT — 1696

Balls are released on four different tracks: straight, broken, circular and cycloidal, as shown. Can you tell which ball will be the first to arrive at the end of the slope? Or, in other words: what is the shape of the "brachistochrone," the curve on which an object descending under gravity will be faster than on any other curve? Or is the straight line the fastest way to travel?

126   CHALLENGE ●●●●   REQUIRES   COMPLETED

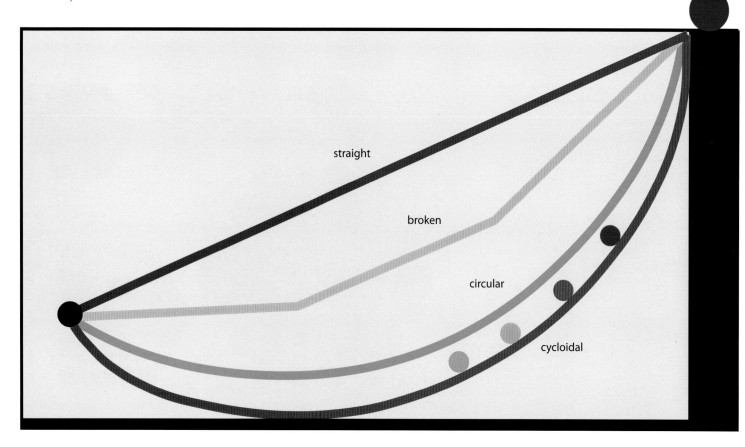

straight

broken

circular

cycloidal

PEG SOLITAIRE — 1697

The origins of the game can be traced back to the year 1697 at the court of Louis XIV, when Claude Auguste Berey made an engraving of Anne de Rohan-Chabot, Princess of Soubise, with the puzzle at her side. Peg solitaire boards feature in a number works of art from that period, showing that the game was extremely fashionable.

In the standard game, the entire board is filled with pegs except for the central hole. The objective is to empty the entire board apart from a single peg in the central hole. The most popular variation of the game is played on a board of 33 cells as shown. Thirty-two pegs are placed on all cells except the cell in the center (cell 17). Easier solitaire puzzles involve starting with a smaller number of pegs, again removing them and finishing in the center cell.

A "jump" consists of moving a peg over any adjacent peg, removing it and landing on the next empty cell. One is allowed to jump vertically and horizontally, but not diagonally. Each move must be a jump, and a chain of continuous jumps counts as one single move.

No one knows how many solutions exist. Obviously, 31 jumps are needed to get a solution, but taking chains into consideration, the number of moves may be less than 31.

The "world record" is a solution involving 18 moves, achieved by Ernest Bergholt in 1912. In how many moves can you get a solution or, how far can you go before reaching a point at which no more jumps are possible?

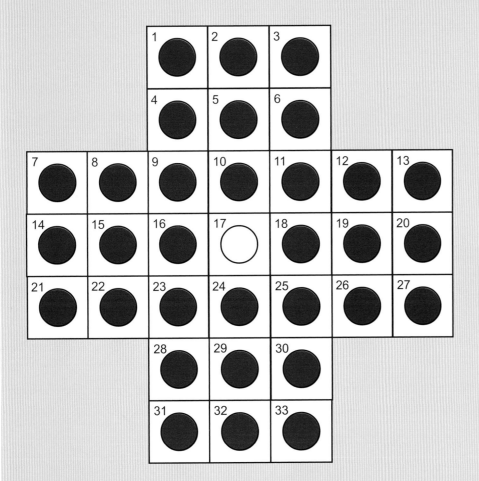

127

CHALLENGE ● ● ● ● ○
REQUIRES
COMPLETED ○

"The game called Solitaire pleases me much. I take it in reverse order. That is to say that instead of making a configuration according to the rules of the game, which is to jump to an empty place and remove the piece which one has jumped, I thought it was better to reconstruct what had been demolished, by filling an empty hole over which one has leaped."

— *Gottfried von Leibniz, 1716*

BINARY MEMORY WHEELS

The possible 3-bit, 4-bit, 5-bit and 6-bit binary numbers can be described by three, four, five or six switches, which may be in either the "on" or "off" position, as shown on the right.

These numbers represent the first 64 numbers (including 0) of the binary numbering system.

Twenty-four switches are necessary to simultaneously express the 3-bit binary numbers; 64 switches for the 4-bit binary numbers; 160 switches for the 5-bit binary numbers; and finally 384 switches for the 6-bit binary numbers.

However in the "binary" wheel above, the same amount of information can be condensed to just 8, 16, 32 and 64 switches respectively, quite an economy. This can be accomplished by having the switches overlap.

Can you find a way to distribute the binary numbers along the binary wheels in such a way that all binary numbers will be represented by a set of adjacent "on" and "off" switches as you go around the wheels clockwise?

Although the switches representing each number must be consecutive, the numbers themselves need not be distributed in a consecutive sequence.

> **"There are 10 kinds of people in the world, those who understand binary math, and those who don't."**
>
> — *Anonymous*

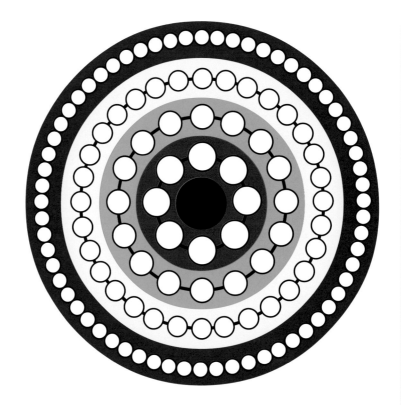

Four wheels:
Red: 3-bit binary numbers
Green: 4-bit binary numbers
Yellow: 5-bit binary numbers
Blue: 6-bit binary numbers

○ = 0
● = 1

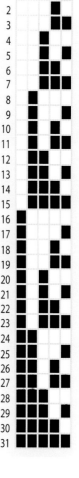

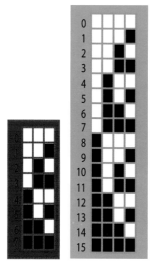

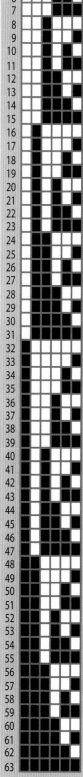

1 meter

? Ping-pong ball

1 meter

? Tennis ball

ROPE AROUND THE EARTH — 1702

One of the most amazing and counterintuitive paradoxes is the "Rope Around the Earth" puzzle of Henry Dudeney, the earliest version of which was published by William Whiston (1667–1752) in 1702.

For this puzzle, one has to assume the Earth is a perfect, smooth sphere and the equator is exactly 40,000 kilometers.

A rope is placed around the equator forming a closed circular loop, tightly around the surface of earth. Then you cut the rope and add 1 meter to its length, and stretch the rope again to form a perfect circle around the earth.

What do you think will be the distance between the rope and the surface of the earth all around? And what if instead of Earth, we do the same experiment with a ping-pong ball, a tennis ball or something else?

129

CHALLENGE ● ● ● ● ● ○
REQUIRES 🧠 ✏️ ✂️ 🔨
COMPLETED ○

?

PARALLELOGRAM OF VARIGNON

Suppose you have five randomly drawn quadrilaterals (i.e. polygons with four sides and four corners). The sides of the one at the top left are bisected and the four points joined forming a parallelogram. Its sides are parallel to the two diagonals of the quadrilateral, therefore they are pairwise equal and parallel.

Can you work out the relationships of the area and the parameter of this parallelogram to the area of the quadrilateral?

Will a similar construction with other quadrilaterals result in a parallelogram? Try it with the other four given quadrilaterals.

130	CHALLENGE	●● ●●●●
	REQUIRES	🧠 ✏️
	COMPLETED	◔

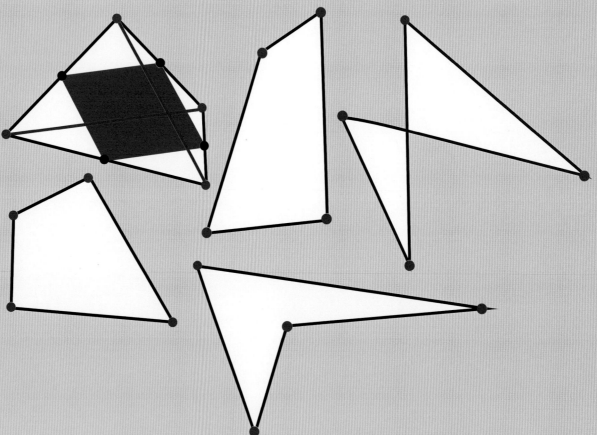

VARIGNON'S THEOREM — 1731

Varignon's theorem is a statement in Euclidean geometry by Pierre Varignon (1654–1722) that was first published in 1731. It concerns the construction of a particular parallelogram (Varignon parallelogram) from an arbitrary quadrangle.

The midpoints of the sides of an arbitrary quadrangle form a parallelogram. If the quadrangle is convex or reentrant, i.e. not a crossing quadrangle, then the area of the parallelogram is half as big as the area of the quadrangle.

In mechanics, the Principle of Moments, also known as Varignon's Theorem, states that the moment of any force is equal to the algebraic sum of the moments of the components of that force. This extremely important principle is often used together with the Principle of Transmissibility to determine systems or forces that are acting upon and/or within a structure.

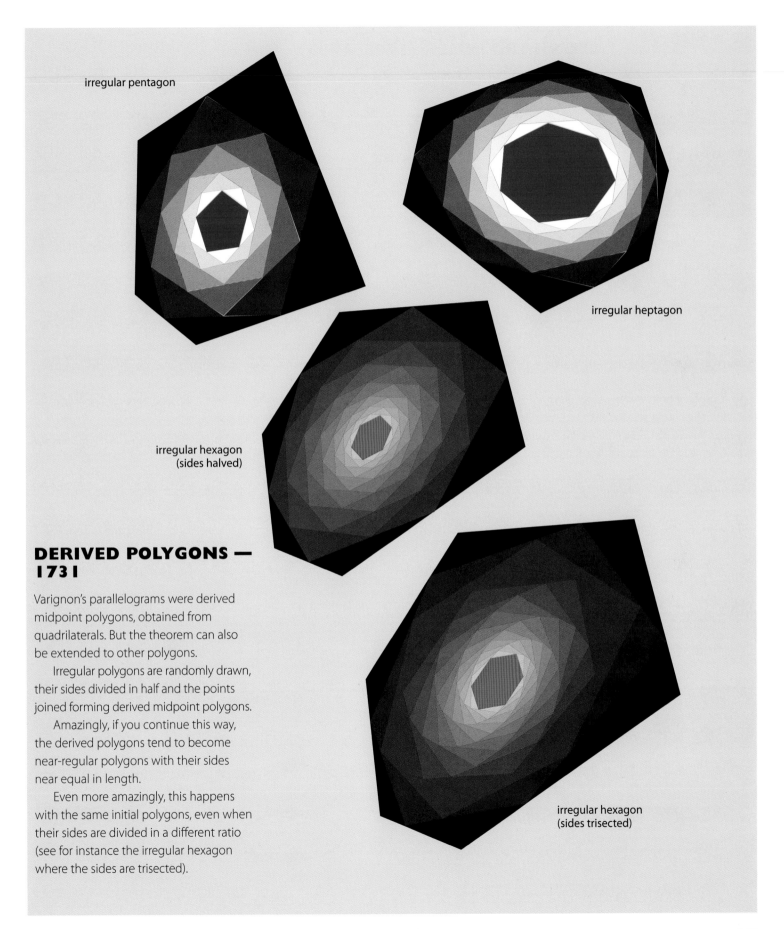

irregular pentagon

irregular heptagon

irregular hexagon
(sides halved)

irregular hexagon
(sides trisected)

DERIVED POLYGONS — 1731

Varignon's parallelograms were derived midpoint polygons, obtained from quadrilaterals. But the theorem can also be extended to other polygons.

Irregular polygons are randomly drawn, their sides divided in half and the points joined forming derived midpoint polygons.

Amazingly, if you continue this way, the derived polygons tend to become near-regular polygons with their sides near equal in length.

Even more amazingly, this happens with the same initial polygons, even when their sides are divided in a different ratio (see for instance the irregular hexagon where the sides are trisected).

LEONHARD EULER (1707–1783)

Leonhard Euler was a Swiss mathematician and one of the most productive mathematicians of all time. He studied at the University of Basel to become a Protestant minister like his father, but his love of mathematics led him to change his studies. Euler had 13 children, and it is often said that he made his greatest mathematical discoveries while holding a baby in his arms.

His contribution to mathematics is enormous, in number theory, differential equations, calculus of variations and other fields. He published more mathematical research than anyone else in the history of mathematics. The Academy continued to publish Euler's unpublished works for 50 years after his death in 1783.

SEVEN BRIDGES OF KÖNIGSBERG — 1735

The following problem dates back to 1735. At that time, the German town Königsberg had seven bridges. The question is very simple (although the answer is not): is it possible to go for a walk, cross each bridge once only, and return home?

It is said that people of the town had tried, but have never been able to solve the problem. Euler solved it in 1735 and laid the foundation of one of the most important branches of mathematics, that of graph theory.

Euler solved the problem in ingenious abstract terms by eliminating many features and replacing them with something that looked simple, only using points and lines. Doing so, he created mathematical structures as shown below, today called "graphs."

The problem then becomes this, given a figure composed of lines joining points, is it possible to traverse the figure in one continuous path, without taking your pencil from the paper and without going over any line twice?

Euler showed that there would have to be at most two places where an odd number of lines meet, and if a return to the start is required, there would have to be no places where an odd number of lines meet. Once seen, the reasoning is easily understood: a continuous journey will enter each junction exactly as often as it leaves — except at the start and finish. The problem of the Konigsberg bridges is then solved by noting that it is equivalent to traversing such

a network of lines, which has four junctions with an odd number of lines.

Conclusion: no solution can exist. Euler's problem is really one of topology, a branch of mathematics that deals with properties of figures that are preserved by continuous deformations. Two networks are topologically equivalent if one can be distorted so that it results in the other one. If a network can be traversed by a single curve, so can any topologically equivalent network. Not bad for a problem that started as an innocent recreational mathematics puzzle and ended as one of the most important branches of modern mathematics.

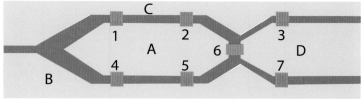

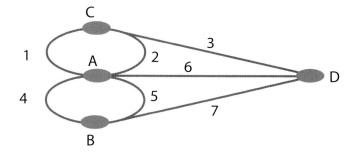

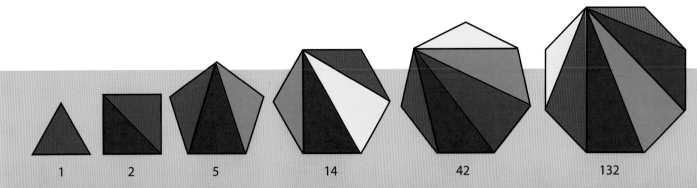

1   2   5   14   42   132

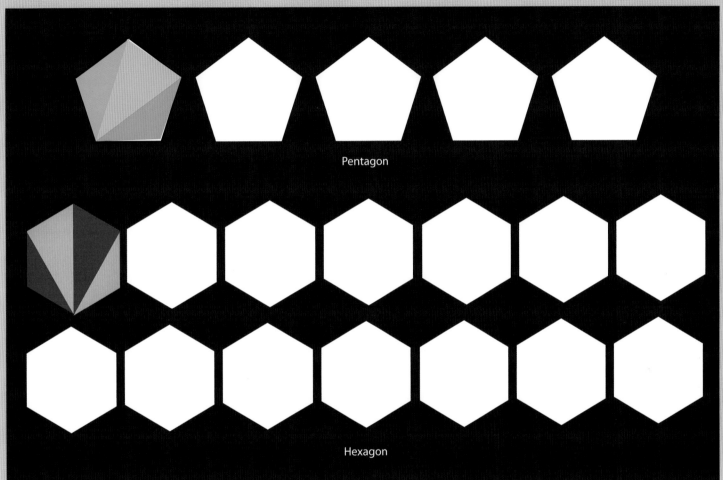

Pentagon

Hexagon

EULER'S POLYGON DIVISION PROBLEM — 1751

Triangulation is the division of a surface or plane polygon into triangles.

One of the basic triangulation problems is the "Euler polygon division" problem that Euler proposed to Christian Goldbach in 1751: in how many different ways can a planar convex regular polygon of n sides be cut into n-2 triangles by its diagonals, if rotations and reflections are considered different? The cuts are not allowed to intersect one another.

Can you find the different triangulations for pentagon and hexagon, shown above? This problem is not as simple as it seems, and has attracted much attention. Can you also figure out how many diagonals and triangles a convex polygon of n sides has?

131   CHALLENGE   ● ● ● ● ○ ○
REQUIRES
COMPLETED   ○

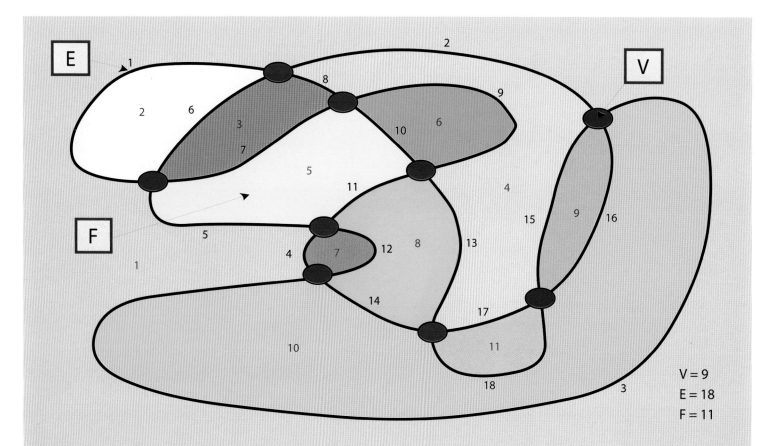

E

V

V = 9
E = 18
F = 11

EULER'S FORMULA — 1752

I have drawn quite a random doodle. To make it really random, I did it by placing the pen on paper, closing my eyes, and without lifting my pen drawing a line, careful not to leave the page, then opening my eyes and joining the beginning and end points. You can try to make and follow your own experiment. The purpose of this exercise is to show that even such random doodles can contain hidden patterns of enormous mathematical significance.

Can you discover any? We are interested in knowing more about our doodle:
1- How many intersection points (V) does it have?
2- How many edges (E) does it have? An edge is a segment connecting two points.
3- How many regions (F) does it have?

We can, of course, count each of these for as long as it takes, but there is

another way. If you know two of the three parameters, the third parameter follows automatically.

Can you figure out the formula, which is called Euler's formula or the Euler characteristic?

It's one of the most beautiful and important expressions of mathematics and a great insight into any connected doodle that we can make in the plane. But there's more.

It can also be shown that all the convex polyhedra have the same relationship of the vertices, edges and faces expressed in the Eular formula.

Although the formula is named after Euler, he did not discover a complete proof himself. It took over two centuries and some of the brightest mathematicians that have ever lived, including Rene Descartes (1596–1650), Euler himself (1707–1783), Adrien-

Marie Legendre (1752–1833) and Augustin-Louis Cauchy (1789–1857) to finally find the complete proof of Euler's Formula.

132 CHALLENGE ● ● ● ● ● ○
REQUIRES 🧠 ✎ ✂
COMPLETED

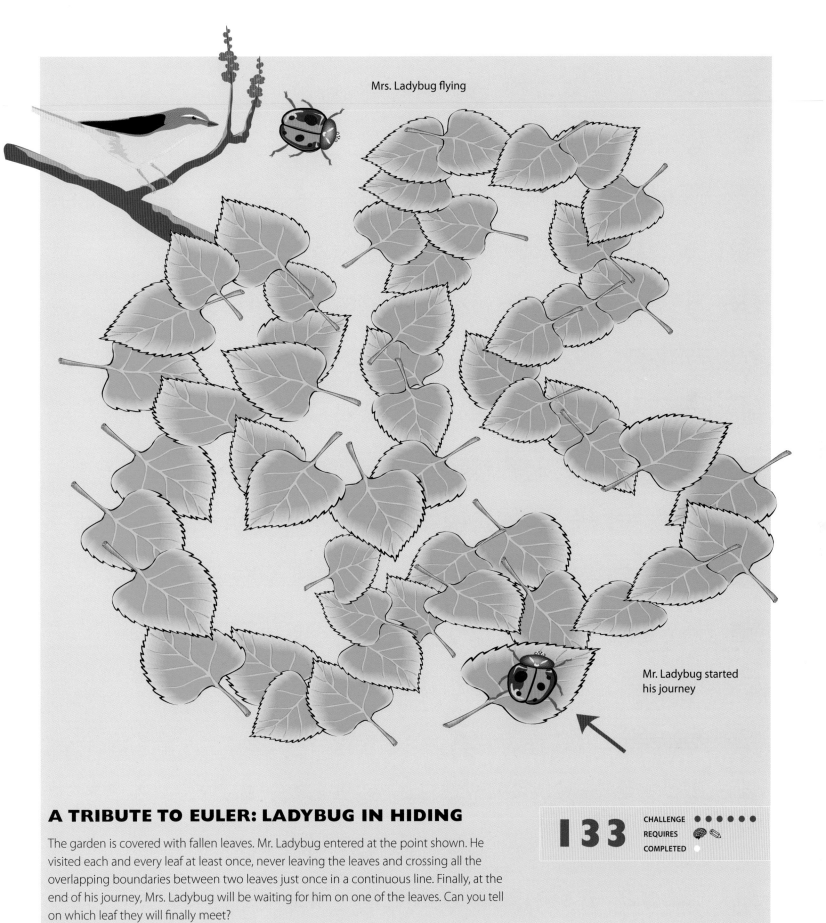

Mrs. Ladybug flying

Mr. Ladybug started
his journey

A TRIBUTE TO EULER: LADYBUG IN HIDING

The garden is covered with fallen leaves. Mr. Ladybug entered at the point shown. He visited each and every leaf at least once, never leaving the leaves and crossing all the overlapping boundaries between two leaves just once in a continuous line. Finally, at the end of his journey, Mrs. Ladybug will be waiting for him on one of the leaves. Can you tell on which leaf they will finally meet?

133

CHALLENGE ● ● ● ● ● ●
REQUIRES 🥔 🥄
COMPLETED ○

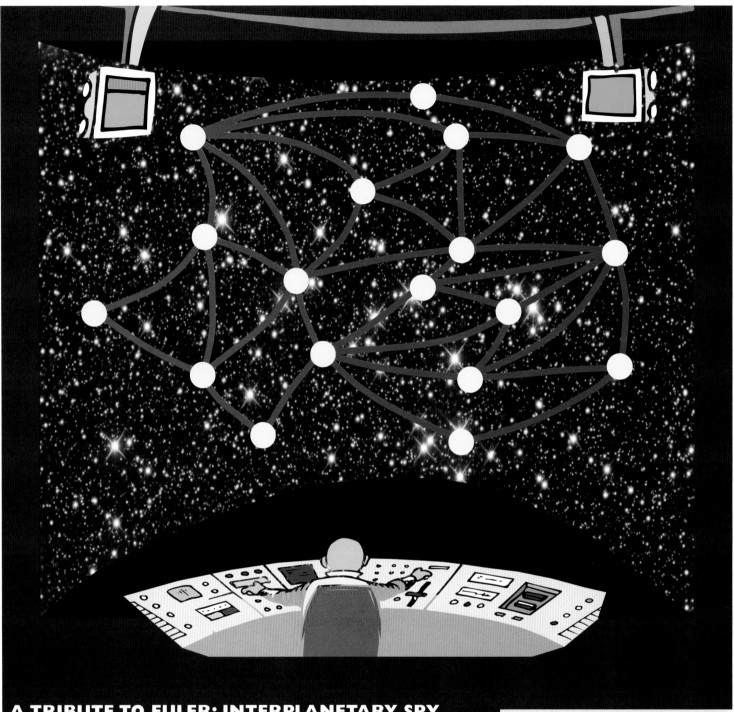

A TRIBUTE TO EULER: INTERPLANETARY SPY

The interplanetary security officer followed the intruder spaceship on his computer screen. The alien spyship entered our planetary system from the north, and in one continuous path crossed all the established routes between the planets, visiting all the planets collecting secret information, never crossing a route more than once, with the obvious intention to leave our system unobserved and as quickly as possible.

But our forces are waiting at the point of its intended departure and its chances of escape are slim.

Can you guess at which exit point our interplanetary defense forces are waiting?

134

CHALLENGE ● ● ● ● ● ●
REQUIRES 🧠 ✏️
COMPLETED ▢

THROWING NEEDLES

If a needle or matchstick is dropped from a considerable height onto our game board, which is a surface on which parallel lines are drawn so that the distance between them is equal to the length of the needle shown, what is the chance that the needle will fall touching a line?

Throw a matchstick 100 times and count the number of times the matchstick falls on a line. Divide 200 by this number. How near will your result be to π (= 3.14)?

135 CHALLENGE ● ● ● ● ●
REQUIRES
COMPLETED ○

BUFFON'S NEEDLE EXPERIMENT — 1750

Buffon's Needle is one of the oldest problems in the field of geometrical probability. It is also one of the most startling examples of the number π showing up unexpectedly in strange places.

Georges-Louis Leclerc (1707–1788), Comte de Buffon, described everything about the natural world in his *Histoire Naturelle*, a 44-volume encyclopedia.

In the appendix he included the problem (completely unrelated to natural history) of the needle experiment. All of a sudden, he became the most important natural historian of his time.

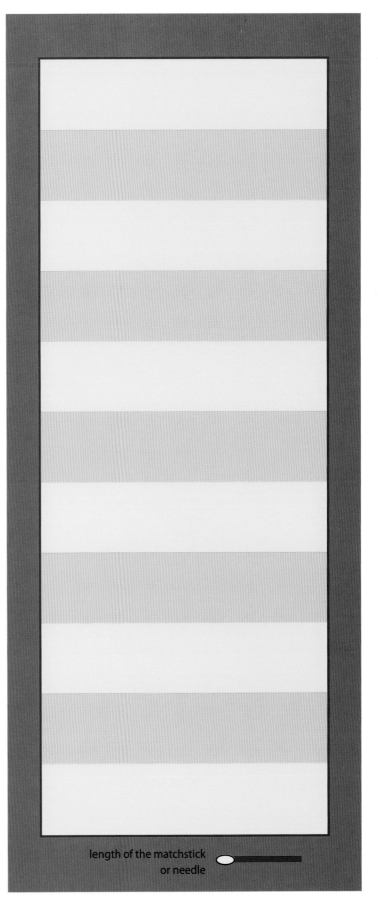

length of the matchstick or needle

COIN TOSSING: D'ALEMBERT'S PARADOX — 1760

Coin tossing reveals many principles of probability. One of the earliest probability paradoxes is called "D'Alembert's Paradox" (after Jean le Rond d'Alembert, 1717–1783).

When tossing two coins there are three possible outcomes. Is the probability of each of these outcomes one out of three? In fact, these results are not equally likely, a fact which escaped the attention of D'Alembert and many others in his time. Actually, there are four possible results when tossing two coins (or two tosses of a single coin), which an average person today is quite well aware of!

A lucky guy traveling back in time with this knowledge could easily become a very lucky gambler in no time at all.

Numbering or coloring of the two sides of the coins helps to show that there are, in fact, four possible results:

1) heads (1) – heads (2)
2) tails (1) – tails (2)
3) heads (1) – tails (2)
4) tails (1) – heads (2)

When a coin is tossed into the air no one can say which way it will land. Yet toss that coin a million times and it will, with increasingly minor variations, come up heads half the time and tails the rest. In essence, this is the basis of the theory of probability.

Basically, two laws underlie probability, a "both-and," to calculate the probability of two events both happening, and an "either-or" law, to calculate the probability of one or the other of two events happening.

The "both-and" law states that the chance of two independent events both happening is equal to the probability of one happening multiplied by the probability of the other happening.

For instance, the chance of one flip of a coin coming up heads is 1/2. The chance of heads landing face up on both the first and second flip is 1/2 x 1/2, or only 1/4.

The "either-or law" states that the probability of one or the other of two mutually exclusive probabilities coming true equals the sum of the separate probability of each coming true individually. For instance, the chance of turning up either heads or tails on a flip of a coin is equal to the probability of throwing heads plus the probability of throwing tails: 1/2 + 1/2 = 1.

But what is the situation when tossing three coins or more? The famous Pascal's triangle (see page 142) gives us all the answers for any number of tosses. In Pascal's triangle, the first number in a line is the number of ways that number of coins can all come up heads; the next number is the number of ways all coins but one can come up heads, and so on.

For example: when tossing four coins, the probability of tossing all heads is 1/16. Looking at Pascal's triangle, can you determine the probability of tossing five heads in a set of 10 tossed coins?

First determine in how many different ways such a throw can happen. The intersection of diagonal 5 and row 10 provides the answer: 252. Now add the numbers of the 10th row to obtain the number of possible throws. This short cut: the sum of the nth row is always 2^n. Thus, the probability of getting five heads is 252/1024.

FOUR-COIN TOSS EXPERIMENT

Probability works! Below you'll find the the results of a statistical experiment in which four coins were tossed 100 times. The number of heads appearing in each toss were recorded, creating a frequency graph of the outcomes, which can be compared to a graph of the outcomes according to the laws of probability.

If we had increased the number of tosses, the outcome would have more closely approached the theoretical curve. But even so, there is a quite good approximation of the probabilities from the fourth row of Pascal's triangle. Try your own experiment and compare the results with those predicted by the laws of probability. Do they match?

Number of Heads	Tally	Frequency	Pascal
0 -	8	8%	6%
1 -	24	24%	25%
2 -	36	36%	37%
3 -	23	23%	25%
4 -	9	9%	6%

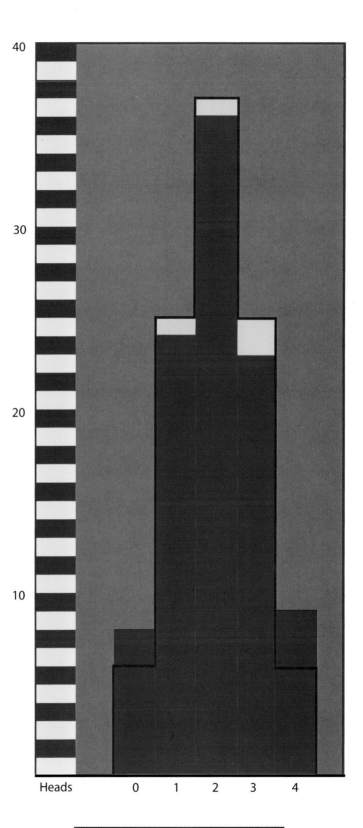

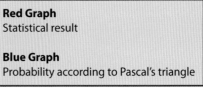

Red Graph
Statistical result

Blue Graph
Probability according to Pascal's triangle

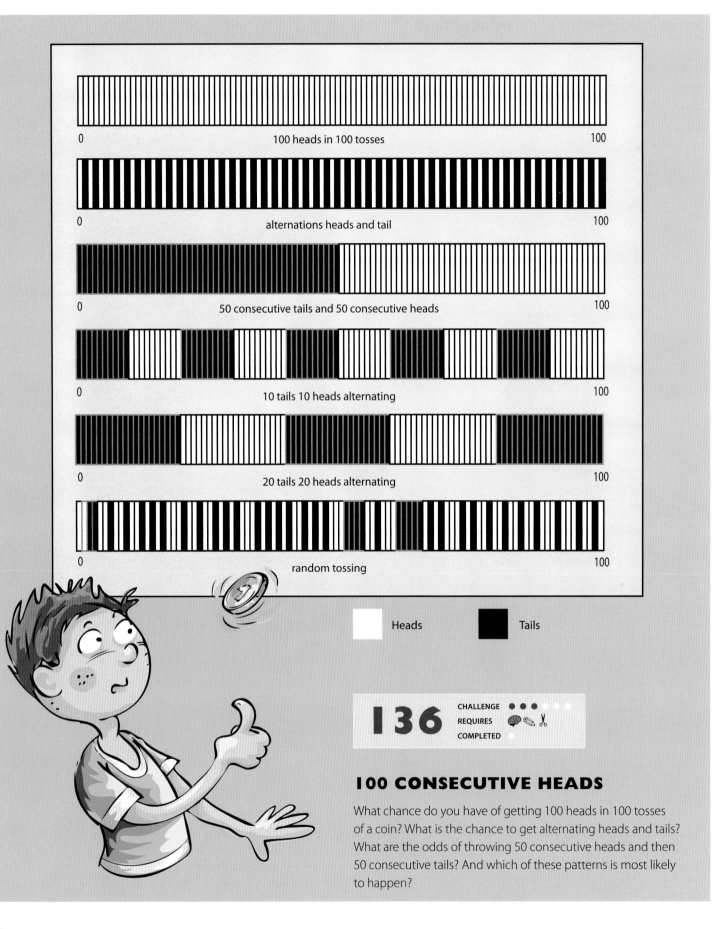

0   100 heads in 100 tosses   100

0   alternations heads and tail   100

0   50 consecutive tails and 50 consecutive heads   100

0   10 tails 10 heads alternating   100

0   20 tails 20 heads alternating   100

0   random tossing   100

Heads   Tails

136   CHALLENGE   ● ● ● ○ ○ ○ ○
REQUIRES
COMPLETED

100 CONSECUTIVE HEADS

What chance do you have of getting 100 heads in 100 tosses of a coin? What is the chance to get alternating heads and tails? What are the odds of throwing 50 consecutive heads and then 50 consecutive tails? And which of these patterns is most likely to happen?

MINIMAL ROUTES

Problems of finding the minimal routes between a number of points is extremely difficult.

For instance, can you guess how the minimal routes will look interconnecting two, three, four, five and six points? The points can represent towns on a map or something else.

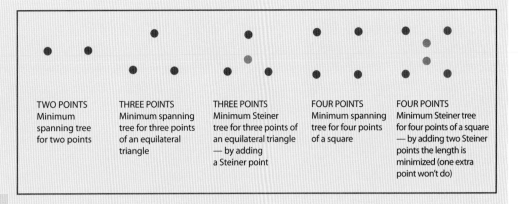

| TWO POINTS Minimum spanning tree for two points | THREE POINTS Minimum spanning tree for three points of an equilateral triangle | THREE POINTS Minimum Steiner tree for three points of an equilateral triangle — by adding a Steiner point | FOUR POINTS Minimum spanning tree for four points of a square | FOUR POINTS Minimum Steiner tree for four points of a square — by adding two Steiner points the length is minimized (one extra point won't do) |

137

CHALLENGE
REQUIRES
COMPLETED

JAKOB STEINER'S MINIMUM TREES

If you have a certain number of points in the plane, another obvious problem arising is how they can be interconnected by straight lines having the minimal possible total length.

In such problems we can distinguish between minimum spanning trees and minimum Steiner trees (which are minimized by adding one or more extra points called Steiner points).

Steinerpoints and trees are named after Jakob Steiner, a Swiss geometer (1796–1863) who was the first to investigate minimal problems.

SOAP BUBBLES AND PLATEAU'S PROBLEM

There is another way to investigate this difficult type of problems: via soap bubbles!

They may seem far removed from serious science and mathematics, but not only small children blow bubbles. Scientists do too, in order to design space stations and many other things and to find answers to some of the most profound questions of nature.

Soap films seem to "know" the principles involved and simple wire models dipped into soap solution often give solutions to complex problems in no time.

When doing such simple experiments we should be aware that we are dealing with problems in the domain of the calculus of variations, a field of mathematics. One of the main general ideas behind it is to learn how to use the least amount of building material to build structures.

Why are soap bubbles round? Because surface tension makes the surface contract as much as possible. Soap bubbles form the shape that encloses a given volume with a minimum amount of surface — the sphere.

First formulated by Joseph-Louis Lagrange in 1760, Plateau's problem involves finding a minimal surface with a given boundary. It is named after the Belgian physicist Joseph Plateau (1801–1883). He was the first to experiment with soap films. (Plateau was also the first to demonstrate the illusion of a moving image in 1832 using a rotating disk he called the phenakistoscope.)

It was not until in 1930 that Jesse Douglas and Tibor Rado independently came up with general solutions using completely different methods.

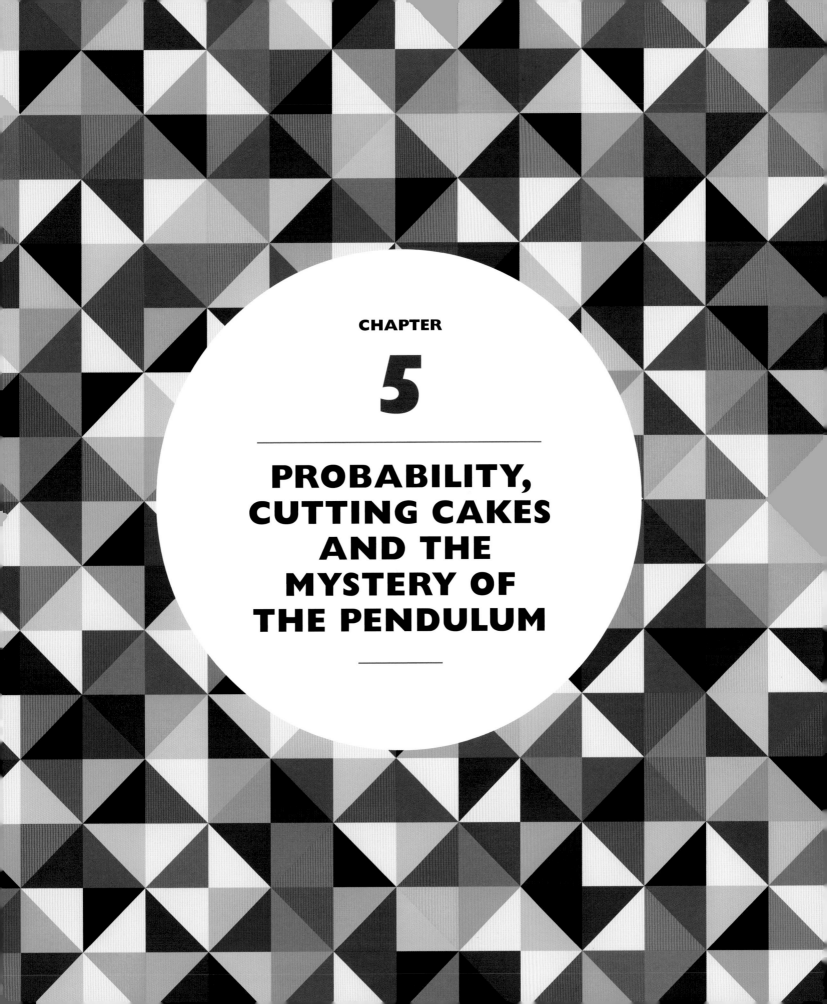

CHAPTER

5

PROBABILITY, CUTTING CAKES AND THE MYSTERY OF THE PENDULUM

KNIGHT'S TOUR

In chess, the knight moves two squares horizontally and one vertically, or two vertically and one horizontally.

One of the oldest and most interesting chessboard puzzles is the knight's tour, originating in the sixth century in India, but Leonhard Euler (1707–1783) made the first serious analysis of the problem.

Can the knight be made to visit every square of the chessboard exactly once?

Mathematically, this is a question about graphs. The problem of finding a closed knight's tour is an instance of finding a Hamiltonian cycle in graph theory (see further in this chapter).

Closed tours can be made only on even-sided boards. Euler found many with unusual symmetries, with visual patterns created in this way, which are aesthetically very pleasing.

Can you find Knight's tours on the smaller boards on the next page?

 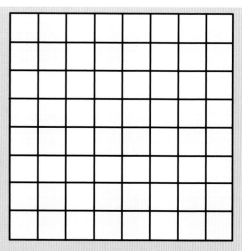

The problem of finding a closed knight's tour is an instance of finding a Hamiltonian cycle in graph theory.

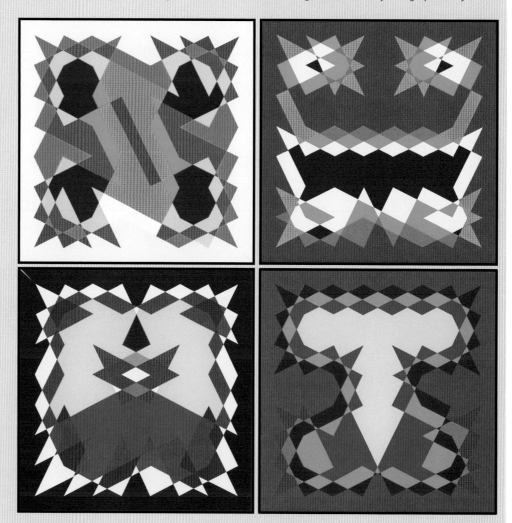

There are 13, 267, 364, 410, 532 closed knight's tours on a chessboard!

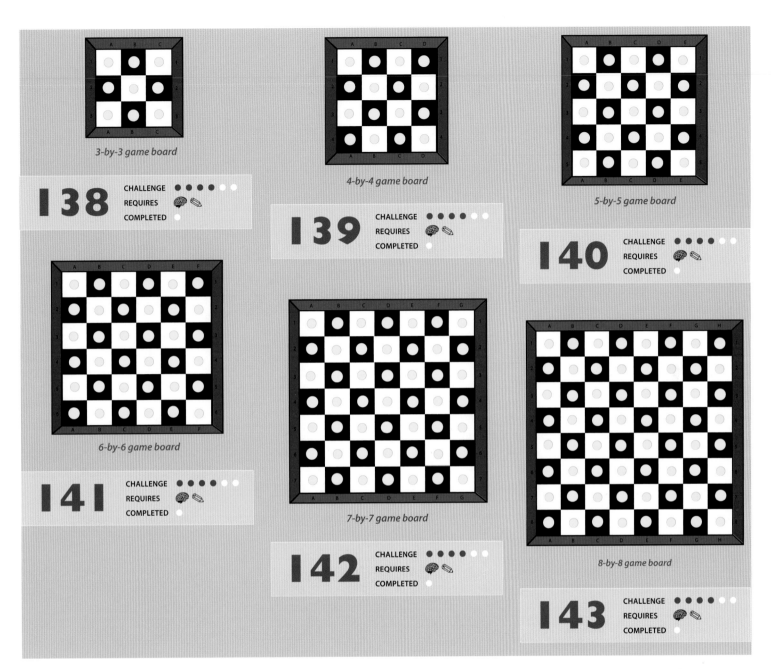

3-by-3 game board

4-by-4 game board

5-by-5 game board

6-by-6 game board

7-by-7 game board

8-by-8 game board

138 CHALLENGE ●●●●○○
REQUIRES 🧠✏️
COMPLETED ○

139 CHALLENGE ●●●●○○
REQUIRES 🧠✏️
COMPLETED ○

140 CHALLENGE ●●●●○○
REQUIRES 🧠✏️
COMPLETED ○

141 CHALLENGE ●●●●○○
REQUIRES 🧠✏️
COMPLETED ○

142 CHALLENGE ●●●●○○
REQUIRES 🧠✏️
COMPLETED ○

143 CHALLENGE ●●●●○○
REQUIRES 🧠✏️
COMPLETED ○

CROSSED AND UNCROSSED KNIGHT'S TOURS

In the *Journal of Recreational Mathematics* of July 1968, L. D. Yarbrough introduced a new variant on the classic problem of the knight's tour. In addition to the rule that a knight touring a chessboard cannot visit the same cell twice (except, in closed tours, for the final jump that can reach the first square, otherwise it's an open tour), the knight is also not permitted to cross its own path (the path consists of straight lines drawn between the starting and finishing square of every jump).

This variant has two names: "non-crossing knight's tour" or "uncrossed knight's tour."

In his *Mathematical Circus*, Martin Gardner presents this problem and explains that the 6x6 tour found by Yarbrough can be improved from 16 to 17 jumps.

Donald Knuth wrote a program to explore all boards up to eight-by-eight. The solutions are 2, 5, 10, 17, 24 and 35.

KNOTS THEORY

The basic problem of knot theory is to recognize the equivalence of two or more knots. This is a difficult problem.

Two knots are equivalent if one can be transformed into the other. Algorithms exist to solve the problem, but they can be time-consuming.

A special case is to recognize the un-knot from a real knot. The figures on the left and in the middle show an unknot and its equivalent. Can you determine whether the figure on the right is an unknot or a real knot?

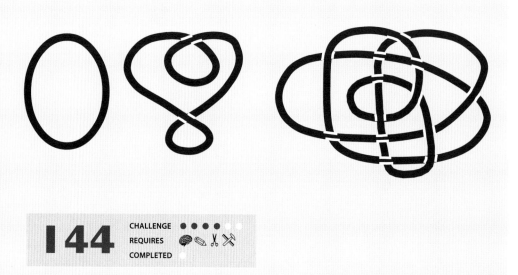

144   CHALLENGE ● ● ● ● ○
REQUIRES 🧠 ✏️ ✂️ 🔨
COMPLETED

TREFOIL KNOTS

The trefoil knot is the simplest example of a knot. It has two forms, both shown, the left-handed trefoil and the right-handed trefoil.

No matter how you try, it is not possible to deform a left-handed trefoil into a right-handed trefoil. The distinction between the two variants depends only on the over and under crossings, not on the orientation of the curve.

When a trefoil is projected on a wall , what is the probability that you can pinpoint the two versions of a trefoil?

145   CHALLENGE ● ● ● ● ○
REQUIRES 🧠 ✏️ ✂️ 🔨
COMPLETED

KNOT TABLES

The above knots are displayed in order of increasing complexity. One frequently used measure of complexity is the crossing number, or the number of double points in the simplest planar projection of the knot. The trefoil or cloverleaf knot is the only one knot with crossing number three (ignoring mirror reflections). The only knot with a crossing number of four is the figure-eight knot.

There are two knots with a crossing number of five, three with a crossing number of six, and seven knots with a crossing number of seven. From this point the numbers increase dramatically. There are 12,965 knots with 13 or fewer crossings in a minimal projection and 1,701, 935 with 16 or fewer crossings. You can see images of the 16 simplest knots above.

CUBIC LATTICE KNOT AND THE TREFOIL KNOT

In topology, the trefoil knot is the simplest example of a non-trivial knot. The trefoil is formed by joining together the two loose ends of a common overhand knot. The result is in a knotted loop. The trefoil, as the simplest knot, is fundamental to the study of mathematical knot theory, which has a wide variety of applications in topology, geometry, physics and chemistry.

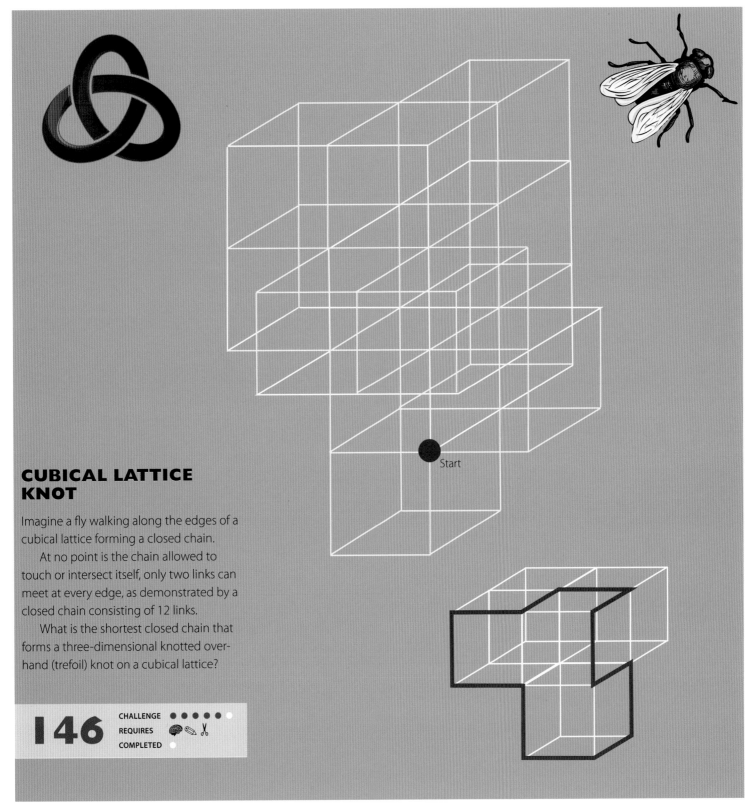

CUBICAL LATTICE KNOT

Imagine a fly walking along the edges of a cubical lattice forming a closed chain.

At no point is the chain allowed to touch or intersect itself, only two links can meet at every edge, as demonstrated by a closed chain consisting of 12 links.

What is the shortest closed chain that forms a three-dimensional knotted overhand (trefoil) knot on a cubical lattice?

146

CHALLENGE	● ● ● ● ●	
REQUIRES	🧠 ✏️ ✂️	
COMPLETED	●	

Start

GAUSS'S HEPTADECAGON — 1796

Euclid provided the solutions to draw regular polygons using only a compass and straightedge: equilateral triangles, squares, pentagons and their derivatives (regular polygons of 6-8-10-12-16-20-24-32-40-48-64, etc. sides).

At the age of 19, Carl Friedrich Gauss (1777–1855) discovered a beautiful construction of a 17-sided regular polygon which convinced him to devote his life to mathematics, a fortunate decision for the advancement of mathematics.

Gauss was so pleased by his discovery of the heptadecagon that he asked that it be drawn on his tombstone. The stonemason objected that the outcome of the difficult construction would look too much like a circle. Eventually a star was inscribed instead.

It required the genius of Gauss to decide which constructions were possible under the Greek constraints and which were not. He showed that the series of constructible polygons was related to numbers called Fermat primes. Pierre de Fermat founded Fermat numbers, from which Fermat primes can be found through this formula (for positive integers): $F_n = 2^{2n} + 1$ where n is a non-negative integer. The first few Fermat numbers are: 3, 5, 17, 257, 65537…, 4294967297.

There are many constructions of the heptadecagon, one of them from 1893 shown below.

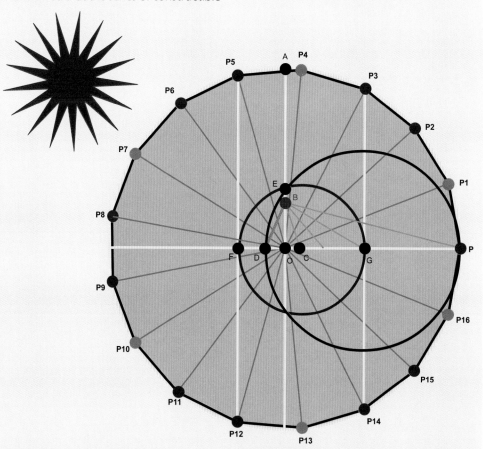

THE MOST BEAUTIFUL CONSTRUCTION OF A 17-SIDED POLYGON — 1893

The most beautiful construction of the heptadecagon is from 1893.

Draw a circle with the center at O, and choose a point P on the circle. Then locate the point A on the circle such that OA is perpendicular to OP, and point B on OA such that OB is 1/4 of OA. Locate point C on OP such that angle OBC is 1/4 of the angle OBP. Find point D on OP (extended) such that DBC is half of a right angle. Let E denote the point where the circle on DP cuts OA.

Now draw a circle centered at C through point E, and let F and G denote the two points where this circle strikes OP. If perpendiculars to OP are drawn at F and G, they strike the main circle at points P5 and P3, as shown. The points P, P3, and P5, are the zero, third, and fifth vertices of the regular heptadecagon, from which the remaining vertices are easily found by bisecting the angles between P3-O-P5 to locate P4, etc.

TANGRAMS — 1802

Originally invented in China at some unknown time in history, one of the oldest known mathematical dissection puzzles is the Tangram, or Chinese Tangram, similar to a Stomachion.

It consists of seven pieces, called tans. A fictitious history of the Tangram described by *The Eighth Book of Tan* claimed that it was invented 4,000 years ago.

The earliest example of a Tangram dates back to 1802. The puzzles was brought to America in 1815. In the years 1817 and 1818 it became the world's first puzzle craze. Its subtlety and the wealth of its combinatorial possibilities will be revealed and appreciated only when playing tangrams.

Truly original inventions in the history of puzzle have always fostered worldwide creativity resulting in novel ideas, variations and new puzzles. Apart from being a basic figure-creating pastime, many variations involve different dissections of not only squares, but also of rectangles, circles, eggs, hearts and other shapes. Today, there are dozens of variations of the tangram, but its original form is probably still the best puzzle in its category.

There are also many intriguing and challenging puzzles and challenges inspired by the tangram, like the Tangram Polygons, Tangram Paradoxes (like the famous Dudeney's Paradox) and many others, examples of which are shown. After you have solved the problems shown here, you can create your own designs — a rewarding and educationally meaningful pastime. Famous devotees of tangrams include Edgar Allan Poe, Lewis Carrol and many others. When in exile, Napoleon spent endless hours inventing and solving tangrams.

CLASSICAL TANGRAM PUZZLES

Copy the seven pieces of the tangram to recreate the black outlines in color.

147

CHALLENGE ● ● ● ● ○
REQUIRES 🧠 ✏️ ✂️
COMPLETED ○

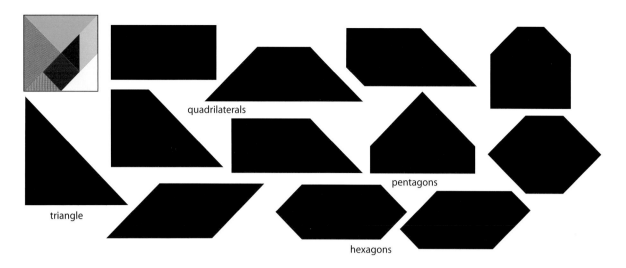

quadrilaterals

triangle

pentagons

hexagons

TANGRAM CONVEX POLYGONS

There is an almost endless number of possible Tangram configurations, but interestingly enough, the number of possible convex polygons is quite limited.

Two Chinese mathematicians, Fu Traing and Chuan-Chih, proved that, using the seven Tangram pieces, only 13 different convex polygons can be formed: one triangle, six quadrilaterals, two pentagons and four hexagons. Try to recreate them.

148 CHALLENGE ●●●●●● REQUIRES 🧠✏️✂️ COMPLETED ○

TANGRAM PARADOXES

All the figures on this page were created using the seven Tangram pieces.

Can you solve the puzzles to explain the small differences between them? (The paradoxical Tangram problems are included in *The Tangram Book*, by Jerry Slocum, collected from Chinese, French, Sam Loyd's, Henry Dudeney's, Gianni Sarcone's and others' books.)

149 CHALLENGE ●●●○○○ REQUIRES 🧠✏️✂️ COMPLETED ○

MALFATTI'S MARBLE PROBLEM — 1803

It happens quite often in mathematics that it is all too easy to jump to wrong conclusions.

The story of Malfatti's problem is a beautiful and convincing example.

In 1803, Italian mathematician Gian Francesco Malfatti (1731–1807) posed the problem of determining the three circular columns of marble (of different sizes if necessary) which, when carved out of a right triangular prism, would have the largest possible total cross section.

This is equivalent to finding the maximum total area of three circles that can be packed inside a triangle of any shape without overlapping.

This problem is now known as the marble problem (Martin 1998). Malfatti thought he knew the answer. He gave the solution as three circles (the Malfatti circles) tangent to each other and to two sides of the triangle.

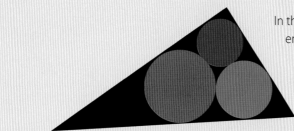

In the particular case of an equilateral triangle, as proven in 1930, Malfatti's "solution" doesn't work and configurations where the big circle touches three sides are better.

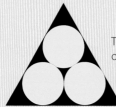

The three circles occupy 0.729 of the triangle area.

These three circles occupy 0.739 of the triangle area.

So it was concluded that the equilateral triangle is an exception to Malfatti's solution. But in 1965, Howard Eves also found that Malfatti's solution was wrong in long and thin right-angled triangles.

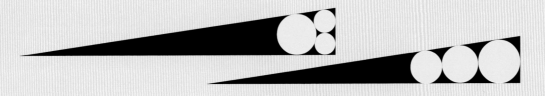

It is obvious that the second triangle gives a better solution than Malfatti's.

Finally, in 1967, Michael Goldberg showed that Malfatti's solution is never correct. The correct solution always has one of the configurations with one of the circles touching all three sides of the triangle.

OVAL TABLES — 1821

In his book *Rational Amusement for Winter Evenings* (1821), John Jackson posed the classic puzzle of how to cut and transform a circular table into two identical oval tables, each with an elongated hole in the middle. His solution is shown on the right, with the table dissected into eight pieces. Sam Loyd in his *Cyclopedia of 5000 puzzles,* solved the puzzle in only six pieces as shown. But he continued the search for a minimal solution and soon found an astonishingly elegant four-piece solution.

150

CHALLENGE ● ● ●
REQUIRES 🧠 ✏️ ✂️
COMPLETED ○ ○

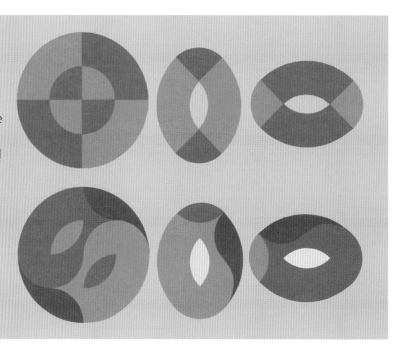

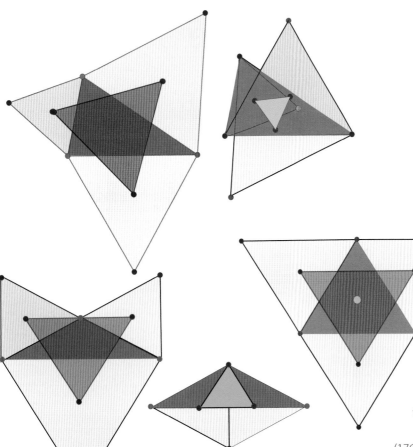

NAPOLEON'S THEOREM — 1825

Napoleon's theorem, states that if equilateral triangles are constructed on the sides of any triangle, either all outward or all inward, the centers of those equilateral triangles themselves form an equilateral triangle. The triangle formed is called the Napoleon Triangle (inner and outer). The difference in area of these two triangles equals the area of the original triangle.

The theorem is often attributed to Napoleon Bonaparte (1769–1821), who was know to be an amateur mathematician, but it is not known whether he either created or solved the problem.

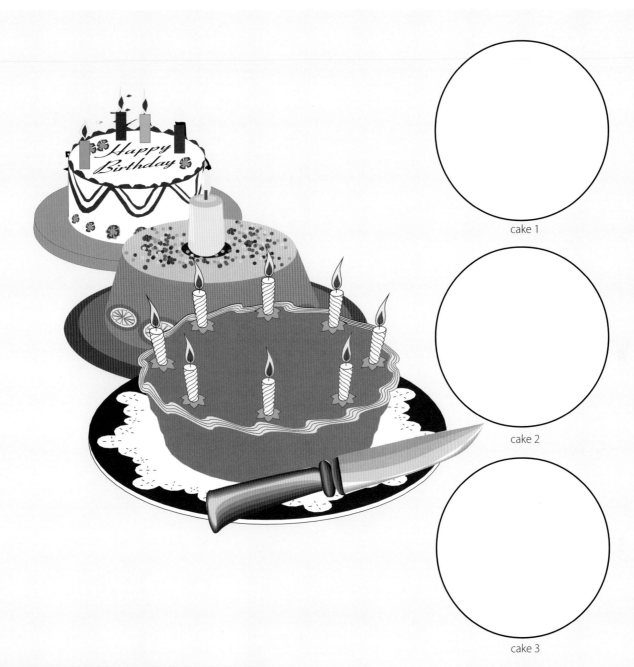

cake 1

cake 2

cake 3

CUTTING THREE CAKES — 1826

Puzzles have infinite variety, but perhaps there is no class more ancient than dissection puzzles. The ancient Chinese dealt with problems involving cutting cakes like this puzzle from the 19th century.

At a birthday party, three cakes have to be divided by straight knife cuts into exactly 34 pieces and divided between 34 children. What is the minimal number of straight line cuts so that each child gets a piece of cake (not necessarily identical)?

There is a condition to the cutting: each cake has to be cut by at least two cuts. Under this condition, can each child get a piece of cake? If we require that each child gets an identical piece of the cake, what would be the minimum number of straight cuts?

151

CHALLENGE
REQUIRES
COMPLETED

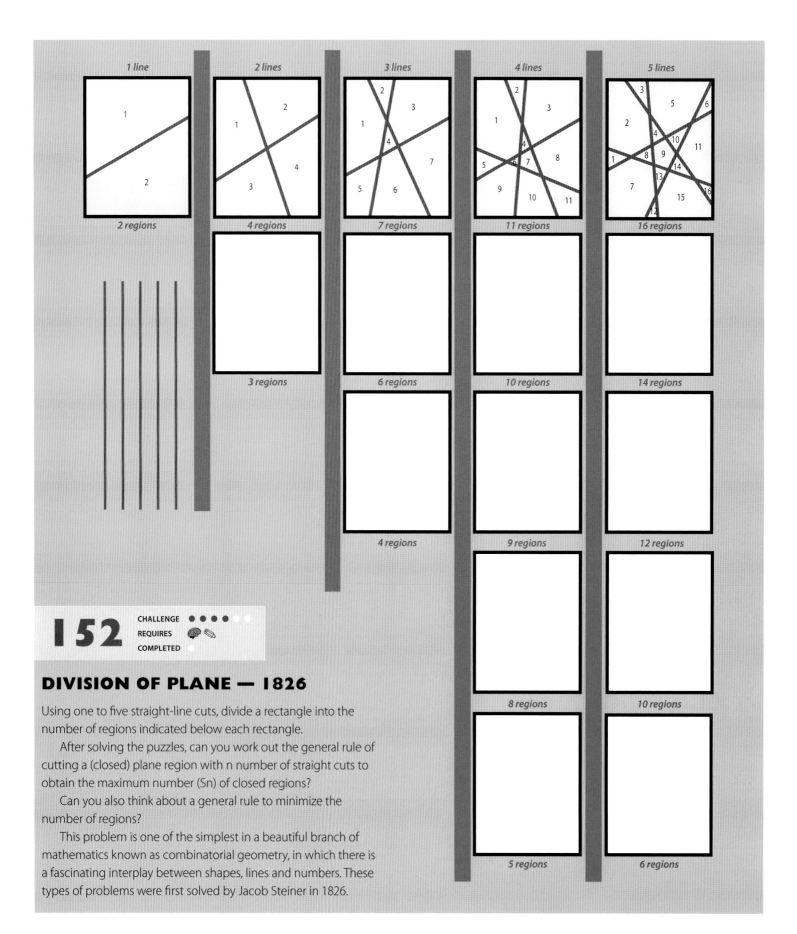

1 line

1

2

2 regions

2 lines

1

2

3

4

4 regions

3 lines

1

2

3

4

5

6

7

7 regions

4 lines

1

2

3

4

5

6

7

8

9

10

11

11 regions

5 lines

1

2

3

4

5

6

7

8

9

10

11

12

13

14

15

16

16 regions

3 regions

6 regions

10 regions

14 regions

4 regions

9 regions

12 regions

8 regions

10 regions

5 regions

6 regions

152

CHALLENGE ● ● ● ● ○ ○

REQUIRES 🧠 ✏️

COMPLETED ○

DIVISION OF PLANE — 1826

Using one to five straight-line cuts, divide a rectangle into the number of regions indicated below each rectangle.

After solving the puzzles, can you work out the general rule of cutting a (closed) plane region with n number of straight cuts to obtain the maximum number (S_n) of closed regions?

Can you also think about a general rule to minimize the number of regions?

This problem is one of the simplest in a beautiful branch of mathematics known as combinatorial geometry, in which there is a fascinating interplay between shapes, lines and numbers. These types of problems were first solved by Jacob Steiner in 1826.

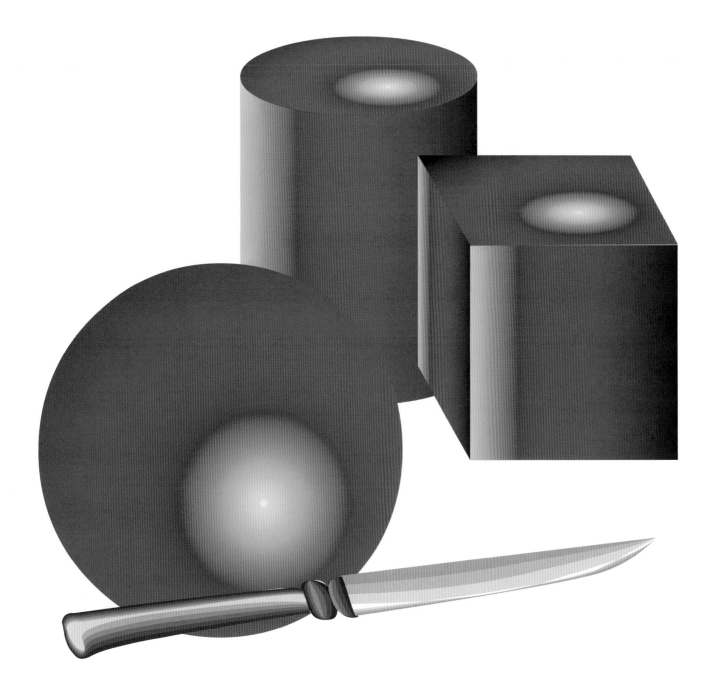

DIVISION OF SPACE — SPHERE — CUBE — CYLINDER — 1826

What is the maximum number of parts into which space, a sphere, a cylinder and a cube can be divided by four plane cuts?

Can you visualize the problem to determine how many separate space regions will result when all four plane cuts go right through the sphere, cylinder or cube?

This division problem was solved by Jacob Steiner in 1826.

As in the problem of dividing the plane, in order to attain the maximum number of partial spaces, it is required that the lines of intersections of no more than two planes are parallel and no more than three planes intersect at one point.

You may easily visualize that one plane cut will produce two space regions, two plane cuts produce four regions, and three plane cuts produce eight regions, but can you try to work out the number of space regions produced by cutting the space with four plane cuts?

153

CHALLENGE ● ● ● ● ● ○
REQUIRES
COMPLETED ○

MATCHSTICK PUZZLES — 1827

Matches were invented by British chemist John Walker in 1827. They quickly took the place of the tinderboxes used until then. It was not long before a new type of recreation emerged in the form of matchstick puzzles, which became popular when several match companies printed them on their boxes. To take advantage of this interest, publishers began to print books of matchstick puzzles.

The puzzles presented here are based on these classical matchstick puzzles.

154

CHALLENGE ● ● ● ● ○
REQUIRES 🧠✏️✂️
COMPLETED

MATCHSTICK TRIANGLES

Can you first move four matchsticks to create two smaller equilateral triangles, and then again four matchsticks to create four smaller equilateral triangles?

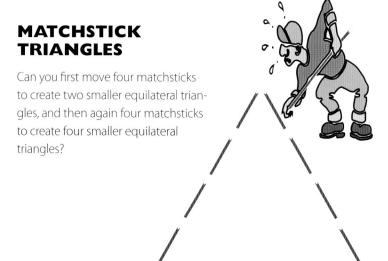

MATCH 4 AND 5

With four matchsticks there are five, and with five matchsticks 12 possible topologically different configurations, taking the following conditions into account:

1. Matchsticks can only touch at their ends.
2. Matchsticks are on a flat surface.

Note: Once a configuration is formed, it can be transformed in an infinite number of ways into topologically equivalent structures, by deforming it without separating its connections at the joints. In each configuration one is missing, can you discover them?

MATCHSTICKS MEETING IN A POINT

An interesting group of matchstick problems involves finding patterns in which a given number of matchsticks meet without crossing at a point. Three sticks forming an equilateral triangle is the smallest pattern in which two sticks meet at every vertex.

Can you find the smallest pattern in which three match sticks meet at every vertex? And four?

MATCH BAMBI

At one of the meetings one morning at breakfast, Martin Gardner challenged me with this tricky puzzle by Mel Stover. Change a single matchstick to make the Bambi look in another direction without changing its shape in any way.

Reflections and rotations are allowed.

MATCH DOG

The playful dog was not careful enough and was run over by a car. Fortunately, he stayed alive and was brought to a vet. By changing the position of only two matchsticks, can you visualize what the dog looked like on the vet's table?

MAXIMUM OVERHANG PROBLEM

A problem from the early 19th century asks how far a stack of identical blocks can be made to hang over the edge of a table?

For example, if the blocks are one unit long, the maximum overhang for three blocks is demonstrated in which the stacking of the blocks is one-on-one fashion, called harmonic stacking. The overhang is almost one unit. The pattern continues. The maximum overhang in harmonic stacking for four blocks has an overhang of just over one unit as shown. With a sufficient supply of blocks, how large an offset can be achieved? For example, in harmonic stacking what will be the maximum overhang for 10 blocks?

In 1955, R. Sutton introduced optimal stacking resulting in larger overhangs, where the one-on-one harmonic stacking restriction is lifted, allowing more than one block in consecutive rows of stacks.

In optimal stacking, as few as three blocks can achieve a one unit overhang. Can you show how? With four blocks in optimal stacking an overhang of over one unit can be achieved as shown.

155 CHALLENGE ● ● ● ● ● ○
REQUIRES 🧠 ✏️ ✂️
COMPLETED ○

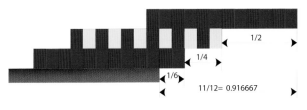

Harmonic stacking of three blocks

1/2
1/4
1/6
11/12= 0.916667

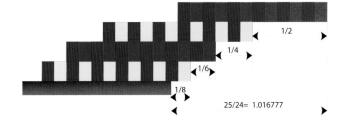

Harmonic stacking of four blocks

1/2
1/4
1/6
1/8
25/24= 1.016777

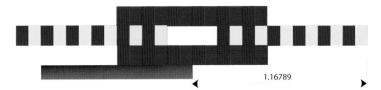

Optimal stacking of four blocks

1.16789

NECKER CUBE — 1832

The Necker cube is an optical illusion that was first published in 1832 by Swiss crystallographer Louis Albert Necker.

It is one of the earliest scientific demonstrations of perceptional ambiguity — a fascinating figure in its simplicity and for the astonishing phenomena it produces when we examine it for a for a few moments. A Necker cube is a line drawing of a wire-frame cube in isometric perspective as illustrated below.

It is a two-dimensional skeleton of a three-dimensional cube in which the front can't be distinguished from the back.

The Necker cube and many other later ambiguous figures demonstrate that we can "see" something in two (or more) different ways, although what we're looking at remains unchanged.

Looking at a Necker cube, we can't tell the front from the back. Back or front is dependent on what you think it is. The

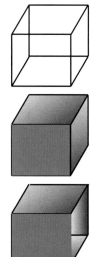

reversal is clearly not in the drawing, it is in you.

Your subjectivity accommodates the object first one way, then another. But strangely enough, the reversal implies your position in space. When you see the orientation in which the red panel is horizontal (see picture below), the whole cube is below your eyes, you are looking down on it. While looking at the reversal in which the red panel is vertical, the cube is above your eyes.

Since you can't be in two places at the same time, it is impossible for you to see both orientations simultanously in the Necker cube. Thus the perceptual structure of the Necker cube is far more complex and ambiguous than initially suggested. We never see both orientations together because our visual system must decide where are we positioned in space. The same reasoning can be applied to drawings of so-called "impossible figures" by Escher, Gregory, Penrose and others.

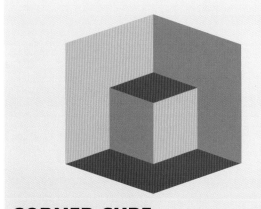

CORNER CUBE

How many different pictures can you see? A small cube in front of a corner of the big cube? Or a small cube inside a corner of the big cube? Or a small cube cut out and missing from the big cube?

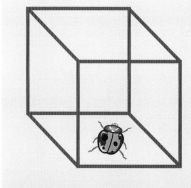

NECKER CUBE LADYBUG

In how many different locations can you see the ladybug?

156 CHALLENGE ● ● ● ● ● ●
REQUIRES 🧠 ✏️
COMPLETED ○

AMBIGUOUS NECKER CUBE

As you look at the cube, it suddenly reverses and what had been the front becomes the back and vice-versa. The Necker cube shows that anything we see is just a "best guess" by our visual system.

The red panels make the Necker cube less ambiguous and you can see each reversal and orientation clearly.

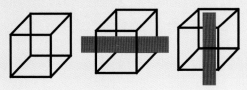

NECKER BOXES

As you look at the wire skeletons of the Necker box, missing the clues about its walls, it can reverse into any of the boxes shown on the right.

PIGEONHOLE PRINCIPLE — 1834

In mathematics, the pigeonhole principle states that if n items are put into pigeonholes with n > m, then at least one pigeonhole must contain more than one item.

Axioms such as "there must be at least two left gloves or two right gloves in a group of three gloves" provide real-life ex-amples of this theorem. It is a counting argument, and although it appears to be self-evident, it can be used to demonstrate results that may be unexpected; for example, that two people in London have the same number of hairs on their heads (see below).

Johann Dirichlet is thought to have been the first to formulate this idea in 1834 under the name Schubfachprinzip ("drawer principle" or "shelf principle"). It is therefore also frequently referred to as Dirichlet's box principle, or the "Dirichlet principle."

50 MAILBOXES PUZZLE

The postman delivers 151 pieces of mail to 50 mailboxes. After all the letters have been equally distributed, one mailbox has more letters than any other mailbox.

What is the smallest number of letters it can contain?

157 CHALLENGE ● ● ● ● ● ○
REQUIRES 🧠 ✏️ ✂️
COMPLETED ○

NUMBER OF HAIRS PUZZLE

Are there two human beings alive today who have precisely the same number of hairs on their bodies?

158 CHALLENGE ● ● ● ● ● ●
REQUIRES 🧠 ✏️
COMPLETED ○

MIQUEL'S FIVE CIRCLES THEOREM — 1836

Five red circles have been drawn with their centers on a fixed black circle. Each circle intersects its neighbor at two points, one of these on the fixed circle (green) and the other within the fixed circle (yellow). By joining adjacent yellow points an irregular pentagram is formed with its five vertices on the five circles. Will this always happen whatever the size of the five circles? Try it!

Auguste Miquel's five circles theorem states that, given five circles centered on a common sixth circle and intersecting each other chainwise on the same circle, the lines joining their second intersection points form a pentagram whose points lie on the circles themselves.

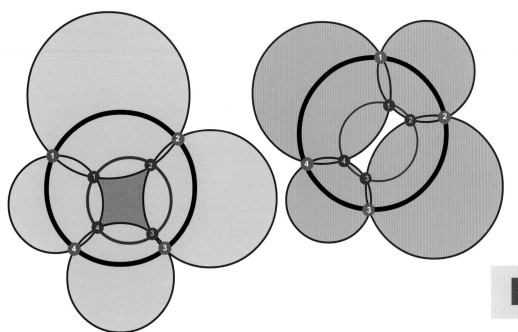

SIX CIRCLES THEOREM (1)

Another interesting theorem is Miquel's six circles theorem. Four points lie on a circle and four circles are drawn through these points as shown. The second intersection points of the four circles will also lie on a circle.

Will this always be the case even if the initial circle becomes a straight line? Try it!

159

CHALLENGE ●● ○○○○
REQUIRES 🧠 ✏️
COMPLETED ○

SIX CIRCLES THEOREM (2)

Starting with a triangle, draw a circle touching two of its sides.

Then draw another circle tangent to the previous circle again touching two sides of the triangle.

Continue in the same way.

The surprising result is a chain of touching circles in which the sixth circle closes the chain as a tangent to the first circle.

Will this always happen, even when some of the circles are outside the triangle? Try it!

160

CHALLENGE ●● ○○○○
REQUIRES 🧠 ✏️
COMPLETED ○

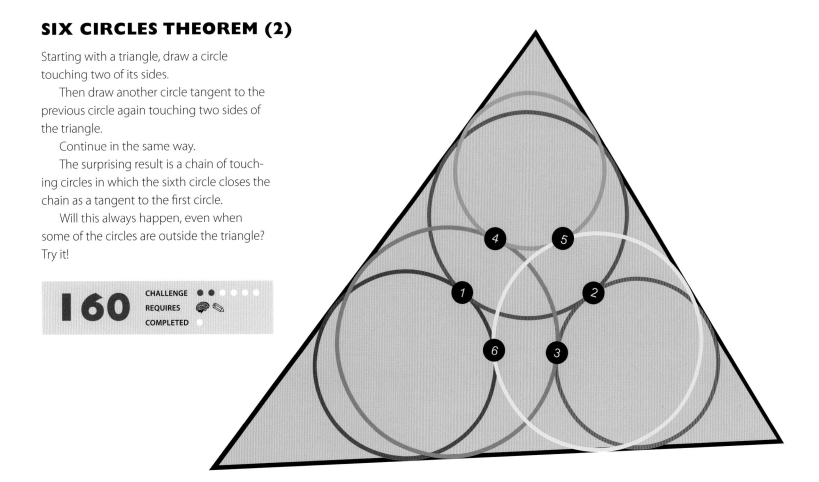

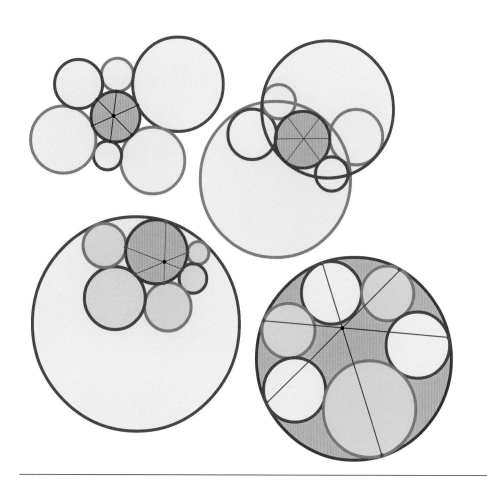

SEVEN CIRCLES THEOREM (1)

Begin with a central (red) circle and arrange six circles tangent to it, touching both the central circle and their two neighbors. The three lines joining opposite points of tangency meet in a point.

Different configurations are possible, four are shown.

Can you imagine what will happen if you let the radii of three of the surrounding circles (blue or green) approach infinity?

(The theorem was discovered in 1974 by Evelyn, Money-Coutts and Tyrrell.)

SEVEN CIRCLES THEOREM (2)

Six congruent circles can be arranged in a familiar configuration as tangents to a seventh circle of the same size as shown.

When we add a circle around the six outer circles, two sets of figures are formed between the circles, six identical figures on the inside and six identical figures on the outside.

The diameters of the smaller circles are 1/3 of the diameter of the big circle.

Can you work out the areas of the two sets of figures, red and yellow?

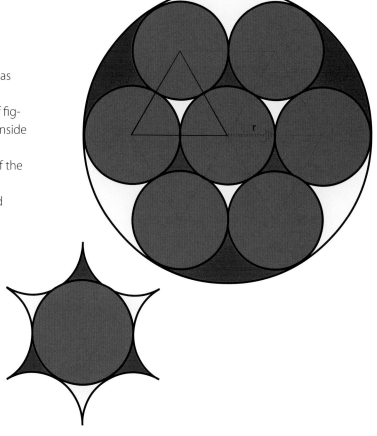

NINE CIRCLES THEOREM

Nine tangent circles can form an unexpected closed chain.

Let 1, 2 and 3 be three circles in the plane, then draw a fourth that is touching 2 and 3. Next, create eight more circles that are tangent to the previous circle, as well as two of the three original circles. Although there are various options for each successive tangent circle in the chain, the ninth and last circle coincides with the first circle of the chain.

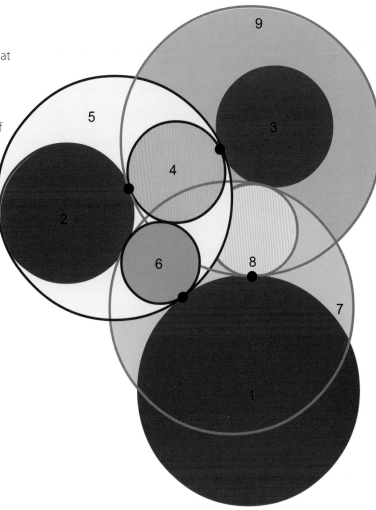

NINE-POINT CIRCLE

It is possible to construct a nine-point circle for any given triangle. It was given this name as it passes through nine significant points defined for every triangle. These points are as follows:
- the midpoint of each side of the triangle (3),
- the foot of each altitude (3),
- the midpoint of the line segment from each vertex of the triangle to the orthocenter (where the three altitudes meet; these line segments lie on their respective altitudes) (3).

In the case of an acute triangle, six of the points (the midpoints and altitude feet) lie on the triangle itself; for an obtuse triangle, two of the altitudes have feet outside the triangle, although they still belong to the nine-point circle.

The nine-point circle is known by several names, including Feuerbach's circle, after the German philosopher Andreas von Feuerbach (1804–1872).

TRIPLES														
DAY 1														
DAY 2														
DAY 3														
DAY 4														
DAY 5														
DAY 6														
DAY 7														

KIRKMAN'S SCHOOLGIRLS PUZZLE — 1848

Reverend Thomas Penyngton Kirkman (1806–1895), an amateur mathematician, posed his puzzle in 1847, which became one of the most famous combinatorial puzzles ever devised: "How can 15 schoolgirls walk in five rows of three girls each for seven days so that no girls walks with any other girl in the same triplet group more than once?"

Or, in a more mathematical language: "How can the first 15 numbers from one to 15 (each number for a specific girl), be distributed into seven matrices, each matrix for a day of the week, consisting of five sets of triplets (groups of three girls in each), so

that no two digits appear in the same row more than once in any of the seven matrices?"

The seven matrices are visualized above. There are seven unique solutions. Can you find one?

How many triplets can be formed from the first 15 numbers? Out of this total, 35 triplets must be selected to provide a solution to the puzzle, and it is a very large number. For this reason it is not an easy puzzle to solve. Kirkman's puzzle has significant relevance to matrix theory.

A matrix in mathematics is an arrangement of numbers or symbols distributed in

rows and columns according to some specific predetermined pattern.

The Kirkman Schoolgirls puzzle is one of several beautiful classic combinatorial problems dealing with different Steiner triple systems. A Steiner triple system is an arrangement of n objects (numbers, symbols or other) in triplets in such a way that each pair of objects appears in a triplet only once. In general, are Steiner triple systems possible for any n number of objects?

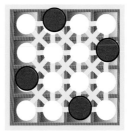

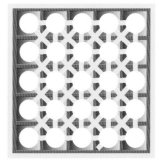

n=4

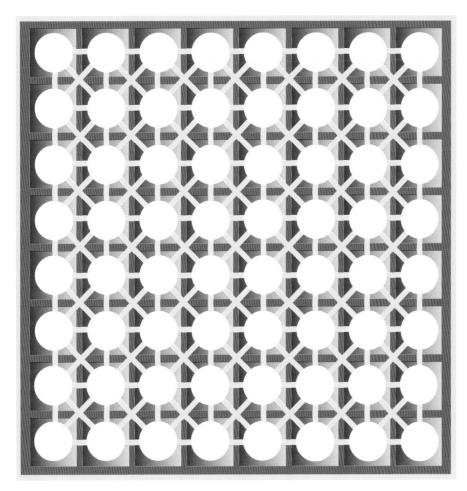

n=8

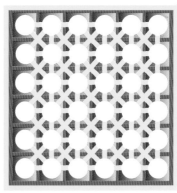

n=5

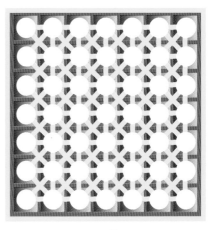

n=6

EIGHT QUEENS PROBLEM — 1848

Problems associated with various chess pieces and their placements have entertained puzzlists for centuries.

Can you place eight chess queens on the chessboard so that no queen can be captured by another? (Remember: a queen can be moved any number of unoccupied squares in a straight line vertically, horizontally or diagonally.)

This problem was first posed in 1848 by Max Bezzel, and is considered one of the gems of recreational mathematics.

There are 12 different solutions. How many can you find? Can you find at least one solution for the mini queens problems from n=5 to n= 7?

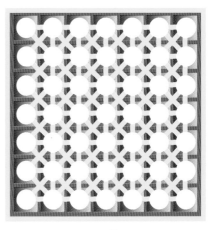

n=7

164 CHALLENGE ● ● ● ● ●
REQUIRES 🧩 ✏️
COMPLETED ○

MÖBIUS MAGIC — 1850

The Möbius strip is a beautiful and mysterious object with a twist. The 19th century German mathematician A.F. Möbius (1790–1868) discovered that it was possible to make a surface that has only one side and one edge and has no "inside" and "outside." You can paint it in one color.

An arrow traveling on the surface of a Möbius band will return after one revolution to the same place from which it started but will be on the "opposite side" of the band, proving that the surface has only one "side."

Although such an object seems impossible to imagine, making a Möbius strip is very simple: take a strip of ordinary paper and give one end a twist, then glue the two ends together.

Variations of Möbius strips are the basis of an endless number of exciting structures and puzzles, with many surprising and paradoxical properties that have led to meaningful developments in topology. Some of these will be demonstrated. It is great fun to play with a Möbius strip, but does it have any practical use? Conveyor belts are sometimes designed as Möbius strips, to give both "sides" equal wear and tear. Some audio cassettes use tape twisted into a Möbius so that continuous playing time is doubled.

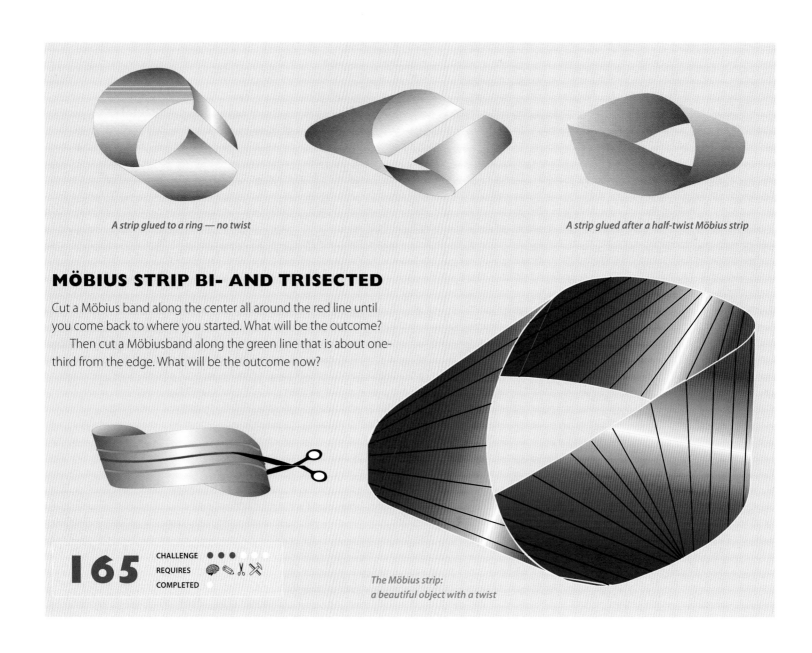

A strip glued to a ring — no twist

A strip glued after a half-twist Möbius strip

MÖBIUS STRIP BI- AND TRISECTED

Cut a Möbius band along the center all around the red line until you come back to where you started. What will be the outcome?

Then cut a Möbiusband along the green line that is about one-third from the edge. What will be the outcome now?

165

CHALLENGE ● ● ● ○ ○ ○

REQUIRES

COMPLETED ○

The Möbius strip: a beautiful object with a twist

KLEIN BOTTLE — 1882

Scientists would define the Klein bottle (first described in 1882 by the German mathematician Felix Klein) as a "closed non-orientable surface." Easier said, it is a surface (a two-dimensional manifold) in which the concept of left and right cannot be consistently defined.

A surface is orientable if it has two sides. Then one can orient the surface by choosing one side to be the positive side and the other to be the negative side.

Any surface that has a Möbius band in its structure is non-orientable. Whereas a Möbius strip is a surface with boundary, a Klein bottle has no boundary. By contrast, a sphere is an orientable surface with no boundary.

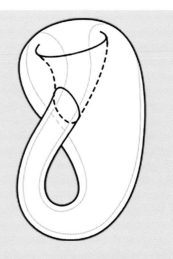

Two Möbius bands result when a Klein bottle is cut along a curve.

THE MÖBIUS BAND

Until you create a Möbius band, it is understandable to be skeptical about whether a one-sided surface can exist.

The Möbius band is the simplest possible one-sided surface. It has a boundary, whereas a sphere has no boundary.

Can a one-sided surface have no edges at all? The answer is yes, but no such surface can exist in three-dimensional space without crossing through itself.

On the right, you can see a beautiful glass Klein bottle blown by Alan Bennett. The glass Klein bottle meets itself in a small, circular curve; topologists ignore that intersection when thinking about an ideal Klein bottle.

> **"A mathematician named Klein
> Thought the Möbius band was divine.
> Said he: 'If you glue
> The edges of two,
> You'll get a weird bottle like mine.'"**
>
> — *limerick by Leo Moser (1921–1970)*

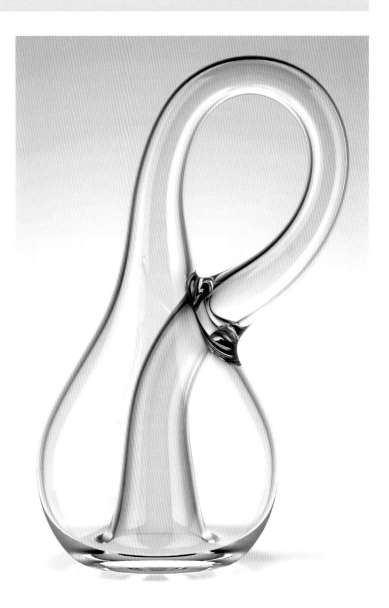

MÖBIUS SIAMESE

Take a strip of paper and cut two longitudinal slots as shown.

Bring the upper pair of ends together and join with a half-twist so that A joins A, and B joins B. Then do the same with the lower pair of ends, but twisting in the opposite direction, joining A to A and B to B.

The result will look like the structure above.

Can you envision what the outcome will be if you cut the structure along the red lines?

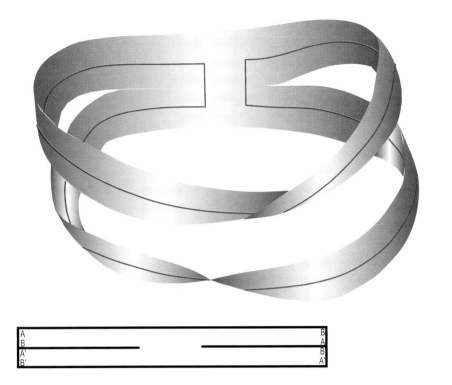

MÖBIUS OR NOT?

Martin Gardner showed the paper structure on the right, sent to him by Josiah Manning of Aurora, Mo., asking his readers whether this surface is topologically equivalent to a Möbius band.

Can you find by working out what the outcome will be if we cut the surface along the red lines?

FOUCAULT PENDULUM — 1851

How do we know that the Earth moves? Astronomers from Plato's time up until the 16th century tended to think that the Earth sat still while everything else rotated around it.

Theories contradicting this view were not lacking but the problem was that convincing evidence was not available. We certainly can't feel that we are on a moving platform, but can we see the Earth moving?

Is it possible to watch the Earth rotate?

In the year 1543, Copernicus sent a copy of his book *On the Revolutions of the Heavenly Spheres* to Pope Paul III with a note containing the famous understatement: "I can easily conceive that as soon as people learn that in this book I ascribe certain motions to the Earth, they will cry out at once that I and my theory should be rejected."

Some still disbelieved the theory when the French physicist Jean-Bernard Foucault was invited to arrange a scientific exhibit as part of the Paris Exhibition in 1851.

From the dome of the Pantheon, Foucault hung a pendulum consisting of 61 meters of piano wire and a 27 kilogram cannonball. On the floor below the ball, he sprinkled a layer of fine sand. A stylus fixed to the bottom of the ball traced the path in the sand, thus recording the movement of the pendulum. After an hour had passed, the line in the sand had moved 11 degrees and 18 minutes. If the pendulum stayed in the same plane, how could it trace different paths in the sand?

The Foucault pendulum demonstration was certainly one of the most beautiful and impressive scientific demonstrations ever, and still is, at science museums and science exhibitions all over the world. Foucault's enormous contribution to science was that with his pendulum he made a complicated idea comprehensible to everybody.

168 CHALLENGE ● ● ● ● ●
REQUIRES 🧠 ✏️ ✂️
COMPLETED ○

Foucault's pendulum in the Panthéon in Paris.

FOUR-COLOR THEOREM — 1852

In 1852, 21-year-old Francis Guthrie stated that what was until recently called the "Four-color problem" was simple enough to state, but not so easy to prove.

How many colors are needed so that any map can be colored in such a way that no adjacent regions (which must touch along an edge, not just at a point) have the same color?

It is not hard to show that at least four colors are needed. In the 19th century a mathematician called Kempe published a proof that no map needed five colors. Ten years later it was noticed that he had made a subtle but crucial mistake, and that his proof showed that no map required six colors. Ever since, that left a tantalizing gap.

For about 100 years people wrestled with the problem. Nobody could find a map that actually needed five colors; but nobody could show conclusively that no such map existed. It became notorious as one of the simplest remaining unsolved classical mathematical problems. To make matters worse, analogous problems dealing with more complicated surfaces could be answered conclusively. For example, a map on a doughnut can always be colored with seven colors; and there exist maps for which six colors do the job. On a strange, one-sided surface called a Klein Bottle, six colors are both necessary and sufficient.

Then, in the late 1970s, two mathematicians at the University of Illinois solved the problem using a supercomputer that finally cracked the four-color problem, so we now have the "Four-Color Theorem," on which many beautiful puzzles are based.

The difficulty of seeing the potential color "cul-de-sacs" that can occur during coloring maps in advance, seemingly requiring a fifth color, makes these problems and games challenging. The solution of the four-color theorem was the first mathematical application of the great computational power of the digital computer by Kenneth Appel and Wolfgang Haken, which required thousand of hours of computer time.

This was the first time that the proof of a mathematical theorem could not be verified by hand.

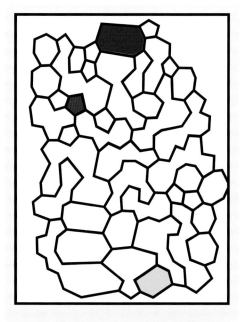

COLORING PATTERN

Color the pattern so that no two regions with a common border are the same color. How many colors will be needed?

169 CHALLENGE ● ● ● ● ○
REQUIRES 🧠✏️✂️
COMPLETED

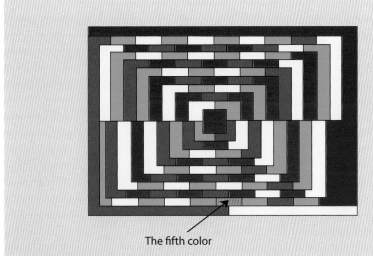

The fifth color

THE FIFTH COLOR

On April 1, 1975, Martin Gardner published the above color map (left), designed by William McGregor, that could not be colored by fewer than five colors. On the right, the map is shown uncolored. Can you do better?

170 CHALLENGE ● ● ● ● ○
REQUIRES 🧠✏️
COMPLETED

COCKTAIL GLASSES

The three cocktail glasses are on a horizontal platform resting on three shapes, one of which is circular and the other two of which are irregular.

What will happen to the glasses when the three shapes start revolving? Will the cocktails spill?

REULEAUX TRIANGLE — REULEAUX POLYGONS — 1854

The circle is the simplest closed curve of "constant width" — which is its diameter. For this reason it was the ideal shape ages ago for cylindrical rollers moving very heavy weights from one point to another.

Are there other curves than the circle that have the same property — i.e. curves of constant width? There is an infinite number of them.

The simplest of non-circular curves of constant width is the Reuleaux Triangle, named after Franz Reuleaux (1829–1905), a German engineer, though the shape was known earlier to Leonard Euler (18th century) and to Leonardo da Vinci (15th century). It can also be found on the windows of the Notre Dame cathedral in Bruges (13th century).

The width of this curve in every direction is equal to the side of the equilateral triangle or the distance across from the triangle's vertex to the opposite arc. This is also the distance between two parallel lines tangent to the figure, which will stay the same as the curves revolve.

The Reuleaux Triangle and its mechanical properties found their application in the original design of the Wankel internal combustion engine in 1957.

Such a figure can easily be constructed. Draw an equilateral triangle. Then, with each of the three corners as centers, draw the circular arc passing through the other two corners. One of the many astonishing properties of the Reuleaux triangle is that the ratio of its perimeter to its width is also equal to π as it is in circles.

A Reuleaux triangle is the simplest and best known Reuleaux polygon of constant width. There is an infinite number of Reuleaux polygons.

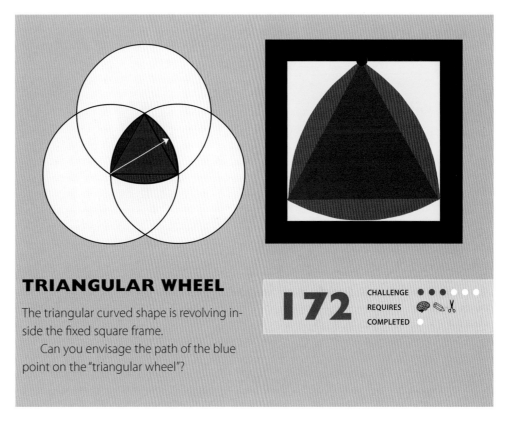

TRIANGULAR WHEEL

The triangular curved shape is revolving inside the fixed square frame.

Can you envisage the path of the blue point on the "triangular wheel"?

WILLIAM ROWAN HAMILTON (1805–1865)

The Irish physicist, astronomer and mathematician Sir William Rowan Hamilton is thought to have been a child prodigy in a number of fields. He made a significant contribution to classical mechanics, optics and algebra, discovering various new mathematical concepts and techniques. In mathematics, he is perhaps best known for discovering quaternions, but is his reformulation of Newtonian mechanics, now called Hamiltonian mechanics, that truly made him famous. His work has become a focal point of the modern study of classical field theories like electromagnetism, and to the development of quantum mechanics.

JOURNEY AROUND A DODECAHEDRON — 1859

We have been traversing graphs on the plane. A more difficult group of problems involves finding routes on three-dimensional objects. One of the first examples of this, and a classic, was invented by W.R. Hamilton in 1859, involving a dodecahedron. His problem asked whether a route along the edges exists, which would come back to the starting point after visiting all 20 vertices (a path called a Hamiltonian circuit) and without retracing an edge.

Note: in a Hamiltonian path or circuit, all the vertices must be visited, while some of the edges can be left untraversed. To make it easier to solve such 3D problems, Hamilton used a two-dimensional diagram of the dodecahedron (so-called Schlegel diagram), which is topologically equivalent to the three-dimensional solid. Hamilton has devised a branch of mathematics to solve similar path-tracing problems on three-dimensional solids, called Icosian calculus. His Icosian Game puzzle was commercialized as a pegboard game with holes at the nodes of the dodecahedral graph. It was later marketed throughout Europe in a variety of forms.

173

CHALLENGE ●●●●●●
REQUIRES
COMPLETED

DIGRAPH TRIANGULATION — 1857

If an arrow is added to each line of a graph giving each line a direction, then the graph becomes a digraph. A complete digraph is a graph in which every pair of points is joined by an arrow. The picture on the right is a complete graph on seven points.

The object of the puzzle is to transform it into a complete digraph by adding an arrow to every line, so that for any two points it is always possible to get to each point in one step from some third point. For example, arrows are added to three lines. For points 1 and 2, we can see that only one step is needed to get to either of them from point 7.

Can you add the rest of the arrows to each line of the graph to fulfill this condition for all points?

The concept of digraphs (or directed graphs) is one of the richest theories in graph theory, mainly because of its applications to physical problems.

174
CHALLENGE ● ● ● ● ○
REQUIRES 🪨 ✏️
COMPLETED ○

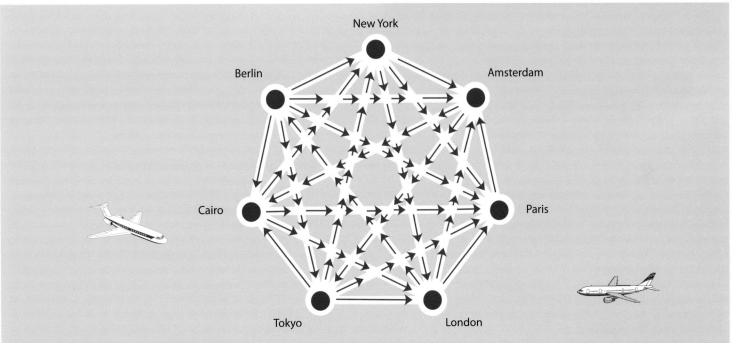

WORLD TRIP DIGRAPH

Can you choose any of the cities and make a trip visiting all the cities following the directions indicated by the arrows on each line and never retracing the lines? For example, what will be the order of the trip from Berlin to London visiting all cities?

A complete digraph, like our graph on seven points, is called a tournament.

An astonishing property of a complete digraph is that no matter how the arrows are drawn, every tournament will have a Hamiltonian path, which is a route that visits every vertex once. It should be noted

that some edges may not be visited during the trip while completing a Hamiltonian path.

175
CHALLENGE ● ● ● ● ○
REQUIRES 🪨 ✏️
COMPLETED ○

TRAVELING SALESMAN PROBLEM — 1859

A related problem to Hamiltonian circuits is the traveling salesman problem.

The traveling salesman problem is the problem of finding a Hamiltonian circuit in a complete weighted graph for which the sum of the weights of the edges is the minimum.

A complete graph is a graph in which every pair of vertices is connected by exactly one edge.

A weighted graph is a graph for which a number, called a weight, is assigned to each edge of the graph (which can be a distance or another value). The sum of all weights is called the weight of the circuit. The solution to the traveling salesman problem is the circuit of minimum weight. In most problems a particular vertex is assigned as the starting vertex.

Can you find the circuit of minimum weight for the weighted graph on five points?

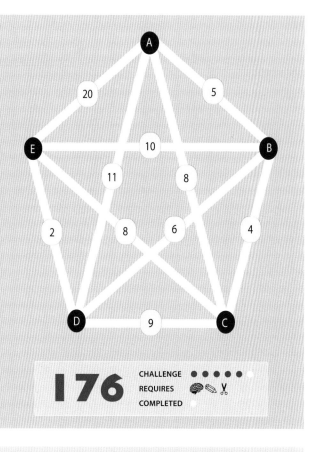

176

CHALLENGE ● ● ● ● ●
REQUIRES
COMPLETED

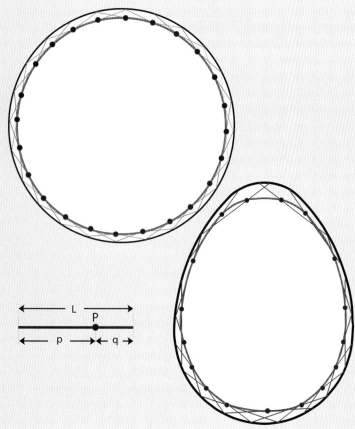

HOLDITCH'S THEOREM — 1858

A chord of constant length is divided by a point P into two segments of lengths p and q, slides along convex curves, with the two ends of the sliding chord touching the curve at all times. Our examples are a circle and an egg-shaped curve. Point π will describe a new curve inside the original curve as shown. The curve inside the circle will be another circle.

The problem is to find the area between the initial curve and the resulting one in each curve (blue areas).

Rev. Hamnet Holditch (1800–1867) published his theorem in 1858. It states that the area between the two curves is π p q, an amazing result, since this area is completely independent of the shapes of the curves involved.

HAMILTONIAN PATHS AND CIRCUITS

Eulerian paths and circuits were concerned with finding paths that cover every edge of a graph. Hamiltonian paths and circuits deal with problems of visiting all of the vertices of a graph, without concern for whether or not all edges have been retraced.

This type of problem was first studied by the Irish mathematician Sir William Rowan Hamilton, who was especially interested in problems of finding a circuit that goes through every vertex exactly once and returns to the starting vertex, which today are called Hamiltonian circuits.

Those paths that don't return to the starting vertex after visiting every vertex are called Hamiltonian paths. Unlike with Eulerian paths and circuits, there is no quick method for determining whether a graph has a Hamiltonian path or circuit in general.

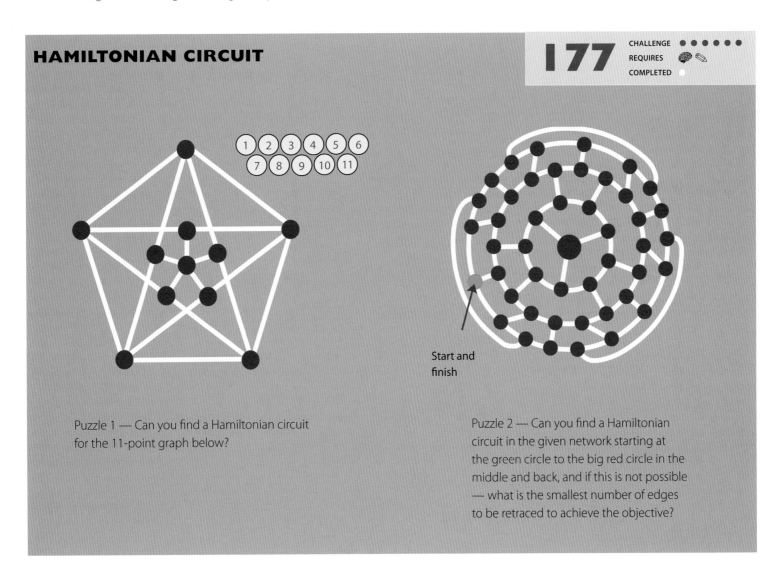

HAMILTONIAN CIRCUIT

Start and finish

Puzzle 1 — Can you find a Hamiltonian circuit for the 11-point graph below?

Puzzle 2 — Can you find a Hamiltonian circuit in the given network starting at the green circle to the big red circle in the middle and back, and if this is not possible — what is the smallest number of edges to be retraced to achieve the objective?

THE PUZZLING WORLD OF ILLUSIONS — 1860

Among the most interesting of perceptual phenomena are optical illusions or, as they are sometimes called, "geometrical paradoxes."

In optical illusions we see things as we think they should be rather than as they are, because of our previous experiences and influences. This visual property of our perceptual system is widely applied in our daily lives, in science, math, art and design.

It should give us a warning as to the reliability of our observations and our senses in general, stressing the importance of measurement.

Optical illusions represent the other way that shapes can seem to change — the way we can see them change. The way we see things — or more precisely, the way we understand what we see — is also based on a set of rules. These are not written rules, but are learned through experience.

Just as logical rules can seem to be broken down into paradoxes, so the rules of perception can also seem to be wrong.

When this happens, an illusion occurs. Understanding what can happen when this occurs helps us understand exactly how important the rules are in the first place, and how much we depend on them.

We can be made to believe that things are larger than they are; we can be made to see depth in a two-dimensional flat surface, see colors where there are none and see motion where there is none.

Much of perception is like a language that has to be learned.

Our contact with the world is 90 percent through the eyes — all day long, until we close that apparatus to the world. The visual system is not simply a camera, a direct receiver and recorder of information.

Together, the eye and the brain are an organizing apparatus that analyzes and processes the large mass of data coming from the outside world. The visual apparatus is not only capable of eliminating the irrelevant and recognizing the unfamiliar, but as we shall see, it is also able to operate with limited information. It "fills in" where there are gaps.

This may be called "the etcetera principle," meaning that when we see a few members of a series and an indication of the rest, it assumes the existence of them all.

Much of art is based on this tendency to fill in, to complete, to organize and so is much of ordinary vision. There is much more wonder about perception in general that you can discover by pursuing the subject further.

In any case, we should be aware that there is a limit to our senses and no amount of practice can ever make them good enough for some special tasks. The solution is to find ways to extend our senses — to invent tools capable of this.

Fortunately, during history, humankind has always been successful in creating such tools whenever the need arose for them.

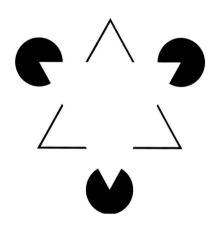

ZOLLNER ILLUSION — 1860

The German astrophysicist Karl Friedrich Zollner (1834–1882) discovered a classic optical illusion that today carries his name: the Zollner Illusion. Zollner wrote of his discovery in a letter in 1860 to Johann Christian Poggendorff, a physicist and scholar and the editor of *Annalen der Physik und Chemie,* who later discovered the related Poggendorff Illusion.

In the Zollner illusion, the black lines appear not to be parallel, although in fact they are. The shorter lines are on an angle to the longer lines thus creating the misleading illusion that one end of the longer lines is closer to the viewer than the other.

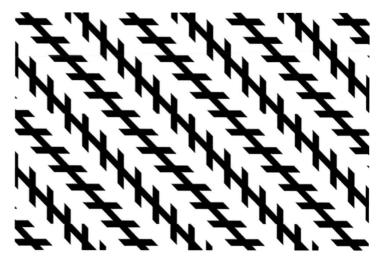

POGGENDORFF ILLUSION

The Poggendorff Illusion is a geometrical optical illusion involving the brain's mistaken perception of the position of one section of a transverse line passing behind an obscuring structure (in this case rectangles).

178 CHALLENGE ● ● ○ ○ ○ ○
REQUIRES 🧠
COMPLETED

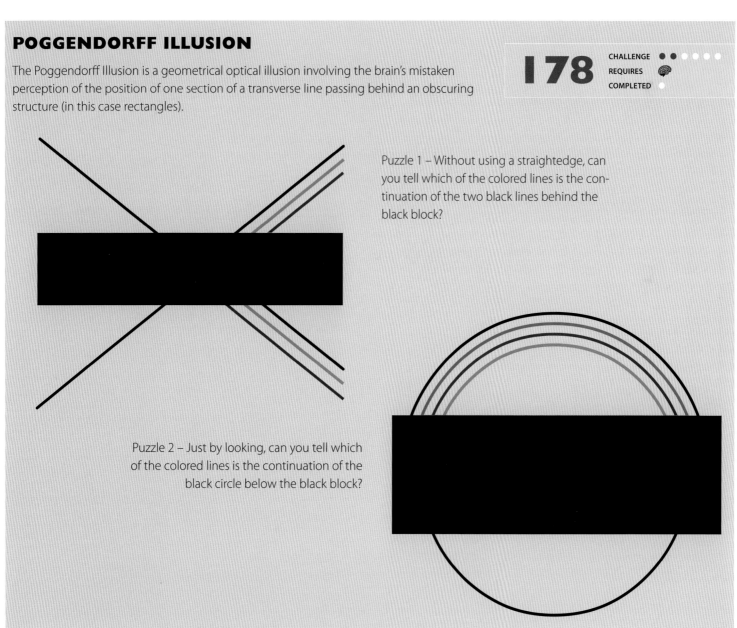

Puzzle 1 – Without using a straightedge, can you tell which of the colored lines is the continuation of the two black lines behind the black block?

Puzzle 2 – Just by looking, can you tell which of the colored lines is the continuation of the black circle below the black block?

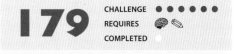

DOGS IN TRIPLES — 1863

Puzzle 1 – Six girls and three boys take turns walking their dogs in triples for 12 days, each pair appearing just once in a triple. Can you find a solution?

Puzzle 2 – A different variation of this problem was also introduced by Steiner.

Suppose the nine children walk the dogs in three triples for four days, again with each pair appearing in a triple only once. How will the solution look in this case?

179

CHALLENGE ● ● ● ● ● ●
REQUIRES 🧠 ✏️
COMPLETED

LINES AND LINKAGES — POINTS AND LINES

There is something fascinating about the motion of linkages. You can easily construct simple linkages from cardboard strips joined by fasteners or eyelets. A linkage in the plane is a system of rods connected to each other by movable joints, or fixed to the plane by pivots about which they can turn freely.

An early problem, long thought impossible exists: can a linkage be found that will produce, by the motion of one of its points, a straight line?

Pivot a single rod at one end: how does the free end move? In a circle. Circular motion is easy and natural for linkages. The trick was how to construct straight-line motion in the absence of a fixed straight line.

This is not just a theoretical problem in geometry. The natural motion produced by a steam engine is rotary. While it can be converted to straight-line motion by a piston, pistons require bearings and bearings are subject to wear.

A linkage would provide a more satisfactory solution. The first practical solution, devised by James Watt (1736–1819), the inventor of steam engines, was only approximate.

> **"Mathematics would certainly have not come into existence if one had known from the beginning that there was in nature no exactly straight line, no actual circle, no absolute magnitude."**
>
> — *Friedrich Nietzsche*

PEAUCELLIER-LIPKIN AND WATT'S LINKAGE — 1864

The Peaucellier-Lipkin linkage (invented in 1864) was the first planar linkage capable of transforming rotary motion into perfect straight-line motion, and vice versa. It is named after Charles-Nicolas Peaucellier (1832–1913), a French army officer, and Yom Tov Lipman Lipkin (1846–1876), son of the famous Lithuanian Rabbi Israel Salanter.

Before the linkage was invented, there was no planar method to produce straight motion without reference guideways, and it therefore became particularly significant as a machine component and for manufacturing.

A specific example is the piston head, which requires an efficient seal with the shaft to maintain the driving (or driven) medium. The Peaucellier linkage was crucial to the development of the steam engine.

The mathematics of the Peaucellier-Lipkin linkage are directly related to the inversion of a circle.

The Sarrus linkage was a slightly earlier straight-line mechanism, though it received little attention at the time. Invented by Pierre Sallus 11 years before the Peaucellier-Lipkin, it comprises a series of hinged rectangular plates, two of which remain parallel but can be moved normally in relation to each other. While the Peaucellier-Lipkin

> **"Mechanics is the paradise of the mathematical sciences because by means of it one comes to the fruits of mathematics."**
>
> — *Leonardo da Vinci*

linkage is a planar mechanism, the Sarrus linkage, also called a space crank, is three-dimensional.

Looking at the two linkages, can you guess what paths the white points will describe when the blue linkages move in a rotary motion along a circular path?

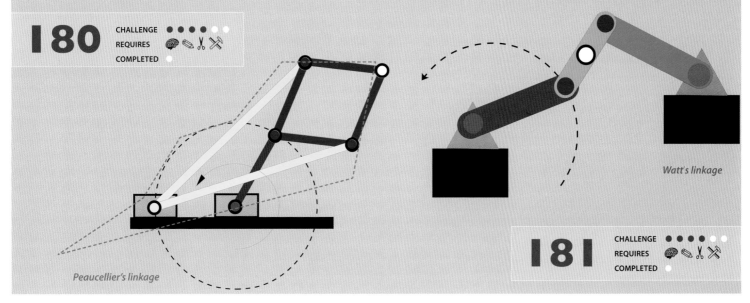

180
CHALLENGE
REQUIRES
COMPLETED

Peaucellier's linkage

Watt's linkage

181
CHALLENGE
REQUIRES
COMPLETED

JOURNEY THROUGH THE EARTH — 1864

Ok, so we have drilled a hole through the Earth.

As a thought experiment, what would happen if, God forbid, a person fell into the hole?

For your journey we shall assume the Earth to be of uniform density, and neglect air friction and the high temperatures inside the Earth, etc.

182 CHALLENGE ● ● ● ● ● ●
REQUIRES
COMPLETED

PROBLEM-SOLVING AND PUZZLES

Marcel Danesi, in his challenging book *The Liar Paradox and the Towers of Hanoi: the Ten Greatest Math Puzzles of All Time,* mentions the Trick Mules puzzle of Sam Loyd as an excellent example of insight thinking, which is so important in solving puzzles and problems.

In general, there are three different types of strategies when it comes to solving puzzles:

1 – Deduction: a strategy requiring previous knowledge related to the problem.

2 – Induction: observing the facts included in the problem, reaching a solution by reasoning and logic.

3 – Insight thinking: a method that may start with a trial-and-error approach, then by guesses and hunches, an intuitive grasp of the hidden answer to the problem is revealed. Insight thinking is the basis of giant strides in the progress of mathematics. Many of the problems of mathematics were originally devised or disguised as challenging puzzles.

TRICK HORSES PUZZLE

The Trick Horses puzzle is based on the original "Trick Mules" advertising puzzle by Sam Loyd, the great American puzzle inventor.

He created it when he was still a teenager. It is one of the most beautiful puzzles ever devised, in a category of its own, and literally a visual masterpiece of lateral thinking.

Using only your imagination, can you mount the riders on their horses when the strip with the two riders is cut out and placed on top of the two horses?

If you can't conceptualize the solution, copy and cut out the strip with the two riders. Arrange the strip so that the two riders properly ride the horses.

Hint: the problem looks deceptively simple until you try it. When the strip is correctly placed, the weary looking horses miraculously break into a frenzied gallop!

No tricks, no bending, folding or cutting is allowed.

Confronted with this puzzle, many have a conceptual "block," and are simply unable to place the riders properly. But the solution is really easy.

Loyd sold his Trick Mules puzzle to P.T. Barnum, who sold millions and earned $10,000 in royalties in just a few weeks — a fortune at the time. Ever since, hundreds of variations of the puzzle have appeared. The Trick Mules puzzle by Sam Loyd may have been inspired by a Persian Horses ink drawing from the 17th century.

> **"It isn't that they can't see the solution. It is that they can't see the problem."**
>
> — *G.K. Chesterton*

> **"A problem adequately stated is a problem well on its way to being solved."**
>
> — *Buckminster Fuller*

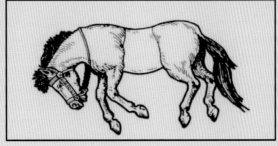

183

CHALLENGE ● ● ● ● ○ ○
REQUIRES 🧠
COMPLETED ○

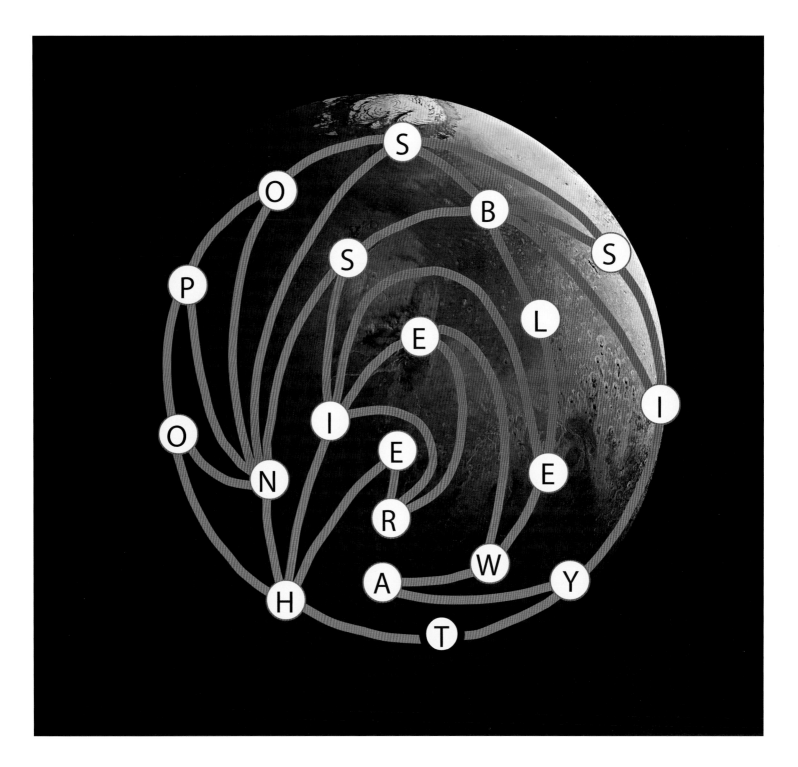

SAM LOYD'S MARS PUZZLE

Start at "T," visit all 20 Mars stations and spell out a complete English sentence. You must travel along the "canals" and you are not allowed to visit any stations more than once.

When Sam Loyd first published his "Mars puzzle," he got over 10,000 letters related to the solution claiming "there is no possible way." Can you solve the puzzle?

184 CHALLENGE ● ● ● ● ● ●
REQUIRES
COMPLETED

GEOMETRICAL VANISHES — 1871

The remarkable images known as "geometrical paradoxes" or "geometrical vanishes" are so subtle that they continue to intrigue and surprise, becoming conversation pieces and causing us to question our sense of perception even after their workings have been explained.

Sam Loyd, the American puzzle creator, was the originator of the most famous puzzle of this kind, the "Get Off the Earth" puzzle.

Mel Stover (1912–1999) and many others have perfected the art, creating subtle variations and perfections of the principle. Geometrical paradoxes involve separating and rearranging parts of a total length or area. After rearrangement, a portion of the figure has somehow disappeared. The explanation lies in the principle of concealed distribution, as Martin Gardner has named it, which depends on the eyes' tolerance for the rearranged version. They often fail to notice a tiny increase in the gaps between the parts or in the lengths of the reassembled pieces, and so believe both have the same length or area.

For instance, in the left picture: 12 vertical lines become 11 when the bottom half is shifted to the right.

In the right picture: 12 radial lines become 11 when the inner wheels rotates one notch counterclockwise. Obviously in both cases nothing has really disappeared.

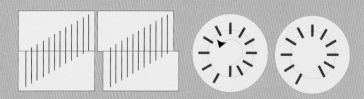

PENCIL MAGIC (1)

When the inside wheel in the top picture is rotated three spaces clockwise, the picture will change from seven blue and six red pencils to six blue and seven red pencils as shown on the right. Can you tell which pencil changed color? The puzzle is a novel design variation of Mel Stover's classic.

185 CHALLENGE ● ●
REQUIRES 🧠 ✏️ ✂️ ✈️
COMPLETED ○

PENCIL MAGIC (2)

Seven red pencils and six blue pencils. Imagine exchanging the two lower parts of the figure. Can you guess what will happen?

186 CHALLENGE ● ● ○ ○ ○
REQUIRES 🧠 ✏️ ✂️ ✈️
COMPLETED ○

BIN PACKING PROBLEM — 1873

A group of problems known as the bin-packing problem is important in industry and technology.

The idea is to pack a set of objects into a number of bins, so that the total weight (length or total volume) doesn't exceed a given number (the size of the packing bins).

The algorithm of packing the bins by taking the objects in any order in the first bin they fit in, etc., is called the "first-fit bin packing." It is not very efficient. Simple experience and logic significantly improved this algorithm, replacing it with the so-called "first-fit, heaviest-to-lightest" algorithm: order the objects from heaviest to lightest, before putting each object into the first bin that can accommodate it. Such an algorithm is never to be off by more than 22 percent.

In 1973, David Johnson, at AT&T, proved that no bin-packing strategy can ever be devised to guarantee to do better than 22 percent.

Ron Graham provided an interesting bin-packing problem associated with a counterintuitive paradox: using the "first-fit, heaviest-to-lightest" algorithm, can you work out the number of bins of 524 kg capacity needed to pack 33 weights in the following sizes (kilograms) shown below: 442, 252, 252, 252, 252, 252, 252, 252, 127, 127, 127, 127, 127, 106, 106, 106, 106, 85, 84, 46, 37, 37, 12, 12, 12, 10, 10, 10, 10, 10, 10, 9, 9.

187 CHALLENGE ●●●● ○ ○
REQUIRES 🧠✏️✂️
COMPLETED

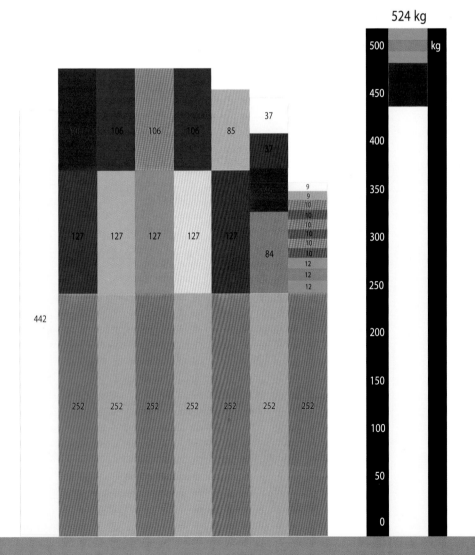

524 kg

BIN PACKING PUZZLES

The first bin packed according to the "first fit, heaviest-to-lightest" algorithm.

How many more bins will be needed to pack all 33 weights?

Now throw out the 46-kilogram weight and repack everything according to the same rules.

How many bins will be needed now?

THE FAMOUS SLIDING BLOCK PUZZLE (AND THE STORY BEHIND IT) — 1880

Sliding the square tiles numbered 1-15, can you change the number configuration into an ordered one, by transforming the incorrect 15-14 sequence into the correct 14-15 sequence?

The forefather of all sliding block puzzles is doubtlessly the famous "Fifteen Puzzle," which is still widely marketed in many different forms and variations.

If you tried to solve the 14-15 puzzle, you may have been disappointed that you were unsuccessful. Don't be! The 14-15 puzzle, thought of long ago by Sam Loyd, cannot be solved.

There were two instances in the history of puzzles and recreational mathematics that involved a world-wide craze: the 14-15 puzzle, more than 120 years ago, and the Rubik's Cube recently (see Chapter 9).

Sam Loyd offered a $1,000 reward to anyone who solved his puzzle and he must have been quite confident that no one would claim the reward. Among over 600 billion possible arrangements of the 14-15 puzzle blocks, 50 percent are impossible to restore to the ordered number sequence, and Loyd's puzzle configuration is just one of these. Loyd knew that restoration of the blocks is possible only in an even number of exchanges.

So a simple parity check will reveal to you whether any arrangement has a solution or not. We switch pairs of numbers until the desired pattern is achieved, counting the number of switches. If the number of switches is even, the change by sliding is possible. Otherwise it is not. In computer language, the 15 Puzzle, as it is usually known, and similar sliding block puzzles are models of sequential machines. Each movement of a block is an input, each arrangement or state of the blocks. The solver will quickly discover that is an al-most hypnotic fascination in pushing the blocks about in search of a minimum chain of inputs that will produce the desired state. It is by no means all trial and error! The mind soon "sees" that certain lines of play lead to blind alleys whereas other lines of play are promising and eventually your intutition will lead you to the solution.

Sam Loyd claimed that he had invented the 15 Puzzle, which was in fact invented by Noyes Palmer Chapman, a postmaster in Canastota, New York under the name "Gem Puzzle" in 1874. He applied for a patent in March 1880, which was rejected because it was not sufficiently different from the 1878 "Puzzle-Blocks" patent (US 207124) granted to Ernest U. Kinsey.

The real story of the 15 Puzzle was revealed by the beautiful ultimate book on the 15 Puzzle by Slocum and Sonneveld.

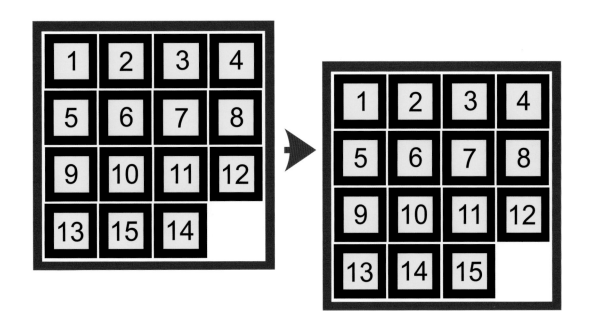

CHAPTER

6

SCIENCE, DIMENSIONS, RANDOMNESS AND THE TOWER OF HANOI PUZZLE

LUCAS' PUZZLE — 1883

For my ninth birthday, I received my first puzzle. It was a wooden base with seven pegs and two sets of three rings placed on the pegs, one set red and the other blue. It looked like the drawing below. The object was to exchange the positions of the two sets of rings, according to simple rules. It looked simple enough to me. I tried to solve it, but after an hour or so I gave up, with the conclusion that it was impossible!

But after a few days, I came back to it, intrigued. It must have a solution, as it said so on the instruction sheet. I decided I had to solve it. I stubbornly renewed the attack on the problem and after a few hours I suddenly came to the solution. I felt great and proud of myself. I decided I liked puzzles. What I didn't know then was that this was the moment when the puzzle bug hit me, and that in solving the puzzle, I had used the method called "brute force."

My first puzzle was the so-called Lucas' puzzle, invented by Edouard Lucas (1842–1891), a famous French mathematician, who invented some of the greatest puzzles of recreational mathematics. Lucas' puzzle was one of the earliest puzzles and games that require arrangement of counters into specified alignments or configurations.

Later, when I started creating and inventing puzzles and games, one of my earliest projects was the Tricky Buttons puzzle range. It was inspired by the basic combinatorial concept of Lucas' puzzle, extending it to any number of playing sets. The four puzzles included here are part of that range. The game board on the next page is designed for playing the puzzles straight out of the book.

TRICKY BUTTON PUZZLES

The object of the four puzzles on the following page is to reverse the pattern by exchanging the two sets of counters (3, 4, 5 and 6 respectively) according to the following simple rules. In the initial configurations, the two sets are placed reds on the left and blues on the right, as demonstrated by the two coin sets sample game, requiring eight moves for the solution:

1 – Only one coin can be moved at a time.

2 – A coin can move into an adjacent empty space.

3 – A coin can jump over one of the opposite colors into a space immediately beyond it.

4 – A coin may not jump over another coin of its own color.

5 – Red coins may only move to the right and blue coins may only move to the left.

What is the minimum number of moves to solve each puzzle? Can you find the general rule, which would give the number of moves required for two sets of any number of play pieces (for example, the minimum number of moves required to solve the puzzle of two sets of 10 coins in each set)?

Two sets of playing pieces: coins or similar

Puzzle 1
Two coins exchange

Puzzle 1
Two coins exchange
RED 1

Puzzle 1
Two coins exchange
BLUE 2

Puzzle 1
Two coins exchange
RED 2

Puzzle 1
Two coins exchange
BLUE 2

Puzzle 1
Two coins exchange
RED 1

Solution of Lucas' puzzle with two sets of two playing pieces solved in eight moves

game board

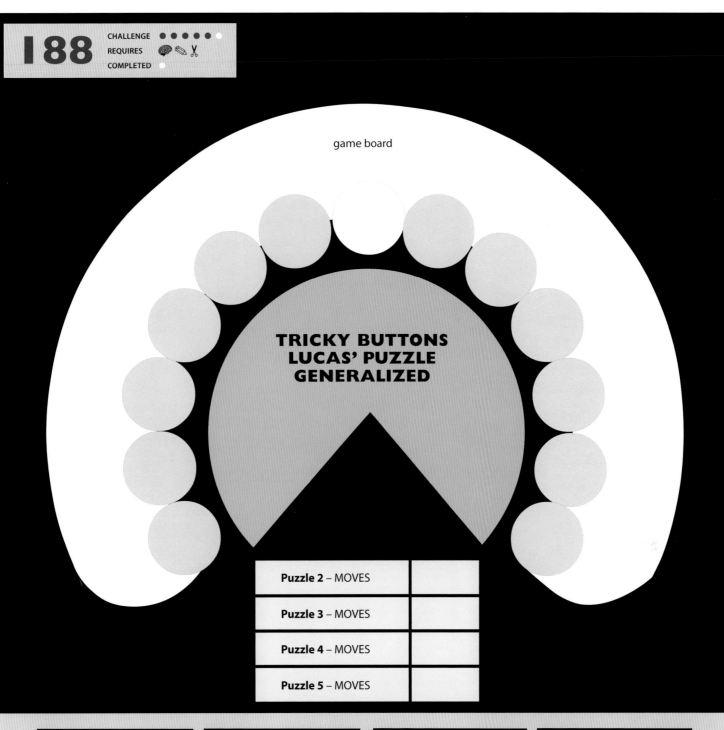

**TRICKY BUTTONS
LUCAS' PUZZLE
GENERALIZED**

Puzzle 2 – MOVES	
Puzzle 3 – MOVES	
Puzzle 4 – MOVES	
Puzzle 5 – MOVES	

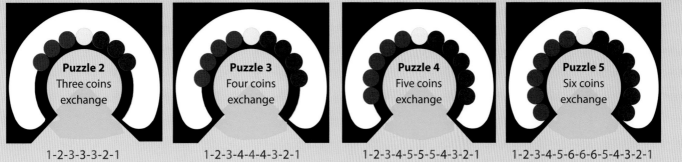

Puzzle 2
Three coins exchange
1-2-3-3-3-2-1

Puzzle 3
Four coins exchange
1-2-3-4-4-4-3-2-1

Puzzle 4
Five coins exchange
1-2-3-4-5-5-5-4-3-2-1

Puzzle 5
Six coins exchange
1-2-3-4-5-6-6-6-5-4-3-2-1

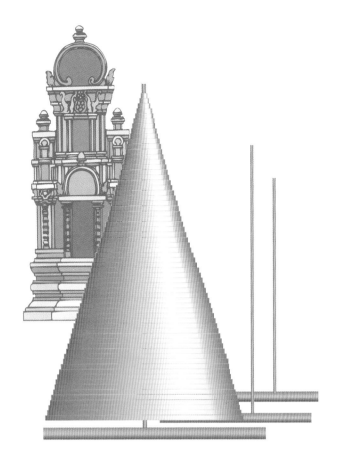

THE TOWER OF HANOI — 1883

Invented in 1883 by French mathematician Edouard Lucas, the Tower of Hanoi is one of the most beautiful puzzles ever created.

The puzzle is accompanied by a legend: At a great temple in Benares, there is a brass plate into which three vertical columns are fixed. At the beginning of time, 64 golden discs were stacked on one pin in decreasing order of size, with the largest resting at the bottom of the brass plate. Day and night, so the legend goes, a priest transfers the discs from one column to another at a constant rate, never allowing any disc to be placed on top of a smaller one. Once the tower is rebuilt on one of the other two pins, the universe will end, or so the legend goes.

Even if the legend were true, there would be no reason to worry. Allowing one second per move of a disc, the task would take about 600 billion years, or about 60 times longer than the lifetime of the sun. The number of moves necessary to complete a Tower of Hanoi of a given smaller number of discs can be calculated as 2n-1. So two discs require three moves, three discs require seven and so on.

BABYLON

189

This puzzle is a variation on the classic known as the Tower of Hanoi.

You can play it on several levels of difficulty, as the sample puzzles below indicate. Initially stack the discs as shown in each puzzle in the left-hand column. Your objective in each puzzle is to transfer the discs to the right-hand column in the same order, with the highest number at bottom.

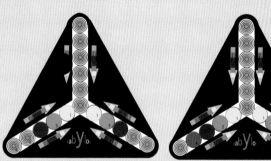

n=3
moves: ?

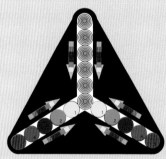

n=4
moves: ?

n=5
moves: ?

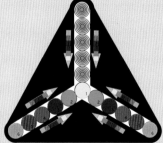

n=6
moves: ?

The object of the four puzzles is to transfer three, four, five, and six discs to the right-hand channel in the same order, the highest number at the bottom of the columns and decreasing numbers toward the top, keeping the following rules:

1 – Move only one disc at a time.
2 – Do not place any disc over another disc of smaller value.
3 – The middle column can be used, but observing rules 1 and 2. How many moves will it take to accomplish the tranfers? Try to solve the first puzzle, with three discs, before you move on to the more difficult ones.

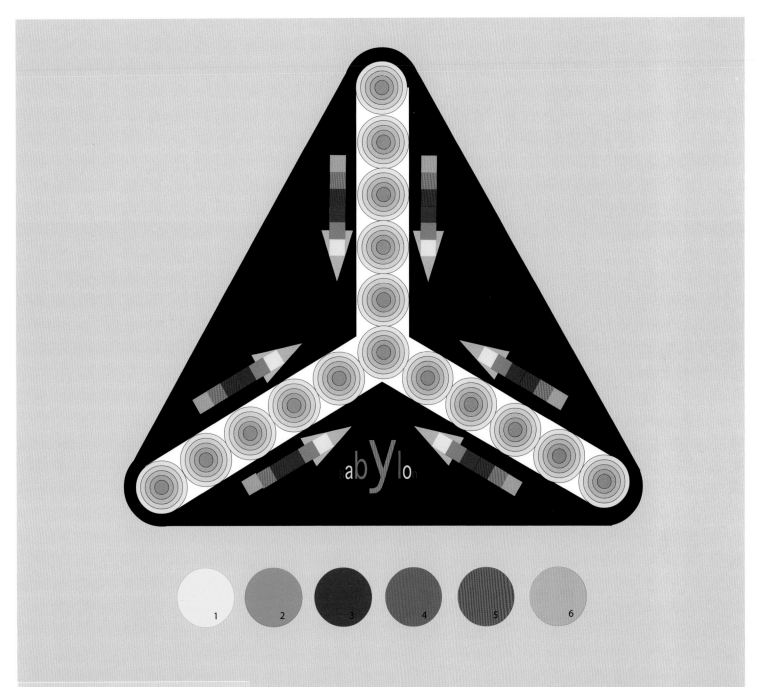

TOWER OF HANOI
GAME BOARD

Use discs or two sets of small coins to play and solve the four puzzles shown on the previous page.

190

CHALLENGE ● ● ● ● ● ○
REQUIRES
COMPLETED ○

SHIPS MEET

This beautiful problem was invented by
Edouard Lucas, the famous French 19th-
century mathematician. Every day at noon,
a ship leaves Le Havre for New York and another ship leaves New York for Le Havre at
exactly the same time. The trip lasts seven
days and seven nights. How many New York
to Le Havre ships will the ship leaving
Le Havre today meet during its journey to
New York?

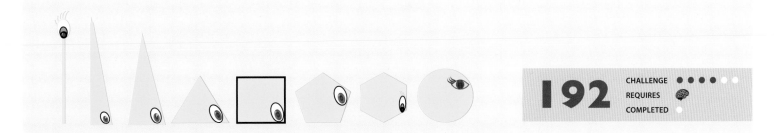

FLATLANDS — TWO-DIMENSIONAL WORLDS — 1884

Astrophysicists say that the universe possesses four dimensions — three of space and one of time — and some recent theories have suggested that there are even higher dimensions.

How can we begin to understand hypothetical higher dimensions? The answer is, by analogy, by getting outside of our normal system: imagining a world that has only two dimensions.

In 1884, Edwin A. Abbott, an English clergyman and popularizer of science, made a fascinating attempt to describe a world made up of only two dimensions. In his satirical novel, called *Flatland,* the characters are basic geometrical figures gliding over the surface of an infinite two-dimensional plane, a vast tabletop. Apart from negligible thickness, Flatlanders have no perception of

the third or any higher dimension.

Although Abbott did not describe any of the physical laws or technological innovations of Flatland, his book spawned sequels that tackle those issues. One such book, *An Episode of Flatland,* written by Charles Howard Hinton in 1907, cleverly extends Abbott's original.

The action in Hinton's book takes place on the apparently two-dimensional planet Astria. Astria is simply a giant circle, and its inhabitants live on the circumference, forever facing in one direction. All males face east, and all females face west. To see what is behind him, an Astrian must bend over backward, stand on his or her head or use a mirror.

Astria is divided between two nations, the civilized Unaeans in the east and the

barbaric Scythians in the west. When the two nations go to war, the Scythians have an enormous advantage: they can strike the Unaeans from the back. The unfortunate and helpless Unaeans are driven to a narrow region bordering the great ocean.

Facing complete extinction, the Unaeans are saved by a scientific advance: their astronomers have discovered that their planet is round. A group of Unaeans cross the ocean and carry out a surprise attack on the Scythians, who have never been attacked from the rear. The Unaeans are thus able to defeat their foes.

Houses in Astria can have only one opening. Tubes or pipes do not exist. Ropes cannot be knotted, although levers, hooks and pendulums can be used.

FLATLANDERS' HIERARCHY

In Abbott's Flatland, a mathematical two-dimensional world:
- ladies are sharp straight lines
- soldiers and workmen are isosceles triangles
- the middle class are regular triangles
- professionals are squares and pentagons
- the upper class starts from hexagons and the ranks go up to circles, who are the high priests of Flatland.

Ladies may be invisible from the back and dangerous to collide with. For these reasons, ladies are required by law to keep themselves visible at all times by executing a kind of perpetual twisting wriggly movement.

FLATLAND CATASTROPHE

Imagine intelligent two-dimensional aliens confined to a two-dimensional surface world called "Flatland." They are confined to Flatland not only physically but also sensuously; they have no faculties to sense anything outside of their surface two-dimensional world.

In an event that occurs every 10,000 years, a three-dimensional giant meteorite cube collides and passes through Flatland. Can you describe how Flatlanders might experience this astronomical catastrophe?

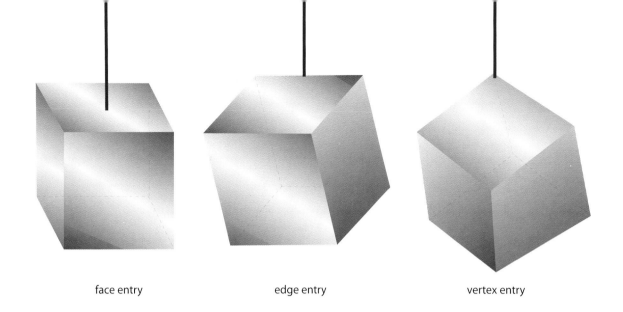

face entry   edge entry   vertex entry

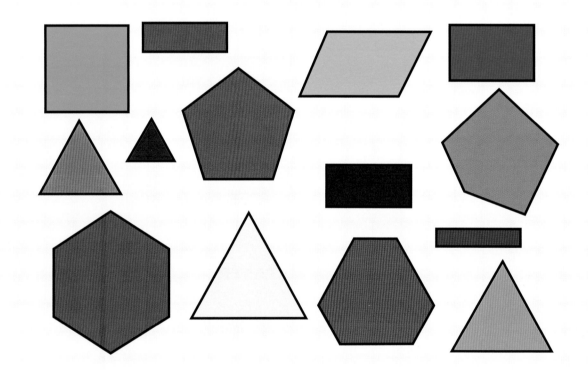

CUBE CUT — 1885

By the 1880s Friedrich Froebel, the inventor of the concept of kindergarten, stressed the importance of introducing children to geometric play and ideas.

When a sphere goes through the plane, like in Abbot's original Flatland, we can easily visualize the sequence: point, increasing circles to a maximum, and then back again, no matter at which points the slicing starts.

But what about a cube? What shapes can be created when a cube goes through the plane? Of the shapes shown above, can all be produced by cutting the plane?

A bonus puzzle: how can you cut a tetrahedron to make a square cross section?

193   CHALLENGE ● ● ● ○ ○ ○
REQUIRES
COMPLETED

JORDAN CURVE THEOREM — 1887

A Jordan curve is another name for a simple closed curve; i.e. a non-self-intersecting continuous loop in the plane.

According to the Jordan curve theorem, every Jordan curve divides the plane into an "interior" region bounded by the curve and an "exterior" region containing all of the nearby and far away exterior points, so that any continuous path connecting a point of one region to a point of the other intersects with that loop somewhere.

The Jordan curve theorem states that a point is inside a simple closed curve if the number of straight-line crossings from an arbitrary direction is odd. Although the statement of this theorem seems to be self-evident, it is no easy feat to prove it by elementary means. More transparent proofs rely on the mathematical machinery of algebraic topology, and these lead to generalizations of higher-dimensional spaces.

The Jordan Curve Theorem is named after the mathematician Camille Jordan (1838–1922), who found its first proof. For a long time it was believed that his proof was invalid, although this belief has been challenged in recent years.

The theorem is indeed obvious for smooth curves. But curves can be fiendishly complicated, and the theorem breaks down for curves like the Koch snowflake.

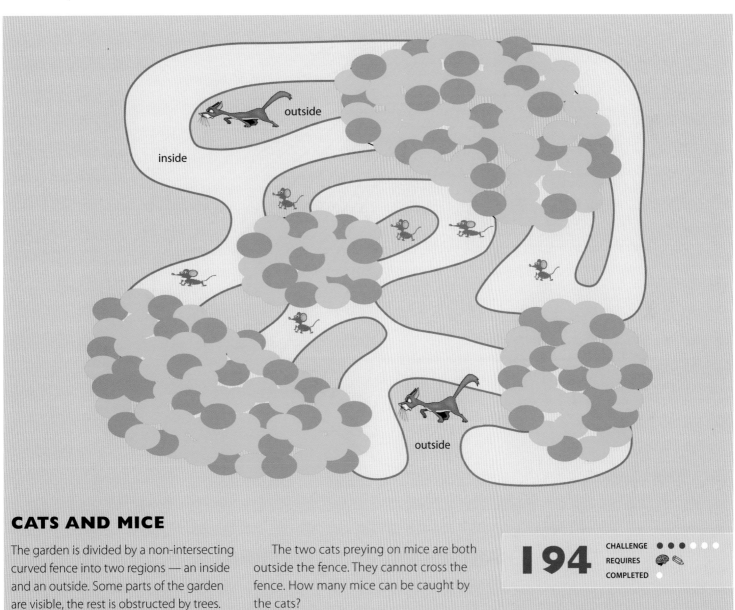

CATS AND MICE

The garden is divided by a non-intersecting curved fence into two regions — an inside and an outside. Some parts of the garden are visible, the rest is obstructed by trees.

Will the Jordan Curve theorem work under such conditions?

The two cats preying on mice are both outside the fence. They cannot cross the fence. How many mice can be caught by the cats?

194
CHALLENGE ● ● ● ○ ○ ○
REQUIRES 🧠 ✎
COMPLETED ○

BERTRAND'S CHORD PARADOX — 1888

In 1888, Joseph Bertrand (1822–1900) introduced a classic and significant problem in probability theory in his work *Calcul des probabilités*, as a way of demonstrating that if the method of producing the random variable is not clearly defined, the resulting probabilities may not be well-defined either.

The problem he posed was: take an equilateral triangle inscribed in a circle. Suppose a chord is randomly chosen. What is the probability that this chord is longer than a side of the triangle?

Bertrand gave three arguments for choosing the random chords, all valid, but all with different results, thus developing the paradox named after him. Since there is no unique selection, there cannot be a unique solution. Only if the method of the random selection is clearly specified can there be a solution to the problem. The three different selections are explained and visualized.

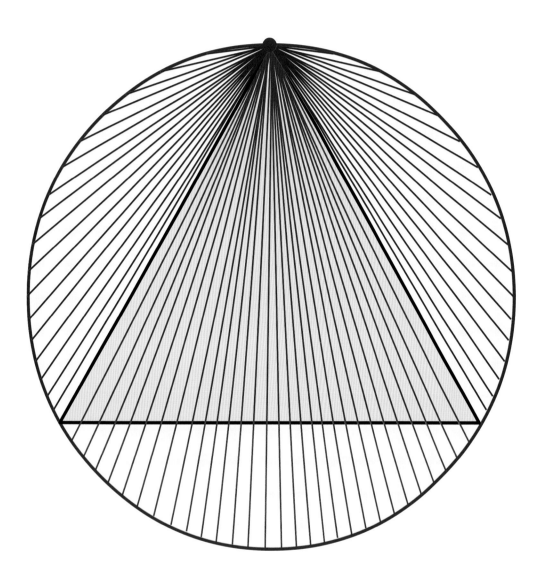

red chords –
longer than
triangle side
Probability: 33%

blue chords –
shorter than
triangle side
Probability: 67%

FIRST SOLUTION: THE RANDOM ENDPOINT METHOD

Choose two random points on the circumference of the circle, one of which coincides with one of the vertices of the triangle. If the other points of the chords lie on the arc between the two other vertices of the triangle, the chords are longer than the side of the triangle. The length of the arc is 1/3 of the circumference, therefore the probability that a random chord is longer than a side of the triangle is 1/3.

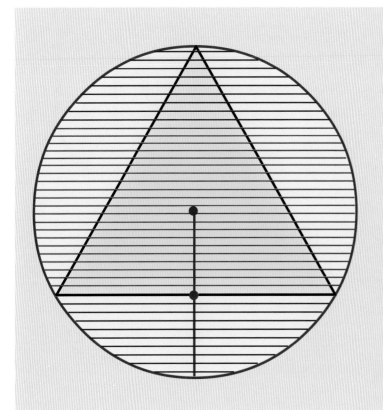

SECOND SOLUTION:
THE RANDOM RADIUS METHOD

Choose a radius bisecting a side of the triangle. Take a point on the radius and construct a chord through this point that is perpendicular to the radius.

The chord is longer than a side of the triangle if the chosen point is nearer the center of the circle than the point where the side of the triangle intersects the radius. Therefore, the probability that a random chord is longer than a side of the triangle is 1/2.

red chords –
longer than
triangle side
Probability: 50%

blue chords –
shorter than
triangle side
Probability: 50%

THIRD SOLUTION:
THE RANDOM MIDPOINT METHOD

Choose a point anywhere within the circle and construct a chord with the chosen point as its midpoint.

The chord is longer than a side of the triangle if the chosen point falls within a concentric circle of a radius 1/2 of the larger circle. The area of the smaller circle is 1/4 of the area of the larger circle and so the probability is 1/4.

red chords –
longer than
triangle side
Probability: 25%

blue chords –
shorter than
triangle side
Probability: 75%

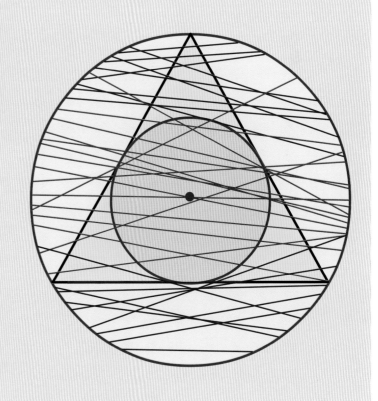

TESSERACT — HYPERCUBE — 1888

The tesseract is the four-dimensional equivalent of the cube, just as the cube is the three-dimensional equivalent of the square. While the surface of the cube has six square faces, the hypersurface of the tesseract has eight cubical cells.

The tesseract is one of the six convex regular 4-polytopes. A generalization of the cube to dimensions greater than three is called a "hypercube," "n-cube" or "n-measure polytope."

The tesseract is the four-dimensional hypercube, or 4-cube.

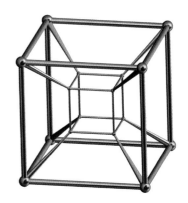

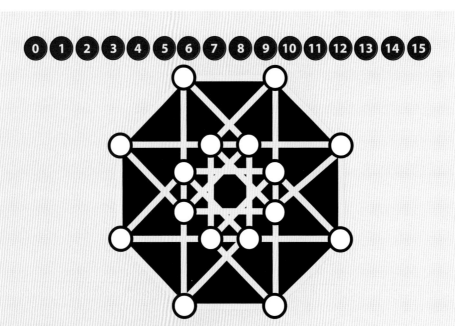

TESSERACT PUZZLES

Puzzle 1 – How many corners, edges, faces and cubes are there in a tesseract?

Puzzle 2 – Fit numbers 0 to 15 in the vertex circles of the tesseract at the top so that the numbers on the square faces of the skeleton cubes in perspective add up to 30.

Puzzle 3 – How many skeleton cubes, like the first shown, can you see in the two-dimensional diagram of the tesseract on the left?

195

CHALLENGE ● ● ● ● ○
REQUIRES 🧠 ✏️
COMPLETED ○

PERIGAL'S REPTILE SQUARE — 1891

Copy and cut out the eight pieces of the truncated triangle and reassemble them into a perfect square.

196   CHALLENGE ●●●●○○
REQUIRES
COMPLETED ○

HENRY PERIGAL (1801–1898)

Henry Perigal was a British amateur mathematician and a member of the London Mathematical Society between 1868 and 1897. He is best known for his cut-and-paste, or dissection, proof of the Pythagorean Theorem. In his *Geometric Dissections and Transpositions* (published in 1891), he included an elegant proof of the Pythagorean Theorem based on the idea of dissecting two smaller squares into a larger square. The five-piece dissection that he found can be produced by overlaying a regular square tiling whose prototile is the larger square with a Pythagorean tiling generated by the two smaller squares.

In *Geometric Dissections and Transpositions* Perigal also wrote of his wish that dissection-based methods could be employed to solve the ancient problem of squaring the circle; the Lindemann—Weierstrass theorem of 1882 had, however, already demonstrated that this problem was impossible to solve.

DOT PORTRAIT

Can you guess how many dots were needed to create a good likeness of Marilyn? Try taking a rough guess, then see if your answer falls within 25 dots either way!

197 CHALLENGE ● ● ●
REQUIRES 🧠 ✏️ ✂️ ⚒️
COMPLETED

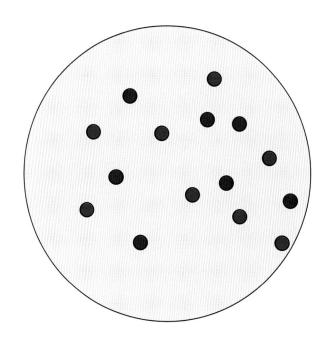

TWO MILLION DOTS IN A CIRCLE

Imagine that inside this circle is a random collection of exactly two million dots but at this magnification you can't see them. Can there exist a straight line crossing the circle having exactly one million dots on each side? Can you find a theoretical procedure, a thought experiment that would solve the problem?

198 CHALLENGE ● ● ●
REQUIRES 🧠 ✏️ ✂️ ⚒️
COMPLETED

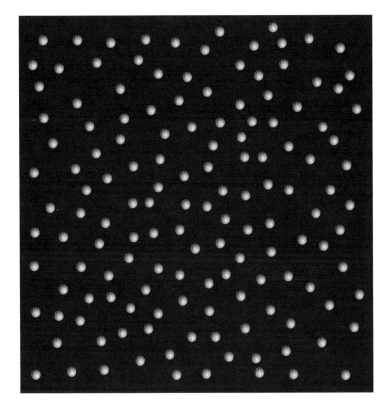

SYLVESTER LINE THEOREM — 1893

Can you find a line that passes through exactly two of the points shown?

James Joseph Sylvester (1814–1897) posed the conjecture in 1893 that given a finite number of points in the plane, there must be at least one line containing exactly two points (otherwise all points are lying on the same straight line), a conjecture proven in 1944 by Hungarian mathematician Tibor Gallai.

The Sylvester's line problem, known today as the Sylvester-Gallai theorem in proved form, states that it is not possible to arrange a finite number of points so that a line through every two of them passes through a third unless they are all on a single line.

199 CHALLENGE ● ● ●
REQUIRES 🧠 ✏️ ✂️ ⚒️
COMPLETED

MAGIC HEXAGON

Volumes were written about magic squares, however the "magic" requirements can be related not only to squares, but also to other polygons, like triangles, hexagons, circles, stars and other polygons.

Puzzle 1 – Is a magic hexagon of order 2 possible? Or, in other words, is it possible to arrange a set of numbers from one to seven in the honeycomb arrangement below with two hexagons on each side, so that all straight rows in any direction have an identical sum? We shall reveal to you that no matter how you place the numbers there won't be a solution. A magic hexagon of order 2 is clearly impossible. Can you find the proof of this impossibility?

Puzzle 2 – On the other hand, a magic hexagon of order 3 is possible as shown below.

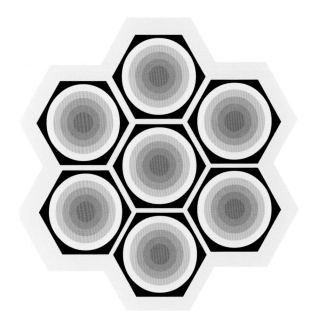

200	CHALLENGE	● ● ● ● ○ ○
	REQUIRES	🧠 ✏️
	COMPLETED	○

MAGIC HEXAGON PUZZLE — 1895

In 1895, William Radcliffe discovered, after much trial and error, that 19 hexagonal pieces numbered from one to 19 can be assembled so that each row of three, four or five hexagons adds up to 38. In 1963, Charles Trigg proved that it is the only magic hexagon of any size!

Radcliffe's Magic Hexagon is a unique and surprising number pattern puzzle: can you distribute the numbers from one to 19 in the hexagonal game board so that the sum of each straight line in any direction is 38?

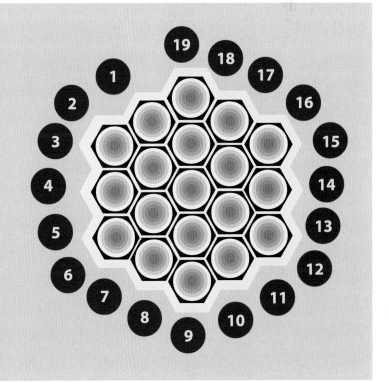

201	CHALLENGE	● ● ● ● ○ ○
	REQUIRES	🧠 ✏️
	COMPLETED	○

PICK'S THEOREM - 1899

Suppose you have a simple lattice polygon (polygon without self-intersection and without "holes") with all the polygon's vertices at points of a square grid lattice, as shown.

The object is to calculate the area enclosed by the polygon. We can do this the hard way, by dividing the interior of the polygon as shown and adding up the areas of each part to get the total area of the polygon (I obtained an area of 84.5 square units in this way).

But there is a beautifully simple way to get the results using Pick's ingenious formula. Pick's theorem provides an elegant shortcut for finding the area of simple lattice polygons.

Pick's theorem – the area of a lattice polygon: $A = i + b/2 - 1$, where i = the number of interior points (blue), and b = the number of boundary points (red). Using Pick's formula the area is 84.5 units, which matches our earlier result.

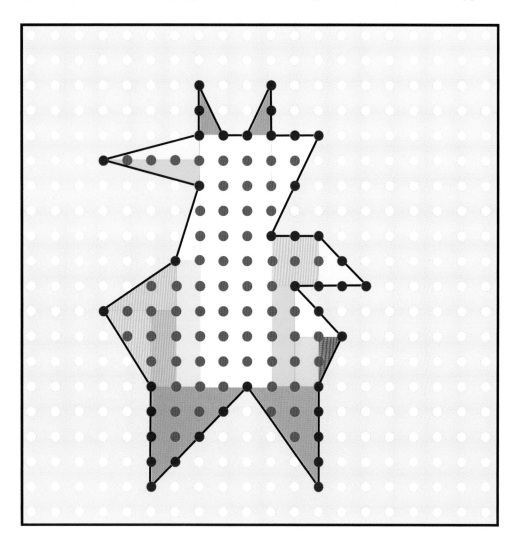

GEORG ALEXANDER PICK (1859–1942)

Georg Alexander Pick was an Austrian mathematician. Victim of the Nazi regime, he died in the Theresienstadt concentration camp in 1942. Pick is best known for his formula for determining the area of lattice polygons. This formula was published in an article in 1899, but only became widely recognized when it was included in Hugo Steinhaus' 1969 edition of his *Mathematical Snapshots*.

"I had been waiting for quite sometime to post something very interesting. Finally, I could find something related to geometry which captured my attention because of its elegance and simplicity. I will not waste any time and will directly delve into the necessary definitions and the theorem itself."

— *George Alexander Pick, about his theorem*

> **"It is indeed wonderful that so simple a figure as a triangle is so inexhaustible in properties."**
>
> *— August Leopold Crelle, 1816*

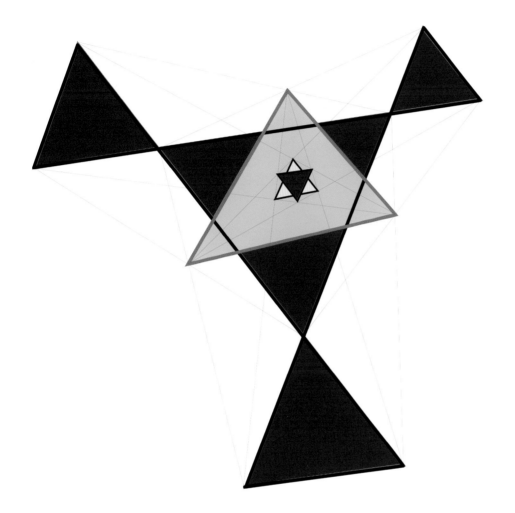

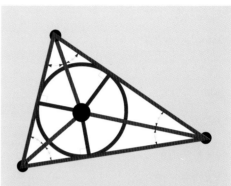

BISECTING ANGLES IN A TRIANGLE

Euclid showed that the bisectors of any two angles in a triangle meet at a point equidistant to the three edges. This point is the center of the inscribed circle called the incenter.

The natural related question would be to ask how the trisectors of a triangle meet. But this question had to wait for over 2000 years, until Morley's trisection theorem.

MORLEY'S TRISECTOR THEOREM — 1899

In 1899, Frank Morley (1860–1937), a British professor of mathematics, discovered a beautiful theorem, one of the most astonishing relationships in geometry.

His theorem states: the adjacent angle trisectors of any triangle intersect at three points, determining an equilateral triangle.

Take any triangle (green). Trisect its angles and join the intersection points of the adjacent trisectors. You will get an equilateral triangle (red) as shown.

Will this always be the case?

Try it with several triangles of your choice. Note that the six trisectors form six inner intersection points. When joining the other three intersections, another triangle is formed, this time not equilateral. This is the second Morley triangle (yellow).

In a generalization of Morley's Theorem, taking into consideration the exterior angle trisectors as well, four additional equilateral triangles are obtained as shown.

M-PIRE PROBLEM — 1890

The famous "four-color" theorem was only recently solved by computer. Herbert Taylor, from the University of Southern California noted that the generalization of the problem of coloring maps is to consider a map in which each country or area to be colored consists of "m" disconnected regions.

With the requirement that all regions of a country must be colored in the same color, what is the smallest number of colors necessary for coloring such maps so that no two regions of the same color touch along a common border? Under this generalization the four-color problem is a special case where m=1, and the number of colors is four.

It is interesting to note that when m=2 ("the 2-pire problem," each country has a colony of the same color), the problem was already proved in 1890 by Percy John Heawood (1861–1955). He first proved that an m-pire map needs no more than 6m colors. He also produced a m=2 map requiring 12 colors. Above is Heawood's 2-pire map. Can you color it in using 12 colors? One 2-pire is already colored in to get you started.

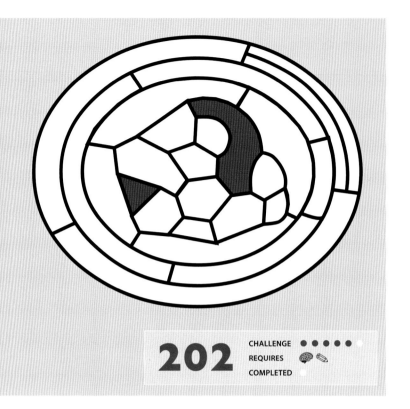

202	CHALLENGE	● ● ● ● ● ○
	REQUIRES	🧠 ✎
	COMPLETED	○

PERCY JOHN HEAWOOD (1861–1655)

Percy John Heawood was a British mathematician educated at Oxford. His lengthy carreer was principally dedicated to the four color theorem. In 1890 he found a mistake in Alfred Kempe's proof, which had been considered valid for 11 years. Now that the four color theorem was once again open, Heawood went on to establish the five color theorem instead. The four color theorem itself was finally established by a computer-based proof in 1976.

In the *Journal of the London Mathematical Society*, G A Dirac wrote: "In his appearance, manners and habits of thought, Heawood was an extravagantly unusual man. He had an immense mustache and a meager, slightly stooping figure. He usually wore an Inverness cape of strange pattern and manifest antiquity, and carried an ancient handbag. His walk was delicate and hasty, and he was often accompanied by a dog, which was admitted to his lectures. His transparent sincerity, piety and goodness of heart, and his eccentricity and extraordinary blend of naiveté and shrewdness secured for him not only the fascinated interest, but also the regard and respect of his colleagues."

DOMINO SETS

Sides, vertices, faces and corners of triangles, squares and cubes are colored in two, three, four and six colors respectively — creating distinct complete color sets of generalized dominoes. The object is to find the number of distinct dominoes in each set, and further, to fit the complete sets in game boards of different shapes and sizes, with the basic condition of the domino principle — each pair of touching edges must be of the same color.

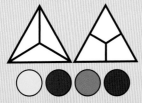

COLOR TRIANGLES

A triangle is divided into three fields as shown. Using four colors to color the sides or vertices, how many different triangles can be created?

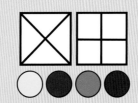

COLOR SQUARES

A square is divided into four fields as shown. Using four colors to color the sides or vertices, how many different squares can be created?

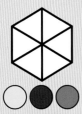

COLOR HEXAGONS

A hexagon is divided into six fields as shown. Using three colors to color the sides, how many different hexagons can be created?

COLOR CUBE

In how many different ways can you place a cube to occupy the same three-dimensional space?

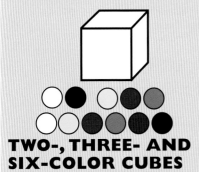

TWO-, THREE- AND SIX-COLOR CUBES

A cube's faces are colored in two, three and six colors. How many distinct two-, three- and six-color cubes can be created?

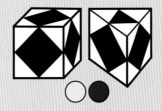

TWO-COLOR CORNER CUBE AND PRISM

In how many distinct ways can the corners of a cube and a triangular prism be colored using only two colors, red and yellow?

MACMAHON'S GENERALIZED DOMINOES — 1900

The classic domino game is basically a linear number game. Adding color and more complex shapes (including three-dimensional cubes), interesting combinatorial games can be created (more on the beauty of combinatorics in Chapter 4).

Alexander MacMahon (1854–1929) devised a number of such generalized domino games, using polygonal shapes that tile the plane and colors in a systematical fashion.

The sets of MacMahon's tiles are not arbitrary: the same basic shape of tile is colored in all possible ways to form a complete set of tiles, no two of which are alike. Reflection of a tile is considered to be different, but rotations are considered to be the same. This is a natural assumption, since the tiles are usually colored on one side only, so they cannot be turned over but can be rotated in the plane without any difficulty. The object of the games is to arrange the complete set of tiles according to the domino principle, in some given geometrical or symmetrical pattern.

MacMahon's mathematical work was based on the theory of symmetric functions — algebraic expressions that remain unchanged if the letters in them are permuted. For example, $a \times b \times c$ and $ab \times bc \times ca$ are symmetric functions of a, b and c. If the colors of a complete set of MacMahon's dominoes are permuted, we can end up with exactly the same set of tiles as before. The beautiful combinatorial properties of these dominoes derive from this deep permutational symmetry. MacMahon's ideas still offer many unexplored areas for new puzzles.

THIRTY COLOR CUBES

Percy Alexander MacMahon (1854–1929) introduced the concept of generalized dominoes, extending the standard dominoes to convex polygons that tile the plane, adding color and restricting such dominoes to complete combinatorial sets.

His classic set of 30 color cubes, introduced in 1893, is one of the real gems of recreational mathematics. It is based on the following problem: if you color each of the six faces of a cube a different color, using the same set of colors for all cubes, how many different cubes can you obtain?

Rotations are not considered as different but reflections are. On the right, nets of 30 cubes are provided. Can you color (or number) the nets using six colors (or numbers from one to six) to create the set of the 30 color cubes?

One tedious way would be to find all the 720 possible permutations of the six colors or numbers. Since you can position a cube in 24 different orientations, there will be 24 appearances of each cube, which gives the number of different cubes as 30. But a better way is to find a way of coloring the cubes systematically.

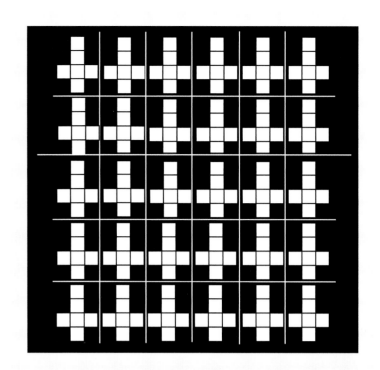

204   CHALLENGE ● ● ● ● ● ●
REQUIRES 🧠 ✏️ ✂️ ⚒️
COMPLETED ○

IVAN'S CUBES

Original design adaptation of MacMahon's cubes, extending its puzzle range to include puzzles for all ages, even for very small children. (Produced by Orda Industries as the "Cu-Zoo," in the '60s.)

HINGED POLYGONS — 1902

"Cut an equilateral triangle into four pieces that can be reassembled to form a square" is also known as the Haberdasher's Puzzle, first published by Henry Dudeney in 1902.

A notable feature of the solution is that each of the pieces can be hinged at one vertex, forming a chain that can be closed clockwise to form the original triangle or counter-clockwise into a square.

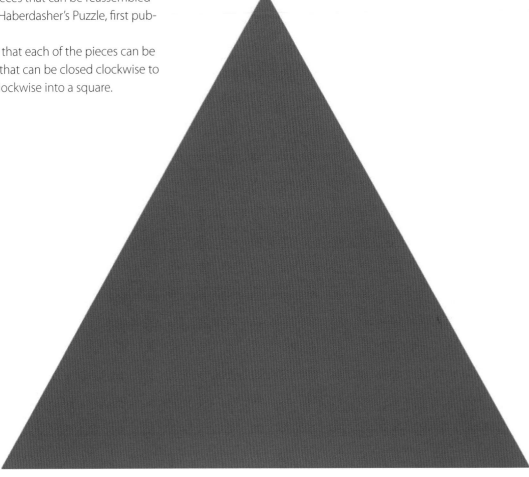

205 CHALLENGE ●●●●●●
REQUIRES 🧠 ✏️ ✂️ ⚒️
COMPLETED ○

HENRY ERNEST DUDENEY (1857–1930)

Henry Ernest Dudeney was an author and mathematician, England's greatest inventor and creator of puzzles. He believed that puzzle-solving is a creative activity of of the highest importance for improving thinking and logical decision-making. His greatest mathematical achievement was the discovery of a dissection puzzle called "The Haberdasher's Puzzle," an equilateral triangle cut into four pieces by straight lines.

STAMP PROBLEM OF DUDENEY — 1903

The classic stamp problem of Henry Dudeney is one of the first problems related to polyominoes and specifically to the five tetrominoes. Stamps can be bought in rectangular blocks of three rows with four stamps in each row. The object is to tear off four stamps joined together along their sides. In how many different ways can you tear off four stamps in the tetromino configurations shown below?

206

CHALLENGE ● ● ● ○ ○
REQUIRES 🧠 ✏️ ✂️
COMPLETED ○

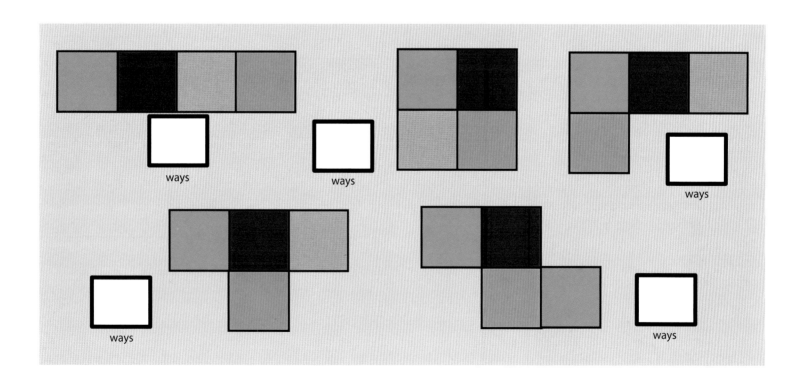

ways

ways

ways

ways

ways

SMART ALEC — 1903

The figure of a concave pentagon formed by removing an isosceles triangle, a quarter of a square, has to be dissected into four parts that could be reassembled into a square. Sam Loyd provided a solu-tion (shown) that was flawed — the "square" is actually a rectangle. No four-piece solution to the puzzle is known. Dudeney found the excellent five-piece solution.

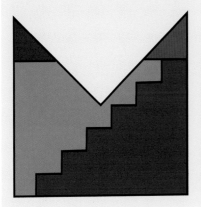 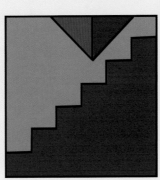 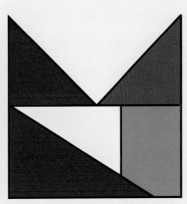 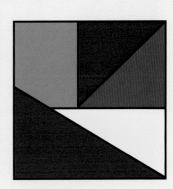

DUDENEY'S QUILT

This beautiful quilt was originally sewn together from unit square pieces. The eight middle square pieces were damaged and had to be removed forming a long hole. Can the quilt be repaired by cutting it along the square grid into two parts that can be sewn together to form a new quilt without holes?

207	CHALLENGE	● ● ● ● ○
	REQUIRES	🧠 ✏️ ✂️ 🔨
	COMPLETED	○

MITRE PUZZLE

In his mitre puzzle, Loyd divided the concave pentagon into 24 identical triangles using four colors as shown.

Can you rearrange the triangles within the pentagon to make four identical connected shapes? Each of the four shapes should be of one color, and the shapes should be considered identical even if they are reflections or rotations of one another.

208	CHALLENGE	● ● ● ● ○
	REQUIRES	🧠 ✏️ ✂️ 🔨
	COMPLETED	○

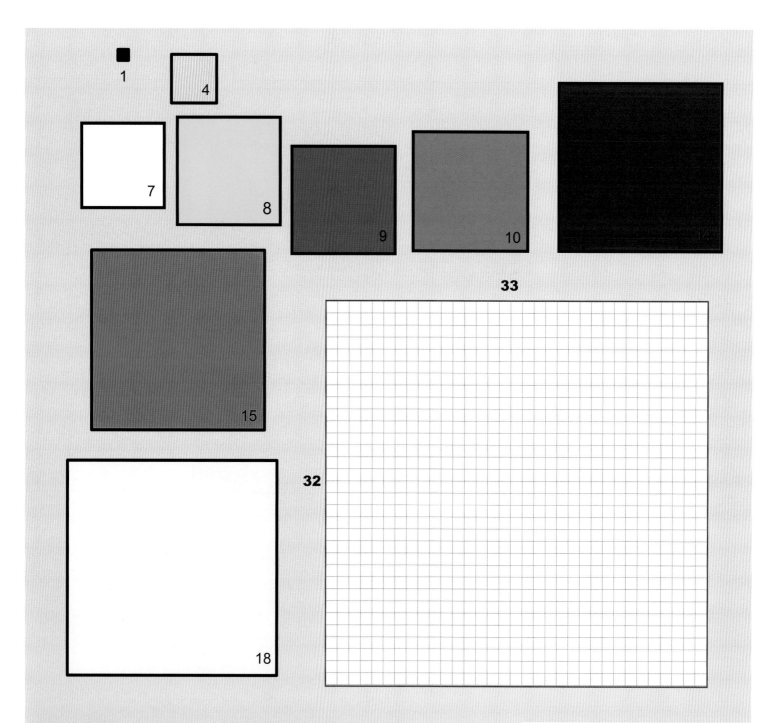

SMALLEST PERFECT RECTANGLE — 1903

Can a rectangle be subdivided into smaller squares no two of which are alike? In 1903, Max Dehn proved the theorem that if a rectangle is dissected into squares, then the sizes of the squares, and the rectangle itself, are commensurable — for integer multiples of a single number.

Choosing a unit of measurement, all sides of the element squares are integers.

In 1909 Z. Moron discovered a rectangle dissectable into nine different squares, which, in 1940, Tutte, Brooks, Smith and Stone proved to be the "smallest," meaning no smaller rectangle can be divided into nine different squares, and no rectangle at all can be divided into eight or less different squares.

209

The smallest perfect rectangle is composed of squares of sides: 1 - 4 - 7- 8 - 9 - 10 - 14 - 15 -18.

It is a 32-by-33 rectangle. Can you use the nine squares without overlap to form the smallest perfect rectangle?

The first four generations

SNOWFLAKE AND ANTI-SNOWFLAKE CURVES — 1904

The black and red figures show the first four generations of the famous snowflake and anti-snowflake curves, also known as Koch fractals. Start with an equilateral triangle whose sides have unit lengths. On the middle third of each of the three sides add (or subtract) an equilateral triangle with sides of 1/3 of the length, and repeat the process.

As the process continues indefinitely, can you tell what the length of the curve will be and what area it will enclose? The curves are basically growth patterns, created as sequences of polygons. Are there three-dimensional analogs of the snowflake curves?

An important principle demonstrated by the snowflake and similar so-called patho-logical curves is that complex shapes can result from repeated applications of very simple rules.

Such shapes are called fractals. The snowflake curve discovered in 1904 by Helge von Koch is one of the early fractals.

210 CHALLENGE ● ● ● ● ● ○
REQUIRES 🧠 ✏️
COMPLETED

RANDOM WALK - 1905

A random walk is a mathematical formalization of a trajectory that consists of taking a series of random steps. Examples are the path traced by a molecule as it travels in a liquid or a gas, the route of a foraging animal and the price of a fluctuating stock.

Karl Pearson first introduced the term random walk in 1905. The concept of random walks has been employed in many fields: ecology, economics, psychology, computer science, physics, chemistry and biology. It explain the observed behaviors of processes in these fields, and therefore serves as a fundamental model for the recorded stochastic activity.

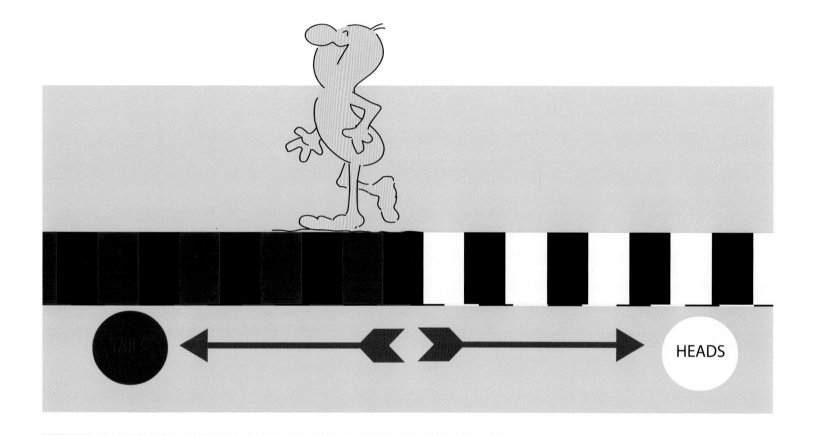

COIN FLIPPING

For this puzzle, flip a coin repeatedly. If it comes down heads, the walker will move one mark to the right; if it's tails, he will move one mark to the left.

Can you guess how far from his starting point our walker will be after 36 flips of the coin? After you've made your guess, flip the coin 36 times to check your prediction.

Can you tell what the chances are that our walker will return to his starting point at some point in his walk? (Assume that the walk continues indefinitely.)

211 CHALLENGE ● ● ● ○
REQUIRES 🧠 ✂️
COMPLETED

> **"Nothing in nature is random... A thing appears random only through the incompleteness of our knowledge."**
>
> – *Baruch Spinoza (1632–1677)*

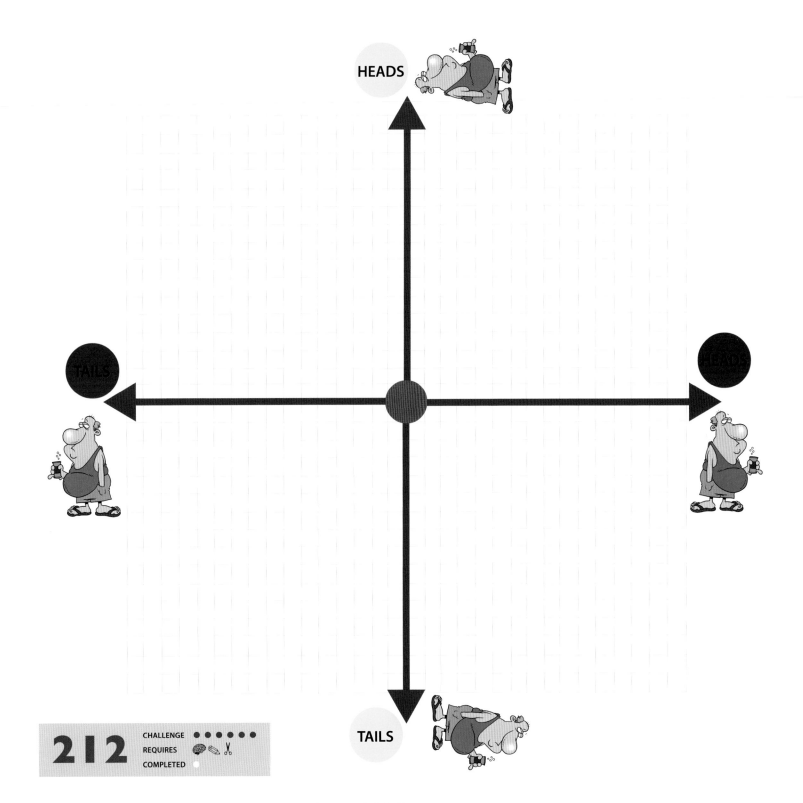

212
CHALLENGE ● ● ● ● ●
REQUIRES 🧠 ✂️
COMPLETED ○

DRUNKARD'S RANDOM WALK — 1905

In the drunkard's random walk, starting from the central lamp post, moves are dictated by flipping two coins (a red and a yellow coin) as shown. This is the simplest demonstration of a stochastic (random) process, and a good analogy for explaining Brownian motion, in which a particle is "kicked around" by molecules of a surrounding liquid or gas.

Where do you think the drunkard will be after a certain number of flips? Can you also guess what the chances are that the drunkard will return to his starting point at the lamp post at some point? Consider the walk as finite by considering the size of the grid as a barrier.

PENTOMINOES — 1907

Dominoes are the playing pieces, or tiles, of a centuries-old game. The tiles are made up of two unit squares joined fully along a common edge. Two identical squares can fit together in only one way (dominoes). But mathematicians, recreational and otherwise, have elaborated on the basic domino shape by successively adding more unit squares.

The results, three-square trominoes, four-square tetrominoes, five-square pentominoes and the like, are collectively known as polyominoes.

If we now take three squares, how many trominoes can we make? How many if we use four squares (tetrominoes), or five (pentominoes)?

The corresponding general problem is: how many possible different shapes can be constructed from a certain number of unit squares or, in generalized form, from any other polygons, called polyforms? No treatise on geometrical combinatorics and puzzles can be tackled without mentioning and devoting parts to polyforms, and especially to pentominoes, on which volumes have been written.

The first polyomino problem appeared in 1907. However, the popularity of these shapes, both as a new form of mathematical recreation and as a form of new enrichment material for schools, owes much to Dr. Solomon Golomb, Donald Knuth and Martin Gardner. It was they who introduced them to wide audiences as puzzles, games and problems.

A domino (two unit squares) is the simplest polyomino. It has only one possible shape: the domino rectangle. A tromino (three unit squares) has two possible shapes. There are five possible tetrominoes (four unit squares), 12 possible pentominoes (five unit squares), 35 possible hexominoes (six unit squares), 108 possible heptominoes (seven unit squares) and 369 possible octominoes (eight unit squares).

There is an enormous accumulation of publications involving polyforms, polyominoes and especially pentominoes.

Polyforms are generalizations of Golomb's polyominoes in which squares are replaced by other polygons. Polyamonds are based on equilateral triangles. Hexiamonds are based on regular hexagons, and so on.

PENTOMINO COLOR PUZZLES

Can you fit the set of 12 colored pentominoes on an eight-by-eight board to solve the six puzzles leaving the four squares uncovered in each?

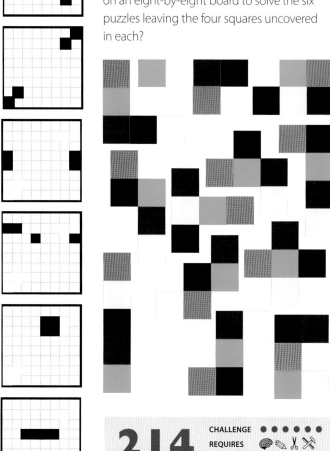

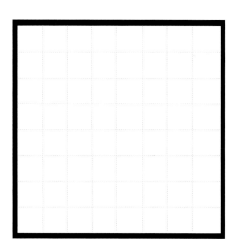

214 CHALLENGE ●●●●●
REQUIRES 🧠✏️✂️🔨
COMPLETED

MINIMAL PENTOMINO GAME

What is the smallest number of pentominoes which can be placed on the eight-by-eight game board, to make it impossible for any further pentominoes to be placed on the board?

213 CHALLENGE ●●●●●
REQUIRES 🧠✏️✂️🔨
COMPLETED

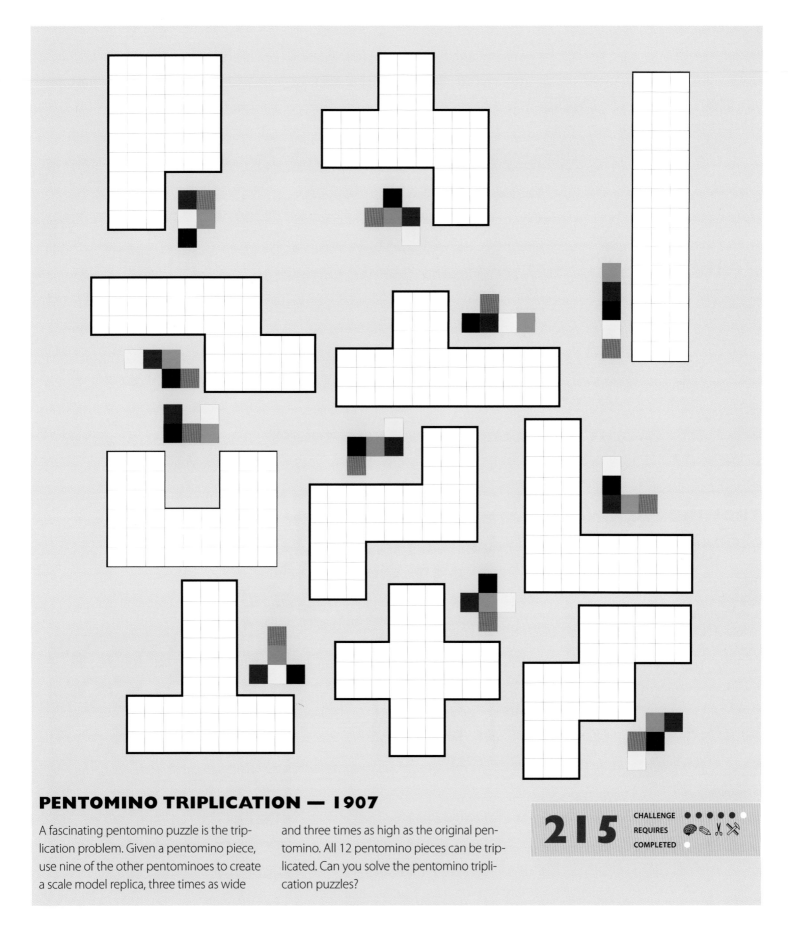

PENTOMINO TRIPLICATION — 1907

A fascinating pentomino puzzle is the triplication problem. Given a pentomino piece, use nine of the other pentominoes to create a scale model replica, three times as wide and three times as high as the original pentomino. All 12 pentomino pieces can be triplicated. Can you solve the pentomino triplication puzzles?

215

CHALLENGE ● ● ● ● ● ○
REQUIRES 🧠 ✎ ✂ ⚒
COMPLETED ○

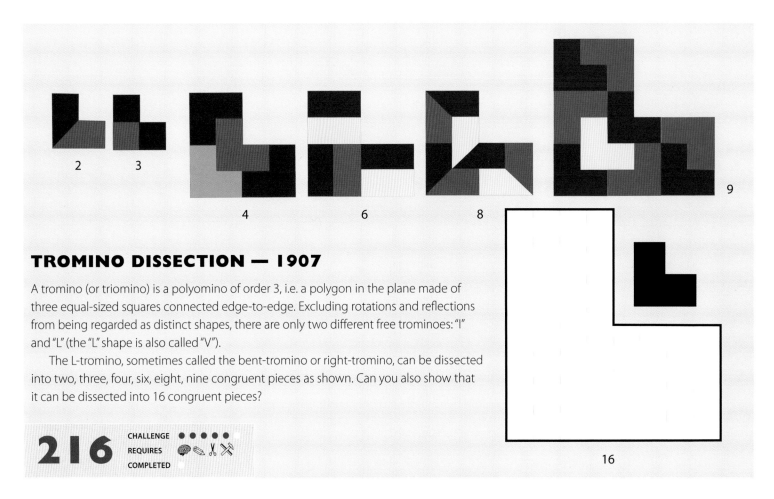

TROMINO DISSECTION — 1907

A tromino (or triomino) is a polyomino of order 3, i.e. a polygon in the plane made of three equal-sized squares connected edge-to-edge. Excluding rotations and reflections from being regarded as distinct shapes, there are only two different free trominoes: "I" and "L" (the "L" shape is also called "V").

The L-tromino, sometimes called the bent-tromino or right-tromino, can be dissected into two, three, four, six, eight, nine congruent pieces as shown. Can you also show that it can be dissected into 16 congruent pieces?

16

216 CHALLENGE ● ● ● ● ●
REQUIRES 🧠 ✎ ✂ ⚒
COMPLETED ☐

TROMINO PACKING

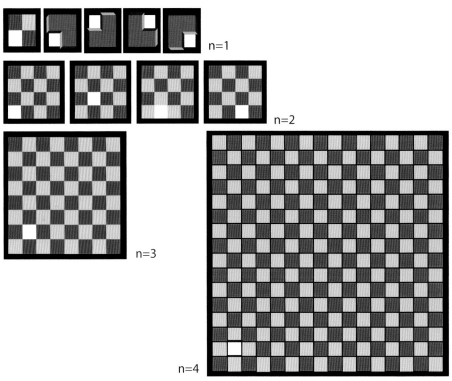

n=1

n=2

n=3

n=4

The L-tromino covers three squares of a chessboard of any size. It is often called the right tromino. We will be dealing with the problem of packing L- trominos into chessboards of dimensions: $2^n \times 2^n$ for any $n > 1$.

Remove a single square from such a chessboard and cover the rest of the board with the appropriate number of L trominoes.

Can this be achieved with every chessboard of the above dimensions, no matter which square is removed? For n=1, we have a two-by-two chessboard as shown above. For n=2, n=3 and n=4, can you cover the chessboards no matter where the missing square is?

217 CHALLENGE ● ● ● ● ●
REQUIRES 🧠 ✎ ✂ ⚒
COMPLETED ☐

REPTILES — 1907

Did you know that some shapes, if they are combined with a specific number of identical copies of themselves, can create larger versions of themselves? And, correspondingly, when such shapes are subdivided appropriately, they also can make smaller versions of themselves?

Reptiles are polygons that can make larger and smaller copies of themselves. Solomon Golomb gave them their name and, by studying them, laid the groundwork for a general theory of polygon replication.

The fish, birds and monuments in the puzzles below are reptiles. How many of each can form bigger replicas of themselves?

FISHNET

Can you fit all 18 fish into the fishnet without overlap?

218   CHALLENGE ● ● ● ○ ○ ○
REQUIRES 🧠 ✏️
COMPLETED

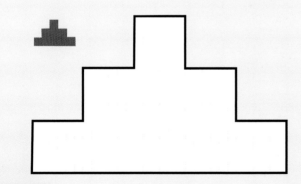

MONUMENT

The monument was built from a number of identical smaller shapes, each one an exact replica of the large monument.

Can you find the number of the smaller monuments and their orientations along the grid lines of the big outline?

219   CHALLENGE ● ● ● ○ ○ ○
REQUIRES 🧠 ✏️
COMPLETED

REPCATS AND REPBIRDS

Nine small birds are in danger of being eaten by the big hungry cat. How many birds can be eaten by the cat or how many birds can be placed in the outline of the cat, without overlap?

220   CHALLENGE ● ● ● ○ ○ ○
REQUIRES 🧠 ✏️
COMPLETED

INFINITE MONKEY THEOREM AND PROBABILITY — 1909

The "Infinite Monkey Theorem," based on an idea first expressed in 1909 in Emile Borel's book on probability, is an intriguing thought experiment about infinity. The Infinite Monkey Theorem asserts that, given an infinite amount of time, a monkey hitting the keys of a typewriter at random will almost certainly type a given text, such as the complete works of Shakespeare, for example.

If such an experiment were conducted, the probability of a monkey typing a given string of text exactly is so remote that the chance of it actually happening over a period of time of the order of the age of the universe is tiny, but it does exist.

The infinite monkey theorem and the images associated with it are considered to be a popular illustration of the mathematics of probability. In 2003 a website called *The Monkey Shakespeare Simula-*tor was launched, containing a Java applet representing a large number of monkeys typing randomly, with the stated intention of finding out how long it would take them to produce a complete Shakespearean play from beginning to end. One of its better results was a partial line from Henry IV, Part 2, which is reported to have taken "2,737,850 million billion billion billion monkey-years" to achieve 24 matching characters.

Perhaps a simpler way to consider this problem is in the selection of lottery numbers. Let's say you are extremely wealthy and very eager to win. With such resources, you have the means to buy all of the lottery tickets, with every possible combination of numbers, thus guaranteeing that you would win.

THROWING A SIX WITH A DIE

Another way to think of the Infinite Monkey Theorem is to imagine rolling a six-sided die numerous times and waiting for a six to come up. It could come up on the first throw, but you might also have to keep trying for a long time without success. Eventually, though, you will roll a six. What is the probability of rolling a six at least once if you roll one die six times?

221

CHALLENGE ● ● ● ● ○
REQUIRES
COMPLETED

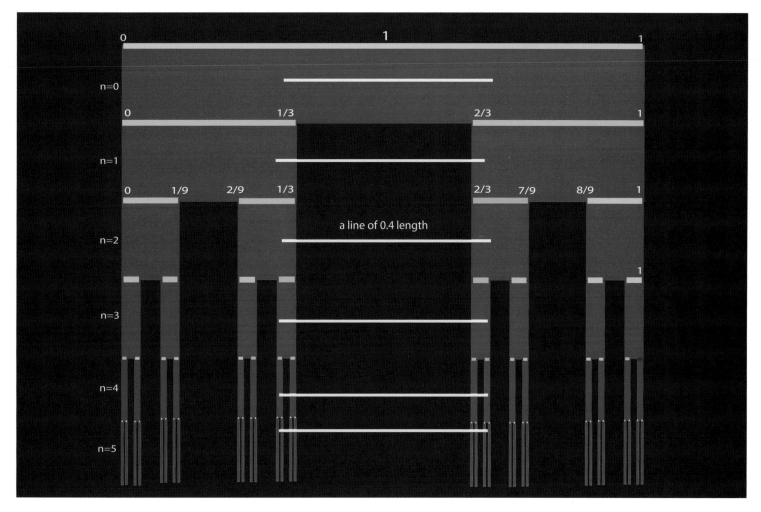

a line of 0.4 length

CANTOR'S COMB — 1910

Start with a line of length one (blue) and remove its middle third. Now remove the middle third of each remaining piece. Repeat indefinitely. What is left is the Cantor's comb. The first five generations are shown. Can you find a formula to determine the total length of Cantor's comb after the n-th stage?

A remarkable property of Cantor's set: no matter how far you go, you will always be able to find two points on the remaining Cantor's set, whose distance apart is any number between 0 and 1, in spite of the fact that great parts of the initial line have been eliminated. A line of 0.4 length on Cantor's

comb is shown in all generations (yellow). Try this with other lengths.

222   CHALLENGE ● ● ● ● ●
REQUIRES 🧠 ✎ ✂
COMPLETED ○

GEORG CANTOR (1845–1918)

Georg Ferdinand Ludwig Philipp Cantor was a German mathematician born in Denmark, renowned for inventing set theory, which has since become a fundamental theory in mathematics. Cantor established the significance of one-to-one correspondence between the elements of two sets, defined infinite and well-ordered sets, and proved that real numbers are "more numerous" than natural numbers. Cantor's method of proof of this theorem implies the existence of an "infinity of infinities." His theory of transfinite numbers was initially strongly criticized, because it was counter-intuitive.

Cantor suffered from severe depressions from 1884 until his death in 1918, and some say the severe criticism of his peers is partly to blame. However, during the last years of his life Cantor received one of the highest honors possible for a mathematician; as in 1904, the Royal Society awarded him its Sylvester Medal. David Hilbert defended Cantor's theory of transfinite numbers from its numerous critics by famously declaring: "No one shall expel us from the Paradise that Cantor has created."

SIERPINSKI FRACTALS — 1915

The Sierpinski triangle (also known as the Sierpinski sieve or the Sierpinski gasket) is a fractal and an attractive fixed set named after the Polish mathematician Waclaw Sierpinski, who first described it in 1915. Similar patterns appear as far back as the 13th century in the Cosmati mosaics in the cathedral of Anagni, Italy and elsewhere, for example the nave of the Roman Basilica of Santa Maria in Cosmedin.

SIERPINSKI CARPET FRACTAL

A unit square is divided into nine squares by dividing its sides into thirds and painting the middle square gold. In the next generation, the remaining blue squares are each similarly divided, with their middle squares painted gold, and so on. If this process goes on indefinitely, can you guess what the proportion of the gold areas in relation to the area of the initial blue square will be?

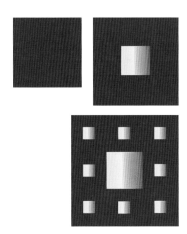
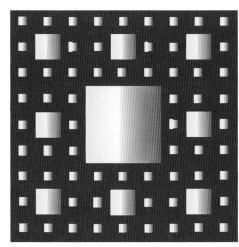
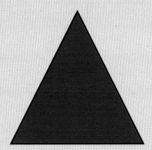
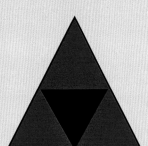
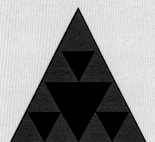
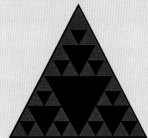

223 CHALLENGE ●●●●○○
REQUIRES 🧠✏️
COMPLETED ○

SIERPINSKI TRIANGLE FRACTAL

Initial triangle	*First generation* 1/4 = 0.25	*Second generation* 7/6 = 0.44	*Third generation* 37/64 = 0.58

Three generations of the Sierpinski triangle are demonstrated. Can you draw the fourth generation in the given triangular grid? Can you also find the number series showing the proportions of the black areas to the area of the whole triangle in each generation? The triangle is obtained by starting with an equilateral triangle, which is then divided into four smaller equilateral triangles, of which the middle one is removed — forming a black triangular hole. The three remaining filled triangles are then divided in the same way, a process that can go on indefinitely. The pattern achieved in this manner is called the Sierpinski fractal.

224 CHALLENGE ●●●●○○
REQUIRES 🧠✏️
COMPLETED ○

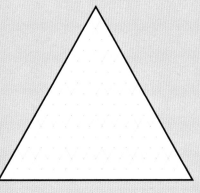

Fourth generation

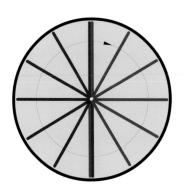

Circle

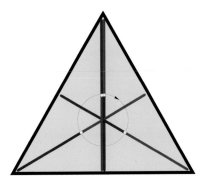

Triangle

Star deltoid 3 cusps

*Star deltoid
5 cusps*

KAKEYA NEEDLE PARADOX — 1917

The famous Kakeya needle problem asks what the minimum area is in the plane, in which a needle (a unit straight line segment) can be turned 180 degrees. The question was first asked by Soichi Kakeya (1886–1947) in 1917.

Obviously, the needle can be revolved in a circle (with diameter = 1 and area = 0.78), or in an equilateral triangle (with height = 1 and area = 0.58). However, Kakeya suggested a better solution: the minimum area would be a deltoid shape, a 3-cusped hypocycloid (with an area of 0.39), which was long believed to be the best solution. At this stage I could ask you whether there is a better solution, but that would be unfair, since there is no best solution.

This conclusion, by Besicovitch in 1928, was a bombshell and a great surprise to the mathematical world, since it is so counterintuitive. Besicovitch showed that the deltoid curves could have as many cusps as one liked and the minimum area can be as small as we want, even zero.

Perron tree

BESICOVITCH PROOF

Besicovitch (1891–1970) proved that there is no answer to the Kakeya Needle Paradox. Or, to put it more accurately, he demonstrated that the answer is that there is no minimum area. The area can be as small as you like. How does this work? You halve the base of an equilateral triangle and then halve it again. You move adjacent triangles towards each other until they overlap slightly. You can repeat this procedure with these pairs of triangles until the area reaches the desired size. Such an iterative construction is called a Perron tree, as shown above. In general, when the figure is restricted to being convex, Kakeya showed that the smallest convex region is an equilateral triangle of unit height.

Besicovitch proved that there is no minimum area for a general shape. You can see this if you rotate a line segment inside a deltoid, star-shaped 5-oid, star-shaped 7-oid and so on.

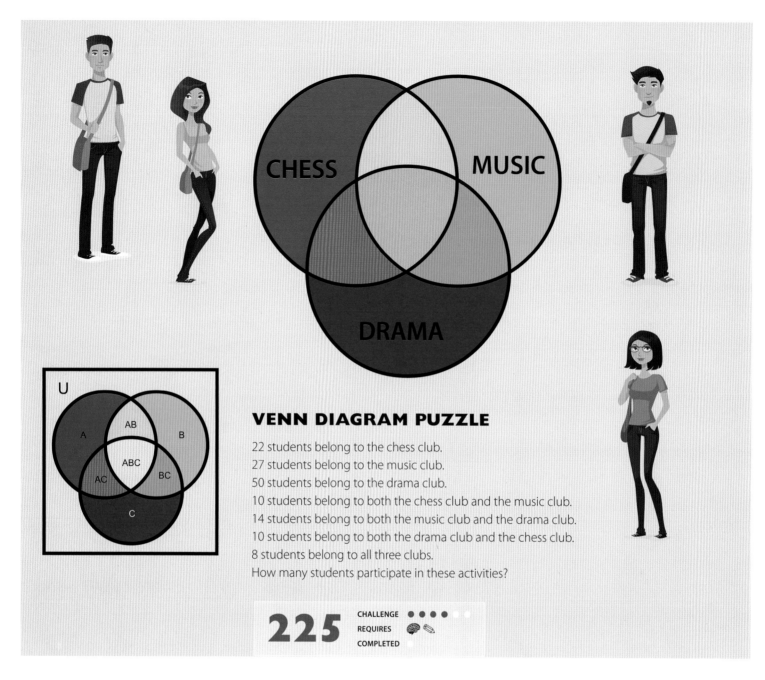

VENN DIAGRAM PUZZLE

22 students belong to the chess club.
27 students belong to the music club.
50 students belong to the drama club.
10 students belong to both the chess club and the music club.
14 students belong to both the music club and the drama club.
10 students belong to both the drama club and the chess club.
8 students belong to all three clubs.
How many students participate in these activities?

225 CHALLENGE ● ● ● ● ○
REQUIRES 🧠 ✏️
COMPLETED

VENN DIAGRAMS — 1920

Mathematical reasoning is based on a system of symbols and ideas that have precise meanings — logic. We all possess an intuitive grasp of many principles of logic. Mathematicians often apply logic to deduce conclusions from more complex premises (starting points of a chain of logically connected ideas) that cannot be worked out with intuition alone.

Such conclusions can be deduced more easily using a "Venn diagram" to simplify relationships between two or more sets, devised by John Venn (1834–1923), a logician and priest who taught at Cambridge.

Venn diagrams are patterns visualizing logical relationships between groups. A Venn diagram helps to describe and compare any number of elements and characteristics of items. To begin a Venn diagram, you must start with the universal set, which is denoted by "U" and represented by a rectangle. Sets are represented by a rectangle with a closed circle drawn inside it. The area inside the circle is associated with the elements in the set.

The area inside the circle is associated with the elements in the set. The overlapping regions imply that sharings are occurring.

The idea of the usefulness of a Venn diagram to obtain hidden logical relationships is demonstrated by the Venn diagram puzzle above. Can you solve it?

A basic Venn diagram consists of a rectangle, the universal set showing the space of all possible things. Each set is represented by a circle. The overlapping regions imply that sharings are occurring. The usefulness of using a Venn diagram to obtain hidden logical relationships is demonstrated by the puzzle above.

CHAPTER

7

INFINITY, IMPOSSIBILITY/ IMPOSSIBLE FIGURES, MIXED HATS AND MIXING TEA WITH MILK

FRANK PLUMPTON RAMSEY (1903–1930)

Frank Plumpton Ramsey was a British mathematician. He died young, at the age of 26, but his theory that any structure will necessarily contain an orderly substructure has become very influential.

The Ramsey Theory aims to figure out just how complex a structure must be to guarantee a certain substructure. Stargazers have experienced the validity of his theory.

They found patterns in the sky. Given a large enough number of stars, a pattern will be produced, from a perfect rectangle to the Big Dipper or something else. The appearance of disorder is really a matter of scale.

The classic example of Ramsey Theory is the famous "Party Puzzle." Ramsey wanted to find the smallest set of objects that would guarantee that some of those objects would share certain properties. For example, the smallest number of people that will always include two people of the same sex is three. If there are only two, you might have a man and a woman. Given that the third person would be either a man or a woman, adding him or her guarantees that there will be at least two of one sex.

Or take this question: Can a complete graph have its edges colored using only two colors, so that no three edges of the same color form a triangle? Ramsey proved a general theorem on this question, but instances with four, five or six nodes are simple enough to analyze using pencil and paper. The Party Puzzle is based on Ramsey's work.

To appreciate how elegant graphs are for solving these kinds of problems, imagine listing all possible combinations of acquaintanceship among six people, a total of 32,768, and having to check if each combination included the desired relationship.

A more advanced Ramsey problem would be to imagine a party where there must be a foursome in which everyone is a mutual friend or everyone is a mutual stranger (or where they either love or hate each other). How large does the party have to be? Ramsey's work demonstrated that this requires eighteen guests. If you draw a complete graph with 18 nodes, no matter how you color the lines using two colors, you will inevitably create a quadrilateral formed by connecting four points (persons) in one of the colors.

The party size required to ensure at least one fivesome of mutual friends or strangers is still unknown. The answer lies between 43 and 49. The beauty of Ramsey theory is its simplicity and the fact that it can be understood intuitively.

> **"Complete disorder is an impossibility."**
>
> — *Frank Ramsey and Paul Erdös*

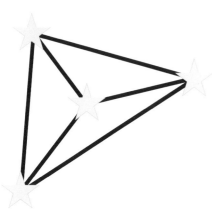

THE HAPPY ENDING PROBLEM

Looking at a starry sky, how many stars do you have to select to guarantee a convex quadrilateral when you connect them with straight lines? Four stars selected as shown won't suffice.

What is the smallest number of n points one can place in the plane in a general direction (no three of which lay on a straight line), which will always result in a convex polygon of n sides?

E. Klein and G. Szekeres proved the theorem for a convex quadrilateral. They got engaged and subsequently married. Consequently, Erdös named the problem as above.

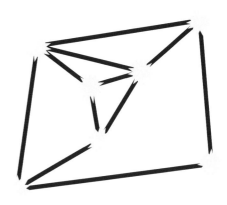

226 CHALLENGE ● ● ● ● ●
REQUIRES 🧠 ✏️ ✂️
COMPLETED

RAMSEY'S GAME

The 15 white lines form a complete hexagonal graph on six points. In our game, the 15 lines can be colored in either of the two colors, red or blue.

Two players alternate coloring lines one by one, red or blue. The first player forced to create a triangle of one of the colors, connecting three points of the graph, loses. There must be a winner in this game. How many different triangles can be colored? How many lines can be colored before there is a winner? What is the total number of games that can be played?

227 CHALLENGE ● ● ● ● ○
REQUIRES 🧠 ✏️ ✂️ ⚒️
COMPLETED ○

PARTY PUZZLE FOR SIX — 1930

Can you invite five friends to a party and avoid having groups of three who all like or dislike each other?

We can simplify the problem representing the six people (you and your five friends) by creating a complete graph of six points. In such a graph, we can clearly distinguish between all possible triplets, which form in-terconnected triangles, including intercon-nected pairs forming lines between points (people). If we create a graph in which the people who like each other are red lines, and the people who dislike each other are blue lines, then we can solve the problem in the brilliant way devised by Paul Erdös.

One by one, color the lines of the graph in either of the two colors, red or blue. If you can choose your coloring so to avoid creating a triangle of one of the two colors formed by connecting three points, you have succeeded to avoid having a group of three who all like or dislike each other. Can you achieve this re-sult in a group of six people?

228 CHALLENGE ● ● ● ● ○
REQUIRES 🧠 ✏️ ✂️ ⚒️
COMPLETED ○

LOVE-HATE RELATIONSHIPS

You can avoid having three people who ei-ther love or hate each other in a group of four or five. This is proven by coloring each line of the graph in either of the two colors, avoiding to create a triangle interconnect-ing any three points in one of the colors as shown on the right.

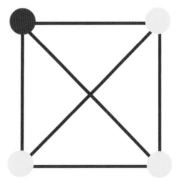

RAMSEY THEORY

The Ramsey theory is actually a generalization of the Party Problem.

The Ramsey number R(n,m) is the smallest possible number of total nodes for which there will be either N red nodes or M blue nodes. Note that the Party Problem states that R(3,3) = 6.

It is known that:

R(3,4) = 9 (our game)
R(5,3) =14
R(4,4) =18
R(6,3) =18
R(7,3) =23
R(5,4) = 25
R(5,5) = 43 or 49

THE PARTY PROBLEM (1)

One by one, color the lines of the graph in either red or blue. How many lines can you color before you are forced to create a red triangle or a blue quadrilateral, formed by connecting three or four of the outside numbered points respectively, or in other words: can you color all the lines so to avoid creating a red triangle or a blue quadrilateral?

229 CHALLENGE ● ● ● ○ ○ ○
REQUIRES
COMPLETED ○

THE PARTY PROBLEM (2)

Below is a sample game in which three lines are left uncolored. Coloring them in any of the two colors would inevitably create either a red triangle or a blue quadrilateral.

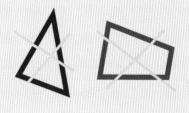

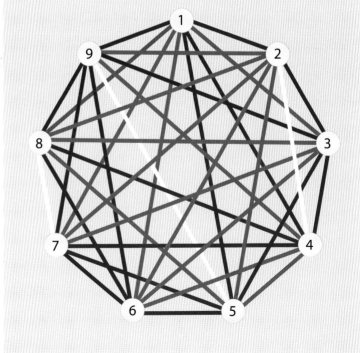

UTILITIES PROBLEM — 1930

Three houses need three supplies: telephone, electricity and water. Each house therefore needs three connections. Can you draw in the connection lines, and connect each house with each utility in such a way that none of the lines intersect?

230

CHALLENGE ● ● ● ○ ○ ○
REQUIRES 🥾 ✏️
COMPLETED ○

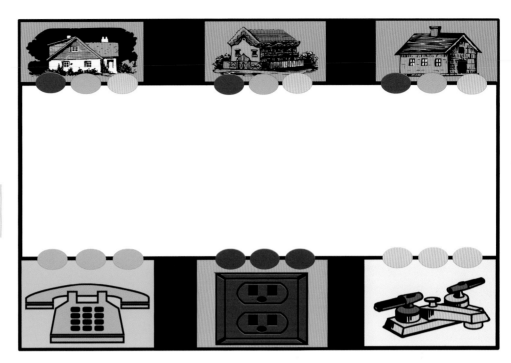

NEIGHBORS CROSSING

Three neighbors in the housing compound want to create separate paths enclosed by fences, which would enable them to leave and enter their houses through their private gates (colored to match their houses) in complete privacy without their paths crossing. Their routes shown do not solve the problem, because their paths cross at the red point shown.

Can you build their paths and do better, so that the three neighbors never cross when leaving their houses?

231

CHALLENGE ● ● ● ○ ○ ○
REQUIRES 🥾 ✏️
COMPLETED ○

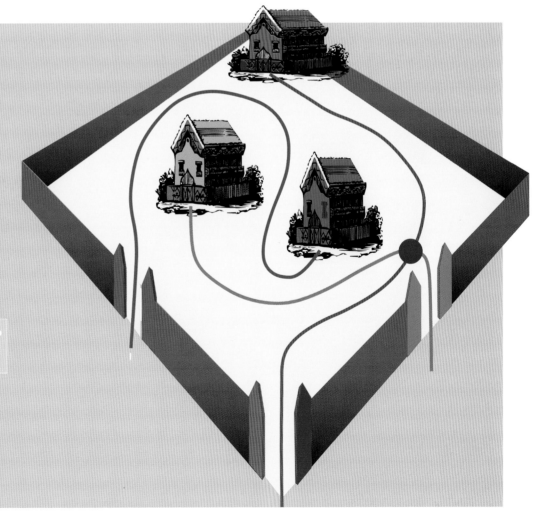

MULTIPARTITE PUZZLE (1)

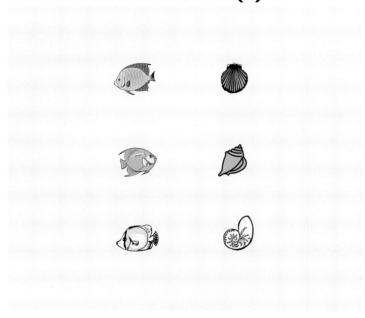

MULTIPARTITE GRAPHS — 1930

In each puzzle, connect groups of animals of different colors, but not those of the same type or group. This means that, for example, in the first puzzle the red fish won't be connected to the red shell or to the other fish, nor will the green fish be connected to the green shell, or the yellow fish to the yellow shell.

Allowing curved lines connections, how many interconnecting lines can you draw in each puzzle without intersection?

Instead of only two sets of points (as in the original Utilities Problem), based on a bipartite graph, this set of puzzles is specifically based on multipartite graphs consisting of three groups of points (K3) or tripartite graphs.

232 CHALLENGE ● ● ● ● ●
REQUIRES 🧠✏️
COMPLETED ○

MULTIPARTITE PUZZLE (2)

233 CHALLENGE ● ● ● ● ○
REQUIRES 🧠✏️
COMPLETED ○

MULTIPARTITE PUZZLE (3)

234 CHALLENGE ● ● ● ● ●
REQUIRES 🧠✏️
COMPLETED ○

SUPERELLIPSE OF PIET HEIN — 1931

You may know that the formula for graphing a unit circle is $x^2 + y^2 = 1$. You can change this into the formula for an ellipse (oval) shape by adding in two extra constant members, a and b: $(x/a)^2 + (y/b)^2 = 1$.

The French mathematician Gabriel Lame (1795–1870) wondered what would happen if we changed the powers from 2 to other values of n: $(x/a)^n + (y/b)^n = 1$.

If $n = 0$, we obtain a pair of crossed lines, if n is less than 1 ($n < 1$), we have a four-pointed star (asteroid), if $n = 1$, we have a diamond, if $n = 2$, we have our ellipse again.

If n is bigger than 2, the ellipse becomes more and more rectangular in nature. For the specific case of $n = 2.5$, Piet Hein described the resulting shape as a "superellipse," which he applied in many ways. For example, he used it to redesign the traffic flow around Sergel's Square in Stockholm as a compromise between the harsh edges of a rectangular layout and the wasted space of a circular ring road. An ellipse would waste much space in the corners and it would endanger smooth traffic flow. A rectangle would slow down traffic flow. His superellipse solved the problem. It's neither circular nor rectangular — but just right.

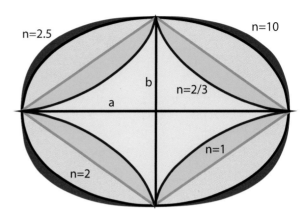

PIET HEIN (1905–1996)

Piet Hein was man of many talents: he was a scientist, mathematician, inventor, designer and poet. During the Nazi occupation of his home country Denmark, he fought in the underground resistance and wrote short poems, known as grooks, that appeared in newspapers after the Second World War. He also invented several games, such as Hex, Tangloids, Morra, Tower, Polytaire, TacTic, Nimbi, Qrazy Qube, Pyramystery and the Soma Cube. He employed his superellipse in city design, furniture making and other uses.

I met Piet several times and admired him enormously. I shall never forget how privileged I was to have him as a friend. To me, he was a genius with a mind equivalent to Leonardo da Vinci.

SUPEREGG

When a superellipse revolves about its long axis we get a surface of revolution — a superegg — an interesting three-dimensional solid, which balances on both ends.

SOMA CUBE

Piet Hein was only 26 years old when he invented his Soma Cube, which is a puzzle game consisting of seven shapes that can be put together to form a cube or a vast number of structures. The Soma concept is more than just a game. It is a sophisticated expression of topological beauty.

Of his Soma Cube, Piet wrote: "It is a beautiful freak of nature that seven simple irregular combinations of cubes can form a cube again. Variety growing out of unity returns to unity. It's the world's smallest philosophical system."

Martin Gardner and John Horton Conway made a detailed analysis of Soma. Conway found that there are 240 distinct solutions of the Soma Cube puzzle, not including rotations and reflections.

PAUL ERDÖS (1913–1996)

Paul Erdös was a famous Hungarian mathematician who worked on problems in combinatorics, graph theory, number theory, classical analysis, approximation and probability theory. He was mathematics' greatest popularizer, regarded by fellow mathematicians as the most brilliant mind in his field. He was indifferent to recognition and material comforts, living out of two suitcases for much of his adult life. "Property is a nuisance," he stated, and generously donated the prize money he was awarded to other mathematicians he believed needed it more.

In spite of, or perhaps because of his eccentricities, mathematicians adored him and found him inspiring to work with. They regarded him as the wit of the mathematical world, one capable of coming up with a short, clever solution to a problem on which others had labored through pages of equations.

Erdös worked with so many mathematicians that the concept of the "Erdös number" came to be born. To qualify for Erdös number 1, a mathematician must have published a paper with Erdös. Erdös number 2 required publication with someone who had published with Erdös and so on for higher Erdös numbers. There are 4,500 mathematicians with an Erdös number of 2.

On Paul's sense of humor and idiosyncratic elements of vocabulary:

Paul Erdös (1913–1996)

– he referred to children as "epsilons" (because in mathematics, particularly calculus, an arbitrarily small positive quantity is commonly denoted by that Greek letter);
– women as "bosses";
– men as "slaves";
– people who had stopped doing math as "dead";
– people who had physically died as having "left";
– alcoholic drinks as "poison";
– music as "noise";
– people who had married as "captured";
– people who had divorced as "liberated";
– giving a mathematical lecture was "to preach" and
– giving an oral exam to a student was "to torture" him/her.

Erdös had a very personal relationship with God. He often spoke of The Book, his imaginary invention, in which God had written down the best and most elegant proof for mathematical theorems and ideas.

In 1985, he said: "You don't have to believe in God, but you should believe in The Book." For his epitaph, he suggested: "I've finally stopped getting dumber."

Erdös died in 1996 from a heart attack at a conference in Warsaw, while he was working on an equation.

ERDÖS-CHEBYSHEV THEOREM

For each number greater than 1, there is always at least one prime number between it and its double. For example, the prime number 3 lies between 2 and 4.

This is knowh as the Bertrand-Chebyshev theorem, named after the Russian mathematician Pafnuty Chebyshev (1821–1894) who proved the theorem in the 19th century. Later, Paul Erdös did the same, but his proof was neater.

ERDÖS' PLUS 1 MINUS 1 SEQUENCES

Equal numbers of n +1 and -1 are lined up in a row. For example for n = 2 and 3: +1 +1 -1 -1 and +1 +1 +1 -1 -1 -1, etc. In how many different ways can you write down two sequences?

235 CHALLENGE ● ● ○ ○ ○ ○
REQUIRES 🧠✏️
COMPLETED

IMPOSSIBLE FIGURES

An impossible figure, also called an impossible object, is an optical illusion involving a two-dimensional image that the brain instantaneously, subconsciously and wrongly perceives as a representation of a three-dimensional object, despite the fact that the existence of such an object is geometrically impossible.

The first deliberate designs of numerous impossible objects were created by Swedish artist Oscar Reutersvärd (1915–2002), who has been referred to as "the father of impossible figures." He produced more than 2,500 of these figures, all in isometric projection, and his work has been translated in several languages.

The impossibility usually becomes clear after focusing on the figure for a few moments, although the initial impression of a three-dimensional object remains even after it has been contradicted. There are other, less obvious, examples of impossible objects whose impossibility is not immediately evident, and it is necessary to analyze the geometry of the suggested object in detail before being able to confirm that it is impossible. What is the greatest illusion of all, if not the illusion of impossibility?

POSSIBLE OR IMPOSSIBLE?

The three gates visualize simple flat planes and the multiple planes on which many impossible figures are based. A multiple plane looks like a flat plane when viewed from one point, when in fact, it consists of two or more planes when seen from another point. Looking at the plane of the base, it is easy to see which gate is impossible.

THE IMPOSSIBLE TRIANGLE — 1934

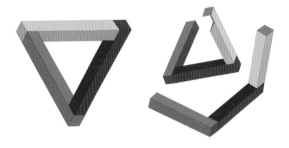

First created by Oscar Reutersvärd in 1934, the Penrose triangle, also called the Penrose tribar, is an good example of an impossible figure. Roger Penrose independently devised and popularized it in the 1950s, describing it as "impossibility in its purest form." It features prominently in the works of artist M. C. Escher. Escher builds inhabited worlds around impossible objects, while Reutersvärd's designs are usually composed of pure geometric forms.

It was long thought that the famous impossible triangle could not exist. However, as you can see on the above photograph, Prof. Richard Gregory built it. It was later also reproduced by John Beetlestone, the director of Techniquest hands-on science center in Bristol, UK. The impossible triangle was of a giant size, large enough for visitors to walk through at the entrance to the museum.

But did Prof. Gregory really build an impossible triangle? Not really, as you can see above. He built only a simple structure which, when viewed from a particular location, looks exactly like the impossible triangle. From this point, when the two ends line up exactly, the perceptual system of the brain assumes that they lie on one plane, a false impression that created the paradoxical perception from the outset.

LOTHAR COLLATZ (1910–1990)

Professor Lothar Collatz was a German mathematician. He did fundamental work in all areas of numerical analysis. The mathematical originality and creativity of Professor Collatz were evident to everyone who knew him.

G Meinardus and G Nürnberger wrote, in *In memoriam: Lothar Collatz*: "He was convinced that mathematics and mathematicians had a responsibility to apply their results to, and be motivated by, real world phenomena. He never wearied of fighting for this conviction."

THE COLLATZ PROBLEM AND HAILSTONE NUMBERS — 1937

Many know the name of Collatz today because of the "Collatz problem," proposed by Collatz in 1937, known also as the Collatz conjecture. The sequence of numbers in this problem is known as the hailstone sequence or hailstone numbers, or as wondrous numbers. The problem has intrigued mathematicians ever since he proposed it.

The Collatz problem is simple to state: take a positive integer "x." If it is even, halve it "x/2."

If it is odd, multiply it by 3, then add 1. Using the resulting integer, start all over again. Continue until you get the number 1, which then produces a never ending loop of 4, 2, 1… Collatz was surprised to find that this always happened, but he could not prove it, nor find a single case that did not eventually end in 1.

The Collatz conjecture asked whether the sequences will always reach 1 for all integers. In the accompanying diagram this is illustrated for the first 14 integers, showing the first 14 hailstone number sequences ending in relatively short sequences. As we can see, in each case the sequence reaches 1 sooner or later, before the endless 4-2-1 loop is encountered. But what happens if you try 15?

The sequences produced by the Collatz problem are known as hailstone numbers, because their values rise and fall like a hailstone from a cloud.

Conway proved that the problem is not decisive as a result of being neither formally provable nor unprovable. Erdős noted that mathematics is not yet ready for such problems.

Today's super-computers tested all numbers up to 27 quadrillion (27,000,000,000,000,000). Not one was found whose hailstone sequence did not eventually end in 1. The longest hailstone sequence to date is a 15-digit number whose hailstone sequence consists of 1,820 numbers.

								1					
								2					
								4					1
						1		8					2
						2		16					4
						4		5		1			8
						8		10		2			16
						16		20		4			5
						5		40		8			10
						10		13		16			20
						20		26		5	1	1	40
						40		52		10	2	2	13
		1			2	13		17		20	4	4	26
		2			4	26		34	1	13	8	8	52
		4		1	8	52		11	2	16	16	16	17
		8		2	16	17		22	4	26	5	5	34
1		16		4	5	34	1	7	8	52	10	10	11
2		5	1	8	10	11	2	14	16	17	3	20	22
4	1	10	2	16	3	22	4	28	5	34	6	40	7
1	2	3	4	5	6	7	8	9	10	11	12	13	14

236

CHALLENGE

REQUIRES

COMPLETED

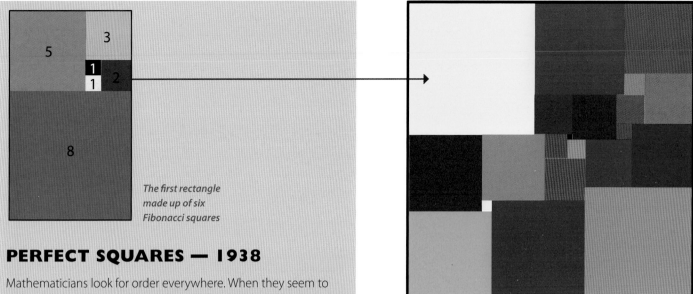

The first rectangle made up of six Fibonacci squares

PERFECT SQUARES — 1938

Mathematicians look for order everywhere. When they seem to discover it, they like to give expression to their enthusiasm by defining numbers, squares, rectangles, triangles and parallelograms as "perfect," "imperfect" and so on.

The problem of squaring the circle goes back to the ancient Greeks, but that of squaring the square is quite recent. In 1934, Paul Erdős, the famous Hungarian mathematician, posed the following dissection problem: Can a square be subdivided into smaller squares of which no two are identical? Such squares are called "perfect" or "squared."

Erdös wrongly concluded that such a square is impossible, probably influenced by the easily proven fact that one cannot dissect a cube into smaller cubes, no two of which are identical. He concluded that the best one could achieve was to dissect a rectangle into smaller squares, no two of which are identical.

For a long time, it was not known whether any squared perfect squares existed. However, in 1938 R. Sprague found a 55-square perfect square. In 1948, a 24-square perfect square was found by Willcocks. For years, it was believed that this square requiring 24 squares (all different) was the smallest perfect square. But in 1978, a better solution was found by A.J.W. Duijvestijn, a Dutch mathematician, requiring only 21 element squares. Presently, this is the smallest known simple squared square (perfect square), and its pattern is unique. If some of the squares used in a dissection are allowed to be identical, the squares or rectangles are called "imperfect" or "Mrs. Perkins's quilts."

SMALLEST PERFECT SQUARE

By substituting the first Fibonacci number square in the Fibonacci rectangle with a perfect square, we have solved the old problem of tessellating the infinite plane with unequal squares.

There is no problem dissecting a square into squares, but the added condition, which made this problem one of the most beautiful and difficult problems for a long time, is the requirement that all the squares must be of different sizes. The lowest-order perfect square, discovered by A.J.W. Duijvestijn, consists of 21 squares of the sizes shown below: 2 - 4 - 6 - 7 - 8 - 9 - 11 - 15 - 16 - 17 - 18 -19 - 24 - 25 - 27 - 29 - 33 - 35 - 37 - 42 - 50.

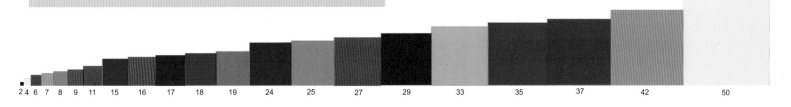

2 4 6   7   8   9   11   15   16   17   18   19   24   25   27   29   33   35   37   42   50

BENFORD'S LAW

On August 4 1998, *The New York Times* published the story of dr. Theodore P. Hill, who asked his mathematics students at the Georgia Institute of Technology to go home and either flip a coin 200 times and record the results, or merely pretend to flip a coin and fake 200 results. "The following day he runs his eye over the homework data, and to the students' amazement," so The New York Times wrote, "he easily fingers nearly all those who faked their tosses. 'The truth is,' he said in an interview, 'most people don't know the real odds of such an exercise, so they can't fake data convincingly.'" There is more to this than a classroom trick. Dr. Hill is one of a growing number of statisticians, accountants and mathematicians who are convinced that an astonishing mathematical theorem known as Benford's Law is a powerful and relatively simple tool for pointing suspicion at frauds, embezzlers, tax evaders, sloppy accountants and even computer bugs.

Benford's Law is named after the late Dr. Frank Benford, a physicist at the General Electric Company. In 1938 he noticed that pages of logarithms corresponding to numbers starting with the number 1 were much dirtier and more worn than other pages. Although Benford's Law applies to a wide variety of data sets, there is no simple explanation for this phenomenon."

FLIPPING FRAUD — 1938

Tell your friend to take a coin, flip it 200 times and record the outcomes, or just pretend to flip it and fake 200 results. Tell him or her that in just a few seconds you will be able to tell whether the results were real or fake. This will be a convincing demonstration of the fact that, if you don't know the real odds of an event, you can't fake it convincingly, and also convincing proof of the workings of laws of probability.

Below are two outcomes of the experiment. Can you tell which one is fake?

237

CHALLENGE
REQUIRES
COMPLETED   TIME

Heads

Tails

Test 1

Test 2

FLEXAGONS — 1939

Flexagons are flat topological structures, usually constructed by folding strips of paper that can be flexed or folded in different ways to reveal faces other than the two that originally appeared on the front and back. They are usually square or rectangular (tetraflexagons) or hexagonal (hexaflexagons).

In 1939, Arthur H. Stone discovered the first flexagon, a trihexaflexagon. The story goes he accidentally stumbled across his trihexaflexagon while was playing with the strips he had cut off his foolscap paper to convert it to letter size. Stone's colleagues Bryant Tuckerman, Richard P. Feynman and John W. Tukey later formed the Princeton Flexagon Committee. They first defined a topological method,

called the "Tuckerman traverse," to reveal all the faces of a flexagon. Flexagons became the very popular on the Princeton campus.

In 1959, Martin Gardner introduced the flexagons to the world in his debut Scientific American article, and flexagons became the rage all around the world.

Flexagons are essentially mathematical curiosities, although there has been a certain amount of commercial interest, with the marketing of some products such as drink coasters, greeting cards and toys. Ever since the 1960s, the author was fascinated by flexagons and paper folding, inventing a considerable number of original puzzles and toys involving paper folding.

ARTHUR HAROLD STONE (1916–2000)

Arthur Harold Stone was one of the foremost topologists of his time, making significant contributions to a number of different areas of general topology. His Jewish parents were immigrants from Romania and he had been a member of the London Mathematical Society since 1948.

In 1935 Stone won a major

scholarship at Trinity College, Cambridge. He excelled at the academic subjects, as well as being an exceptional violinist and a formidable chess player. He achieved his BA in 1938, before moving on to Princeton to study for a PhD under S. Lefschetz.

Although a single-minded mathematician, he had wide-

ranging interests, and this combination often showed up in unexpected ways, like the discovery of the famous flexagons.

He was also intrigued by how one can dissect a square into unequal smaller squares (see Perfect Squares). He managed a dissection with 69 squares, which was later improved by others.

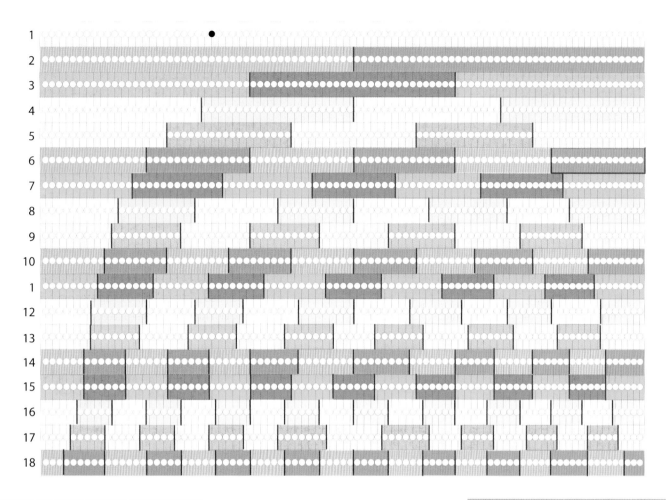

18-POINT PLACEMENT GAME

Imagine you have a strip of land on which there is a tree, represented by a point that can be placed anywhere (row 1). Dividing the land into two halves, you plant another tree in the second (empty) half (row 2).

Then you decide to divide your land into three equal parts and plant another tree (in the empty third). And again, and again. Each time, the trees already planted turn out, luckily, to be in their separate plots.

Can you be foresighted and farsighted enough to plant your trees where they will have their own space, no matter how often you divide your land into a growing number of equal parts? How many points will you be able to place before two of the trees land in the same compartment? This is the end of the game.

In the Solutions chapter at the back of this book, you can find a solution that ends in the sixth generation. Remember: the sequence of generations represents the same length of land divided into an increasing number of equal divisions. A challenging two-person game can also be played while solving the problem. Players alternate by placing their trees. The loser is the player who on his or her turn is forced to plant a tree in an area already occupied by a tree planted earlier.

18-POINT PROBLEM

The 18-point placement game is a simplified variation of the famous 18-point problem, in which a line with an infinite number of points is to be divided in a similar way with the objective of placing 18 points on the line.

One might expect that one could place an infinite number of points on the line, but this is wrong. No matter how the points are placed, one cannot go beyond 17 points. The 18th point will always end the game.

This elegant problem was first seen in 1939 in Polish mathematician Hugo Steinhaus' publication *One Hundred Problems in Elementary Mathematics*.

Later, it was extensively covered by Gardner, Conway, Warmus, Berlekamp, Baxter and others.

There are 768 different 17-point solutions.

THE BIRTHDAY PARADOX — 1939

You want to have a party at which at least two people share the same birthday. Their birthdays must be on the same day of the same month, but they do not necessarily have to have been born in the same year. If you don't know the birthdays of any of your guests, how many people do you have to invite so that the probability of two people sharing the same birthday is more than 50 percent? How many people do you need to invite for birthday sharing to be a practical certainty?

Remarkably, the probability of two people sharing a birthday in a group of just 23 people is about 50 percent. To calculate this, you have to look at the probability that everyone has a different birthday. The probability of a group of two people having different birthdays is extremely high, at about 364/365. The probability is slightly lower for a group of three — 363/365 — and as the group of three still includes the group of two, the probabilities are multiplied.

Continue in this way until the probability of every person in the group having different birthdays drops below 0.5, which means the probability of two people sharing a birthday is now more than 0.5.

The phenomenon is known as the Birthday Paradox, because of the fact that the probability nears certainty at 99 percent, with just 57 people or more, and not with 366 people, as intuition would dictate.

These conclusions are based on the assumption that there is an equal probability of a birthday falling on any day of the year (except February 29).

RICHARD VON MISES (1883–1953)

Richard von Mises was a Jewish scientist and mathematician who worked on solid mechanics, fluid mechanics, aerodynamics, aeronautics, statistics and probability theory. Van Mises was born in Austria but fled to Turkey to escape the increasing threat of Nazi policies after they came to power in 1933. He moved to the United States in 1939 and was appointed Gordon-McKay Professor of Aerodynamics and Applied Mathematics at Harvard University in 1944.

He was the first to propose the now famous "birthday problem" in probability theory. The Birthday Paradox is a problem in which two randomly chosen samples may have the same value, such as the birthdays of people at a party. The paradox lies in the probability of this coincidence actually happening being much greater than expected.

MIXING MARBLES AND TEA WITH MILK

One of the most counterintuitive and interesting puzzles I have encountered is the classic problem about cups of tea and milk, with a teaspoon of milk being added to the tea, and then another tea spoon of the mixture returned to the glass of milk.

The question is whether there is more milk in the tea than there is tea in the milk or vice versa?

As tricky as the problem seems, the counterintuitive answer is that there is exactly the same amount of milk in the tea as there is tea in the milk. The explanation is that the total volume in each glass is unchanged by the transfers; the net volume transferred from glass A to glass B cancels out the volume that went from glass B to glass A.

Initially I was a bit skeptical about this answer until, many years later, I performed an analogous experiment involving marbles in two colors instead of tea and milk, which you can easily try out yourself.

Take two boxes filled with marbles, let's say 50 in each box, one box with red marbles and the other with green marbles as shown.

Take five marbles from the red box and transfer them into the green box. Mix the green box well and return five randomly selected marbles into the red box. Which box will contain the wrong color marbles?

The experiment is visualized. There are six possible ways to return the marbles back into the red box. In each case there is the same number of marbles of the wrong color in each box. The same will always happen no matter how many marbles are transferred.

1. *Transferring five marbles into the green box*
2. *Mixing the green box*
3. *There are six different possible ways to return five marbles into the red box as shown.*

HEX GAME — 1942

One of the most beautiful topological games is Hex, invented in 1942 by Piet Hein, a remarkable Danish inventor and freedom fighter. He invented Hex while analyzing the four-color problem of topology.

In 1948, John F. Nash, a Nobel prize winner at MIT, independently reinvented the game. The game was the first prototype of a game introducing connectivity, the board-crossing principle on which many later games like Twixt, Bridge-It and others were based. Although it is a very simple game to learn and play, it offers surprising mathematical subtleties.

Hex is played on a hexagonal gameboard of 11-by-11 hexagons shown on the next page, but the size of the board in later variations varies. One player has a supply of red pieces; the other a supply of green pieces. It can also be played as a paper-and-pencil game, using colors or just marks like O and X, on game pads or using a set of small coins, heads or tails up, for the two players. Players alternate by placing their pieces on an unoccupied hexagonal cell. The object of the game is to complete an unbroken chain of a solid color from one side of the board to the other. The corner hexagons may belong to either of the two colors or marks. No draw is possible; one player will always win.

On a two-by-two board the player who makes the first move easily wins.

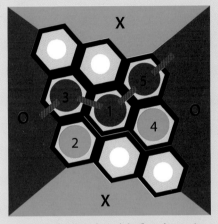

On a three-by-three board the first player wins when his first move is the center cell.

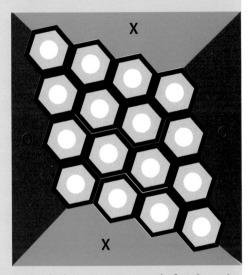

How and in how many moves can the first player win on a four-by-four gameboard?

How can the first player win on a five-by-five gameboard?

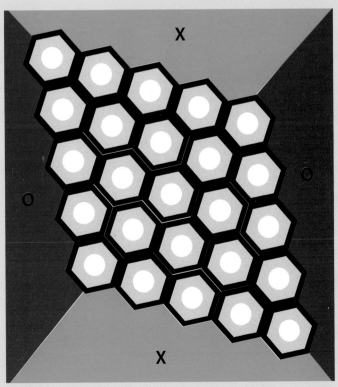

239 CHALLENGE ● ● ● ● ● ○
REQUIRES 🧠 ✏️ ✂️
COMPLETED

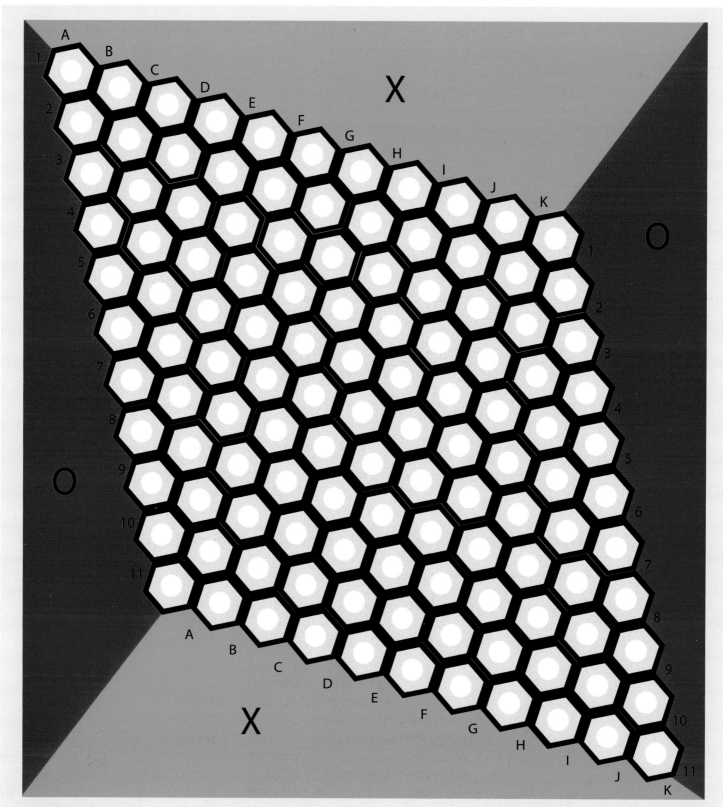

HEX

The standard game board for a two-person game. With a supply of small coins you can
use it to play the game right away.

HAROLD SCOTT MACDONALD COXETER (1907–2003)

Harold Scott MacDonald "Donald" Coxeter was a Canadian geometer. He was born in London but spent most of his life in Canada. He is considered to be one of the greatest geometers of the 20th century and was often referred to as the "king of geometry." He made a huge contribution to the field of geometry.

Coxeter accepted a position at the University of Toronto in 1936 and continued to work there for 60 years, publishing 12 books. He was best known for his work on regular polytopes and higher-dimensional geometries. Against the trend, he championed the classical approach to geometry, at a time when an algebraic approach to geometry was increasingly popular.

REGULAR POLYGONS AND STARS — 1950

As we've seen already, a polygon is called regular if it has the following two properties:
– All of its sides are equal.
– All of its angles are equal.

A circle may be thought of as a regular polygon with an infinite number of sides.

There is also an infinite number of regular polygons that can be subdivided into the following subgroups as demonstrated:
– Simple regular polygons (in red);
– Regular star polygons;
– Regular compound polygons.

Stars (blue) and compounds (green) of the first seven regular polygons are shown.

A star polygon is a figure formed by connecting with straight lines every q-th point out of P regularly spaced points on the circumference of a circle. P and q are positive integers.

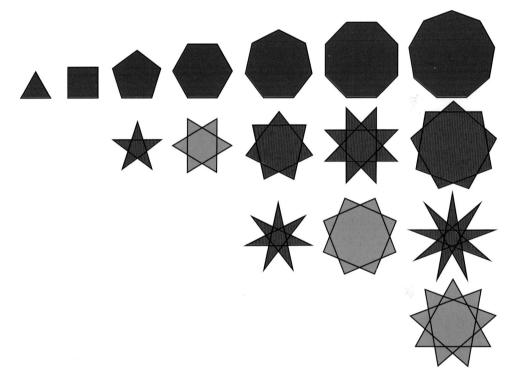

STARS AND COMPOUNDS

How many stars and compounds have regular decagons, undecagons and dodecagons (regular polygons with 10, 11 and 12 sides)?

Regular decagon   Regular undecagon   Regular dodecagon

240 CHALLENGE ●●●●○
REQUIRES
COMPLETED

GOLOMB RULERS: PERFECT AND OPTIMAL GOLOMB RULERS — 1952

The concept of the Golomb ruler is an unusual measurement idea. It was introduced in 1952 by W.C. Babcock.

Today, they are called Golomb rulers after Solomon W. Golomb, a professor of mathematics and electrical engineering at the University of Southern California who extensively analyzed and extended the concept into new and unexpected directions.

A Golomb ruler is a ruler constructed so that no two pairs of marks can measure the same distance. The markers on a Golomb ruler must be placed at integer multiples of fixed spacings. The aim is to place the markers so as to achieve as many distinct measures of distance between two markers as possible with a given number of markers. In order to achieve this, the markers must be placed very efficiently, avoiding redundant distances between markers.

In a perfect Golomb ruler of length n, all of the distances from 1 – 2 – 3 – … n, can be measured exactly once. Perfect Golomb rulers exist only for ruler lengths up to four marks. The condition for an optimal Golomb ruler — in other words the shortest possible Golomb ruler for a specific set of marks — remains that no two pairs of marks may measure the same distance. However, it may not have all consecutive distances from zero to the ruler's length.

It becomes more difficult to find and prove optimal Golomb rulers as the number of marks rises.

Today, optimal Golomb rulers up to 24 marks are known, and there is now a search for the 25 and 26 versions.

The Golomb rulers problem is one of the most beautiful problems in recreational mathematics, but such rulers are also needed in a variety of scientific and technical disciplines and are at the forefront of mathematical research, proving the relevance of recreational problems to pure math. Golomb rulers provide a general spacing principle applied in astronomy (placement of antennas), x-ray sensing devices (placement of sensors), and many other fields.

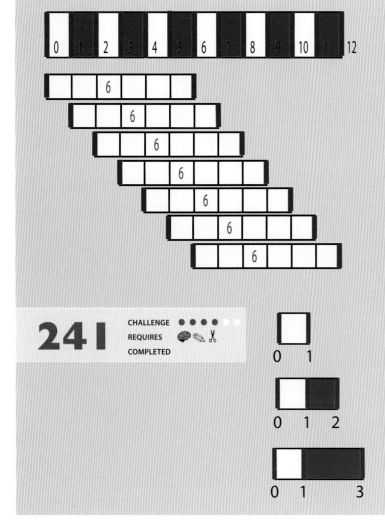

241 CHALLENGE ● ● ● ● ○ ○
REQUIRES 🧠✏️✂️
COMPLETED

A 12-UNIT LENGTH RULER WITH 13 MARKS

The 12-unit length ruler shown with 13 markers (at every unit), allows us to measure any integer distance between 1 and 12 units. From a mathematical viewpoint, this is not a very economical arrangement. Using 13 markers to measure lengths from 1 to 12 is not very efficient. For example, we can measure unit lengths in 12 different ways, lengths of six units in seven different ways (as shown), etc. It obviously isn't a Golomb ruler. Can we reduce such redundancies and create Golomb rulers for different numbers of markers, optimal or perfect?

PERFECT GOLOMB RULERS

A one-unit length ruler with two markers is "perfect" but trivial. A two-unit length ruler with three markers is not "perfect," since it can measure a one-unit distance in two ways.

A three-unit length ruler with three markers is in fact the first "perfect" ruler.

The definition for a perfect ruler of length n units is one which is capable of measuring all integer lengths from 1 to n in one way only. Can you find the next perfect Golomb ruler?

OPTIMAL GOLOMB RULER

Can you place six markers on the ruler to create an "optimal" Golomb ruler in which the maximal number of distances can be measured between two markers, without measuring any distance more than once? Which will be the distances that can't be measured?

six markers

A 17-unit length ruler

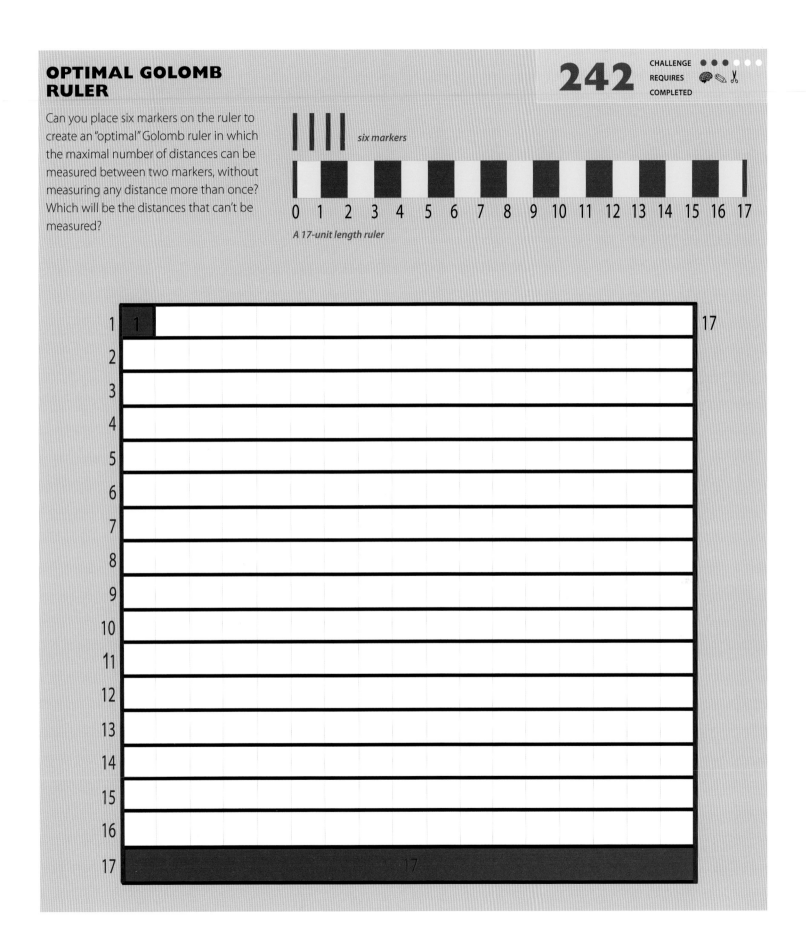

DISAPPEARING CUBE

A chessboard is cut into two parts along its diagonal as shown below. The lower part is shifted along the diagonal one square to the left and the remaining triangle (blue) is fit into the triangular space at the bottom left, as shown below.

A seven-by-nine rectangle is formed of an area of 63 unit squares, one less than the original chessboard. Can you explain the paradox?

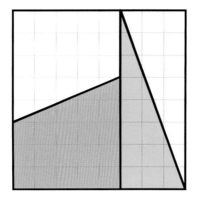

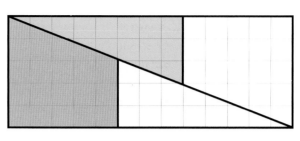

243	CHALLENGE ● ● ● ● ●
	REQUIRES 🧠 ✏️ ✂️ ⚒️
	COMPLETED

CURRY'S CHESSBOARD PARADOX — 1953

An eight-by-eight chessboard is cut into four pieces along its unit squares gridlines — two trapezoids and two triangles, as shown. The four parts can be put together into a five-by-13 rectangle containing 65 unit

244	CHALLENGE ● ● ● ● ○
	REQUIRES 🧠 ✏️ ✂️ ⚒️
	COMPLETED

squares (one more unit square than the original square. How can you explain the gaining of a unit square?

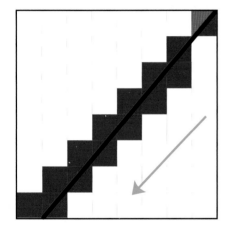

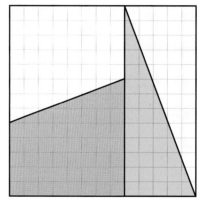

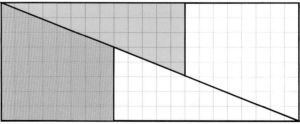

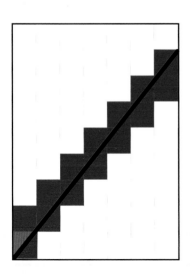

CURRY'S SQUARE PARADOX

Similarly a 13-by-13 chessboard (169 unit squares) is cut into four parts, rearranged into an eight-by-21 (168 unit squares) rectangle — one unit square less than the original square. How can you explain the loss of a unit square?

245	CHALLENGE ● ● ● ● ●
	REQUIRES 🧠 ✏️ ✂️ ⚒️
	COMPLETED

The earliest references of dissection paradoxes, often called "geometrical vanishes," can be found in *Rational Recreations* by William Hooper, from 1774.

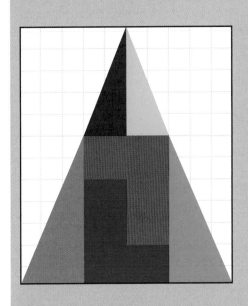

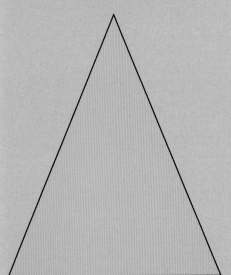

CURRY'S TRIANGLE PARADOX — 1953

The Curry triangle, also known as the missing squares puzzle, is a dissection fallacy that was created by American neuropsychiatrist L. Vosburgh Lions as an example of a phenomenon discovered by New York magician Paul Curry in 1953. The filled triangle has an area of 60 unit squares. The six pieces also have a total area of 60 unit squares.

Copy the six pieces and rearrange them into the triangular grey outline, which also has an area of 60 unit squares, but this time with a hole in the middle. How can this be done? How can you get nothing from something? Since Curry was a magician, you might expect that magic is involved. But no magic is needed, as shown in the figures below.

The figure appears to show that a triangle of area 60, a triangle of area 58 containing a rectangular hole, and a broken rectangle of area 59 can all be created from an identical set of six polygonal pieces.

This can be explained by the inaccuracy of the initial subdivision. The combined hypotenuses of the big and small triangles are bent, producing the illusion of a paradox.

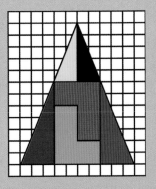
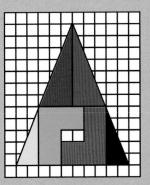
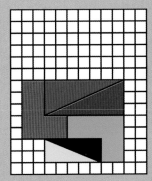

TIED DOG

Fido is tied to a thick tree of 2 meters in diameter by a rope that is 20 meters long. He could run and cover a circle with a 21-meter radius, but he is obstructed by a shed built in the circle. This reduces the area that Fido could cover.

Can you work out the reduced size of the area he could cover compared to the big circle area and also whether he could reach the inviting bone positioned as shown?

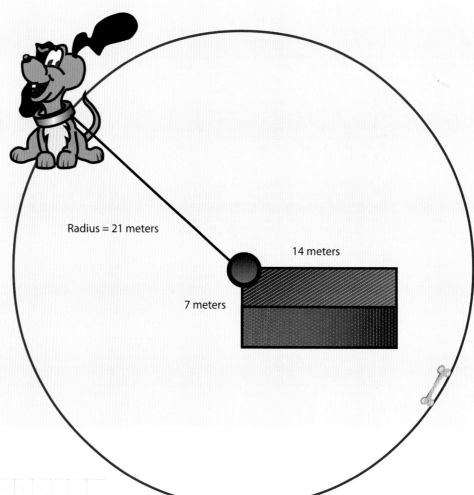

Radius = 21 meters

14 meters

7 meters

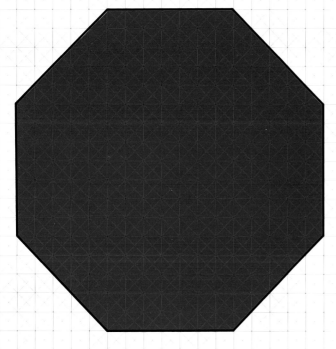

REGULAR OCTAGON

Are you sure that the pink octagon is the area of a regular octagon?

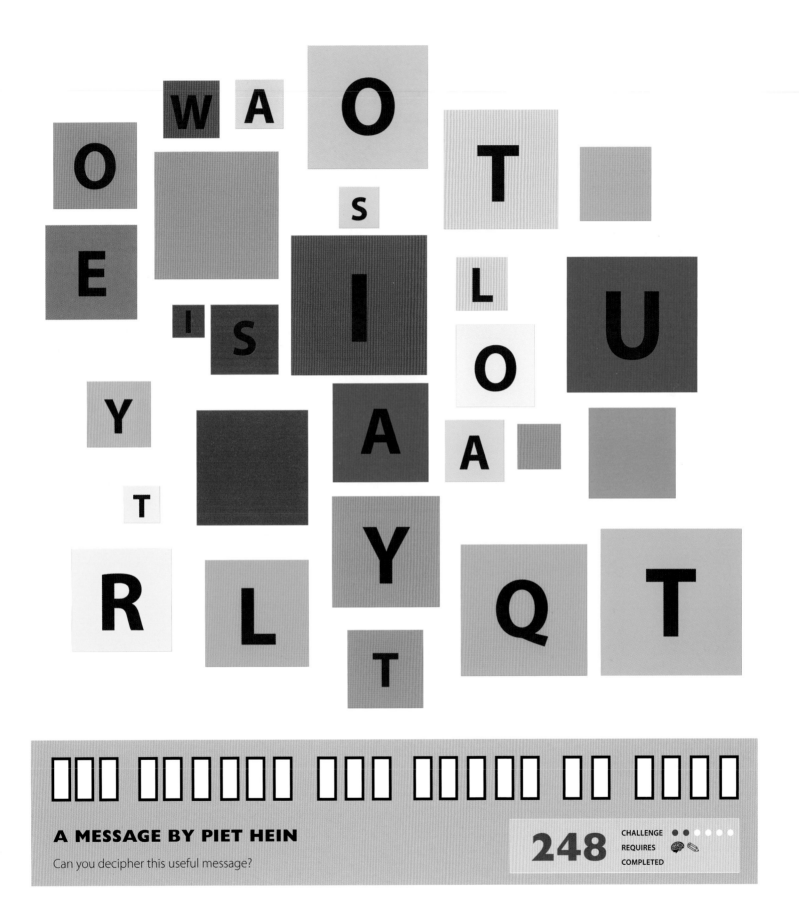

A MESSAGE BY PIET HEIN

Can you decipher this useful message?

248

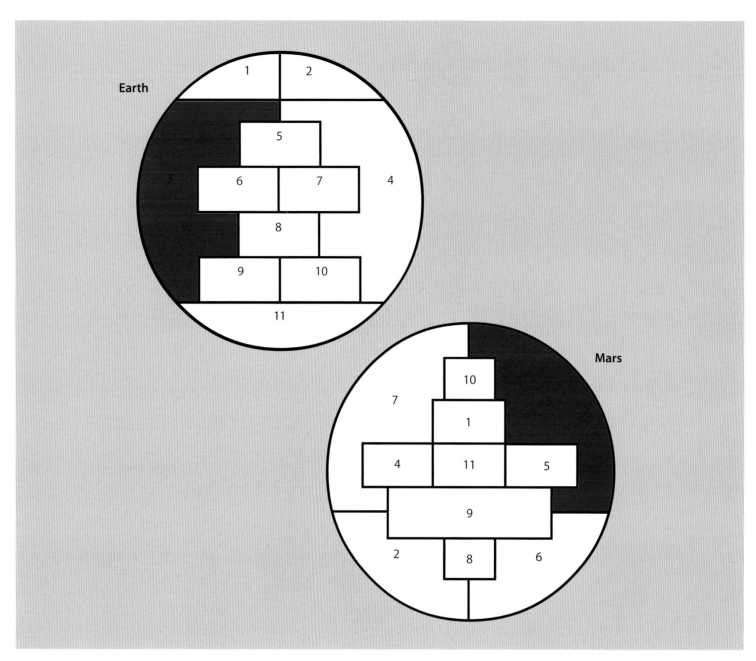

MARS COLONY - 1959

An interesting map-coloring problem was proposed by Gerhard Ringel in 1959. Assume that the nations of the Earth have colonized Mars and that there is one region for each nation on each planet (the home country and its colony).

Naturally, the countries will insist that Mars maps use the same colors for the colonies as Earth maps do for the home countries. Color the eleven 2-pires of the two spheres so that both regions sharing the same number are given the same color and, of course, no neighboring regions match in colors. One region on both planets has been colored. What is the minimum number of colors needed?

249 CHALLENGE ●●●●●
REQUIRES 🧠✎✂
COMPLETED

GERHARD RINGEL (1919–2008)

Ringel was a German mathematician at the Free University of Berlin and a pioneer in the field of graph theory. He made a significant contribution to the proof of the Heawood conjecture (now the Ringel-Youngs theorem), a problem closely associated with the Four Color Theorem. He later accepted a position at the University of California.

THE MINIMAL SPANNING TREE PROBLEM — 1956

In graph theory, a spanning tree is a subset of a graph that has no circuits (closed loops), and includes all of the vertices, but usually not all of the edges. When the edges are weighted the problem is to find the spanning tree that has the lowest cost.

Kruskal's algorithm solves this problem. The procedure is listing the weights in increasing value, and then choosing them in order, avoiding those that complete a circuit.

Can you find the Minimal Spanning Tree for the graph on your right, which has 10 vertices and 21 weighted edges?

Using Kruskal's algorithm, a sample solution is given below.

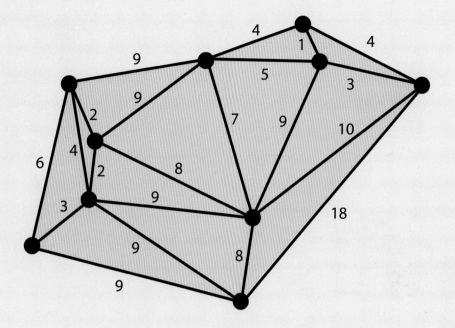

1 – 2 – 2 – 3 – 3 – 4 – 4 – 4 – 5 – 6 – 7 – 8 – 8 – 9 – 9 – 9 – 9 – 9 – 9 – 10 – 18

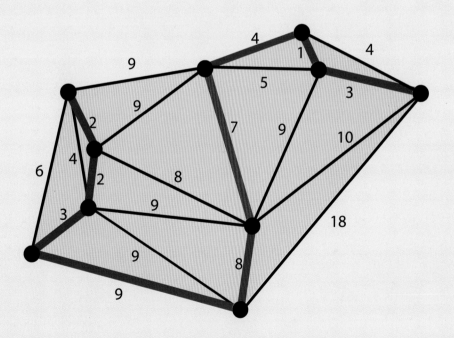

Solution: 1 – 2 – 2 – 3 – 3 – 4 – 4 – 4 – 5 – 6 – 7 – 8 – 8 – 9 – 9 – 9 – 9 – 9 – 9 – 10 – 18 = 39

CHESSBOARD SQUARES — 1956

How many squares of different sizes can you find along the grid of a chessboard? Offhand, you might say there are 64 squares. But there are more than 64 unit squares in the square matrix. Can you find the total number of squares of different sizes? Can you generalize a way to find the number of squares of different sizes there are in a square grid with n unit squares on a side?

250 CHALLENGE ● ● ● ● ●
REQUIRES
COMPLETED

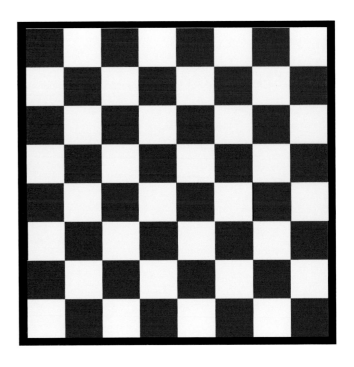

CHESS CUBE

How many cubes of different sizes composed of unit cubes are there in a 3D chess cube?

251 CHALLENGE ● ● ● ● ○
REQUIRES
COMPLETED

SQUARE LATTICE NUMBERS

If we extend the Chessboard Squares problem to include rectangles of different sizes, we obtain the lattice number for a square lattice.
Puzzle 1 – Can you work out what this number L(n) is for square lattices from n=2 to n=8 ?
Puzzle 2 – On the chessboard, an eight-by-eight square lattice, how many different sizes of squares and rectangles are there?

252 CHALLENGE ● ● ● ● ○
REQUIRES
COMPLETED

LAMP IN THE ATTIC

Puzzle 1 – This is an old castle with black curtains on the windows and a single lamp in the attic. At the entrance gate there are three light switches. One of them turns on the lamp in the attic. Your job is to find out which of the three switches turns on the lamp, but you are allowed only one trip to the attic to check out the light. Can you figure it out how to tell which light switch works?

Puzzle 2 – In the previous puzzle there were two switches that had no use. Now we have the same switches and three lamps in the attic, each lamp turned on by one of the switches. As before, you are only allowed to enter the castle to check the lights once. How can you figure out now which lamp is switched on by which switch?

253   CHALLENGE ● ● ● ● ● ○
REQUIRES 🧠 ✏️ ✂️
COMPLETED

OLD FRIENDS' MEETING

Two Russian mathematicians meet on a plane. "If I remember correctly, you have three sons," says Ivan. "What are their ages today?"

"The product of their ages is 36," says Igor, "and the sum of their ages is exactly today's date."

"I'm sorry, Igor," Ivan says after a minute, "but that doesn't tell me the ages of your boys."

"Oh, I forgot to tell you, my youngest son has red hair."

"Ah, now it's clear," Ivan says. "I now know exactly how old your three sons are."

How did Ivan figure out the ages of the boys?

254 CHALLENGE ●●●●●● REQUIRES 🧠✏️ COMPLETED

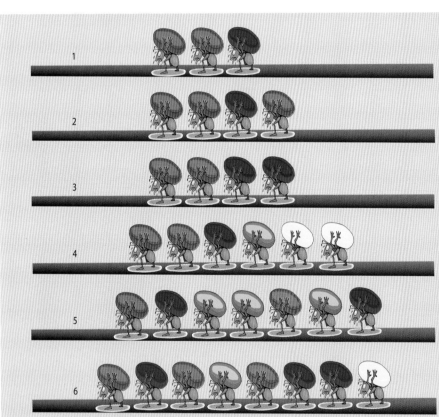

ANT PROCESSIONS — 1958

Dennis E. Shasha, a professor of computer science at New York University, defined a sequence of symbols as being "surprising" if there are no two symbols x and y, such that there are two or more pairs of occurrences of the two symbols where x precedes y by the same distance. For example, procession number 3 is not surprising since there are two occurrences of the symbols x and y where x precedes y by two symbols. In our puzzles, the symbols are ants carrying eggs of different colors. Can you work out which of the six ant processions are surprising and which are not?

255 CHALLENGE ●●●● REQUIRES 🧠 COMPLETED

LANGFORD'S PROBLEMS — 1958

Langford's Problem is named after Scottish mathematician C. Dudley Langford, who proposed it in 1958 after watching his young son play with colored cubes. The boy arranged three pairs of blocks in a row with one block between the two red blocks, two blocks between the two blue blocks and three between the two yellow blocks. He also managed to add two green blocks with four blocks in between and retain the above properties after making some adjustments.

In combinatorial mathematics, a Langford pairing, also known as a Langford sequence, is a permutation of the sequence of pairs of identical numbers in which the ones are one unit apart, the twos are two units apart, and in more general terms the two copies of each number k are k units apart. Langford's problem is the task of finding Langford pairings for a given value of n.

RUNNERS PUZZLES

Puzzle 1 – We have four teams of runners, with two runners in each team, shown in a configuration during the race. At the finish line, their configuration changes, so that one runner is between the red pair, two runners between the blue pair, three runners between the green pair, and finally, four runners are between the yellow pair. All we can be sure of is that a yellow runner will be the last. Can you work out what colors are the first three winners?

Puzzle 2 – We have nine teams of runners, with three runners in each team. Each team is numbered in nine consecutive numbers and nine colors. At the finish line, their configuration changes, so that each numbered middle element in a triplet is separated from its outer triplets on both sides by their number. This is demonstrated, for example, for the number 2 triplet. Can you find the runners' configuration at the finish line?

MIXED HATS — 1959

There are many puzzles and problems dealing with coincidence. One of the most famous is the mixed hats or mixed letters problem, sometimes also called the "Montfort problem."

Suppose n people attend a party, checking in their hats with a checkroom girl, who mixes up the hats. What is the probability that, in spite of the mix-ups, at least one man will get back his own hat? Do you think any of them has a better than 50-50 chance of getting his hat?

The surprising conclusion is that, as n grows, the chance that any given man will get his own hat gets smaller and smaller, but on the other hand there is a growing chance that at least one man will get his own hat. The two effects cancel each other out. The probability that at least one man gets his own hat is about 63 percent.

You can check the validity of this result with a deck of cards. Shuffle it, then turn up the cards one at a time, counting: "ace, two, three, four… ten, jack, queen, king, ace, two, three…" What will be the chance that your count will identify at least one card? This is, in fact, the same problem as that of the mixed hats. The odds are quite good that you will have at least one match, and possibly more. Try it!

MIXED HATS PUZZLE (1)

Three men check in their hats.

A careless hatcheck girl mixes up the checks before she hands them out. When the three men later call for their hats, what do you think the chances are that the right hats will be returned to all of them?

257 CHALLENGE ●●●●○○ REQUIRES 🧠✏️ COMPLETED

MIXED HATS PUZZLE (2)

Six men check in their hats as before. What now are the chances that the right hat will be returned to at least one of them?

258 CHALLENGE ●●●●○○ REQUIRES 🧠✏️ COMPLETED

EARLY COMPUTER ART

The first exhibition of computer graphics took place in 1965. Many followed. The most noteworthy was the Cybernetic Serendipity Exhibition in London in 1968. The catalog of this exhibition has been published in book form and it is still the richest and most comprehensive survey of information on the state of the new art form. Ours is an age of technology, in which human labor is increasingly taken over by the computer, not all of it purely utilitarian. The earliest computer drawings were those of mathematical curves and figures, a beauty that has been explored in art from the earliest times.

THE MACHINE THAT MADE SCIENCE INTO ART — 1951

The French physicist Jules Antoine Lissajous (1822–1880) discovered the Lissajous figures that are named after him. He used sounds of different frequencies to vibrate mirrors attached to tuning forks. A beam of light reflected from the mirrors would trace pleasing patterns based on the sound frequencies. A similar setup is used today in laser shows.

Classical Victorian harmonograph toys drawing Lissajous figures usually consisted of two coupled pendulums, oscillating at right angles to each other, one pendulum carrying a pen and the other the paper. The resulting families of Lissajous curves ended in a point when the pendulums were damped by friction. Dozens of early harmonographs were patented. Their designs limited the size and quality of their drawings and did not allow any artistic expression, but were fascinating early science toys.

In the late 1950s, I patented a novel entitled *Harmonograph* based on a completely new design concept that enabled beautiful artistic creations of giant size. The "Harmonograph of Moscovich," as it was patented

worldwide, and its creations the Harmonograms, aroused enormous interest at the Cybernetic Serendipity exhibition in London in 1968.

As a consequence of its acclaim at the Cybernetic Serendipity exhibition, and winning medals at the Geneva Inventions fair during the '70s and '80s, I was invited to participate in many art exhibitions and several one-man shows worldwide (including the International Design Center in Berlin, Museum of Modern Art in Mexico City, Didacta exhibitions in Basel and Hannover and the Israel Museum in Jerusalem).

In 1980, Peter Pan Playthings, a UK toy company launched a toy version of the Harmonograph, which was marketed successfully in the '80s. Today, the harmonograph of Moscovich is one of the major interactive exhibits at the Technorama, a science center in Winterthur, Switzerland. My daughter Hila is presently creating unique original harmonograms, which are highly acclaimed by art lovers and collectors worldwide.

LISSAJOUS' OPTICAL SETUP

Lissajous' setup projecting moving Lissajous curves and patterns by vibrating tuning forks, mirrors and pin-point light source.

VICTORIAN HARMONOGRAPH

The basic design of harmonographs since Victorian times, until the 1951 invention by Moscovich.

HARMONO-GRAPH OF MOSCOVICH

The first prototype of the world patented harmonograph from 1951 with a harmonogram on its revolving board.

HARMONOGRAMS

Harmonograms of Moscovich at the Cybernetic Serendipity exhibition in London 1968.

POLYABOLOES OF O'BEIRNE — 1959

Polyaboloes are analogs of polyominoes. They are composed of a number of n isosceles right triangles (half-squares) joined along edges of the same length. Polyaboloes are identical if they have the same boundaries. The inner structure of the triangles does not make them different.

There is one monoabolo, three diaboloes and four triaboloes.

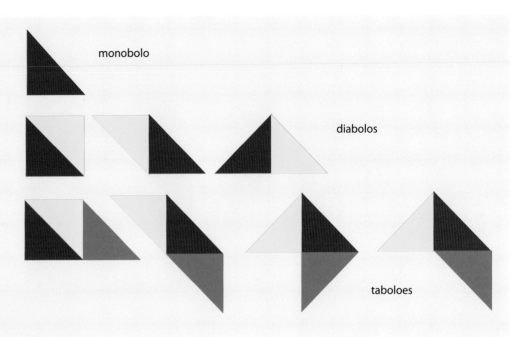

monobolo

diabolos

taboloes

TETRABOLOES OF O'BEIRNE

On the right, you can see 14 tetraboloes. The number of polyaboloes composed of n triangles creates the following number sequence: 1, 3, 4, 14, 30, 107, 318, 1106…

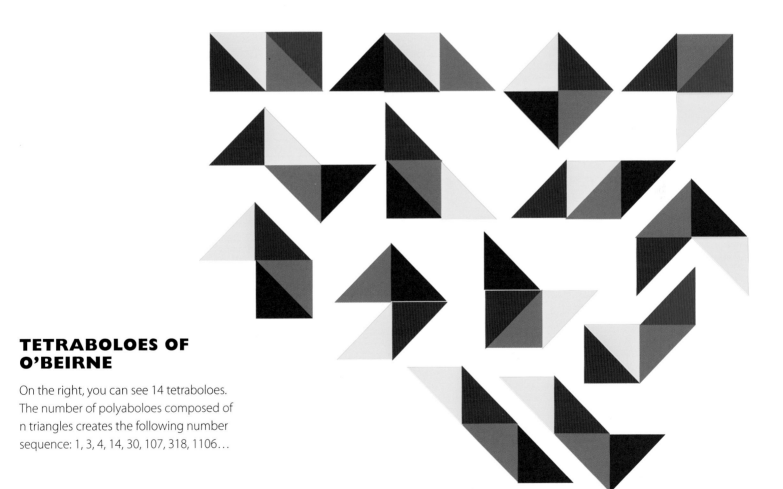

THE HEXIADIMONS OF O'BEIRNE — 1959

In 1959, Thomas O'Beirne noticed that among the polyominoes that can be formed by joining six equilateral triangles, five are symmetrical while seven are not. If we count the reflections of the unsymmetrical hexiamonds, we have 19 shapes, which cover the same area as a game board consisting of a three-by-three configuration of regular hexagons. He posed the following problem: can the 19 shapes cover the game board of 19 regular hexagons?

O'Beirne's hexiamonds problem is one of the most challenging two-dimensional puzzles ever devised.

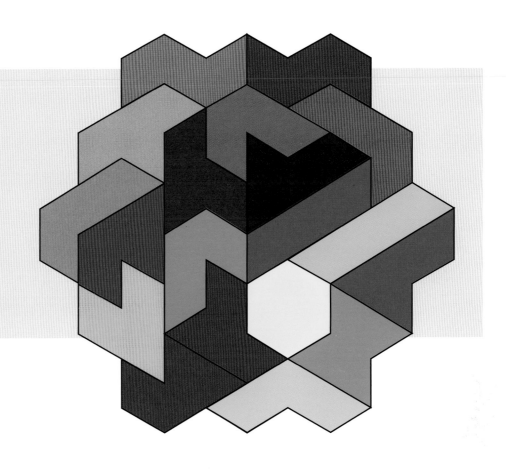

O'BEIRNE'S SOLUTION

It took O'Beirne a number of months to find a solution. His first solution is shown.

Richard K. Guy classified the solutions. According to his guess, there are about 50,000 solutions, and there are already more than 4,200 of them in his collection. Can you find a different solution?

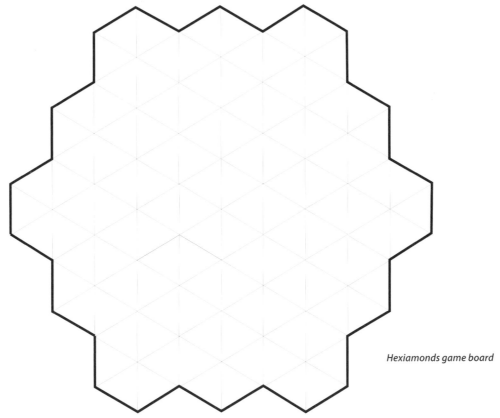

Hexiamonds game board

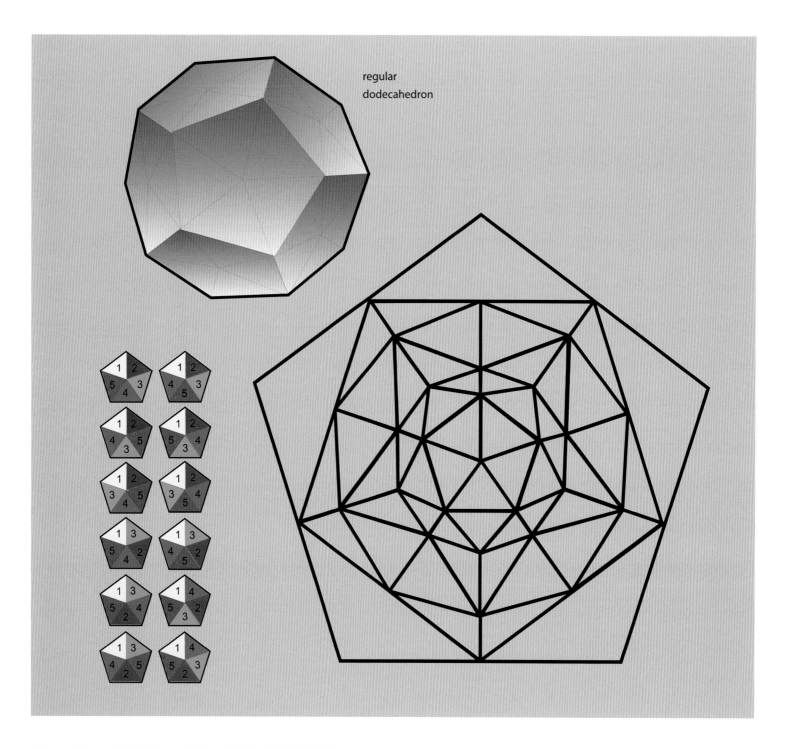

regular
dodecahedron

QUINTOMINO DODECAHEDRON

A regular dodecahedron is a three-dimensional solid whose faces consist of 12 regular pentagons. Conway wondered if it was possible to color the edges of a dodecahedron so that each of the 12 quintominoes would appear on one of its faces.

Can you find a way to place the 12 quintominoes on the sides of a dodecahedron? You can try to solve the puzzle by constructing a three-dimensional solid or through a Schlegel diagram of dodecahedron in the plane, which is topologically identical to the three-dimensional solid and more conveni-

ent for solving the problem. In the distorted diagram, note that the back face is stretched to become the outer edge of the diagram.

259 CHALLENGE ● ● ● ● ●
REQUIRES
COMPLETED

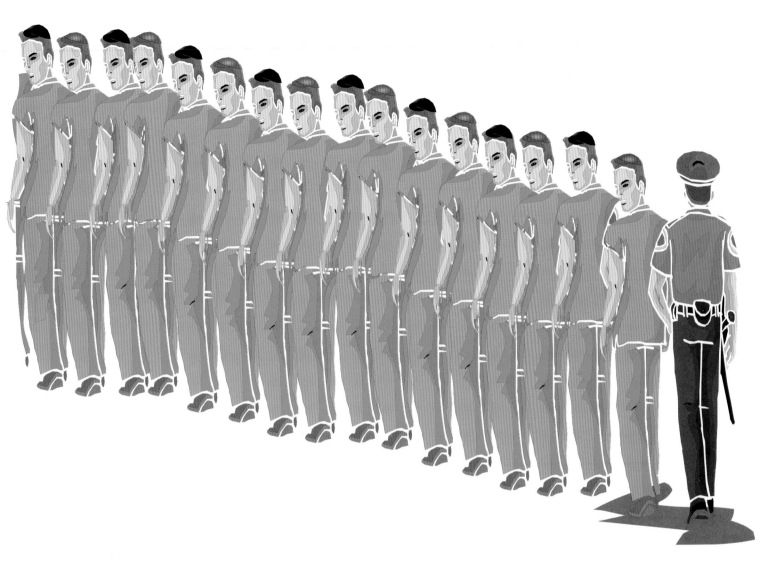

HATS AND PRISONERS

The latest variation of logical problems involving the classic category of hats puzzles, gives it an original twist: during the World War, there were 100 captured soldiers in a prisoner-of-war camp. The guards wanted to take a vacation, and one of their ideas was to get rid of the prisoners by shooting all of them. The camp commander, of a fairer mind, agreed to the idea, but decided that he was going to tell the prisoners that they were going to be shot, unless they could answer one question.

So all the prisoners were gathered and he said: "You are all dirty dogs, so I shall have all of you shot. But being a fair hunter, I am going to give you one last chance. You will be taken to the dining hall where you will find your last drink on the tables. You will be able to talk to each other while I arrange for a large crate containing the same amount of red and black hats to be placed in front of the hall. You will leave one by one. A hat, randomly picked from the crate will be placed on your head. You will not be able to see the color of your hat, but you will be able to see those of the others. You will form a single line and you will be shot on the spot if you utter a word or communicate with each other in any way. After that, I shall pass from one of you to the other asking each of you for the color of the hat you are wearing. If you answer correctly, you will be set free. If not, you will be shot."

The soldiers were ushered into a big hall, where they could discuss the situation and come up with some kind of a strategy to cope with the situation. Some time later, each man was given a hat. The camp commander, expecting that he would be able to shoot at least 50 percent of the prisoners, started asking them for the color of their hats.

"How many prisoners do you think he had to set free?"

260 CHALLENGE ●●●●○
REQUIRES 🧠✏️
COMPLETED

THE SHORTEST PATH PROBLEM

We show a simple weighted graph on seven points (vertices, nodes). Our problem, the shortest path problem, is to find a path between two vertices A and G, such that the sum of the weights of the interconnecting edges is minimized.

In a simple graph like this, you may conclude by trial and error that the solution is 16. But in more complex graphs you can never be sure without mathematical proof, like Dijkstra's algorithm demonstrated below.

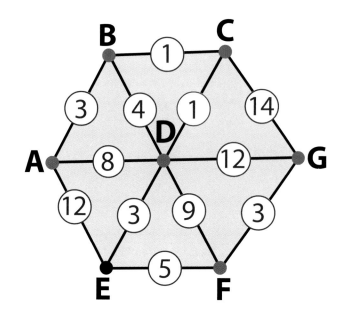

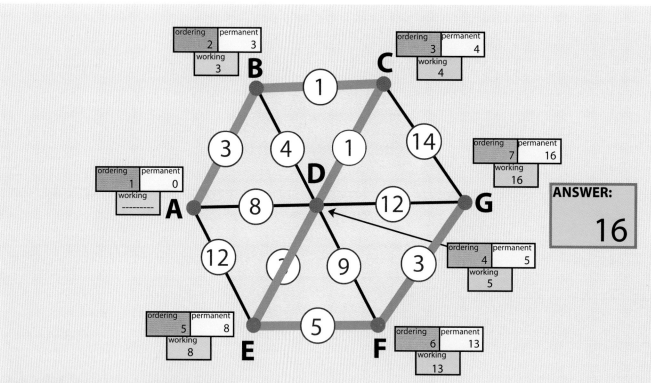

ANSWER: 16

DIJKSTRA'S ALGORITHM

Can you find the shortest path between points A to G, using Dijkstra's Algorithm, as as explained below?

Dijkstra's Algorithm requires you to assign (node) labels to each point: an ordering, permanent, and working label. These labels are assigned in the following way:

1. Give the start point the permanent label of 0, and the ordering label 1.

2. Any vertex directly connected to that last vertex given, a permanent label is assigned a working label equal to the weight of the connecting edge added to the permanent label you are coming from. If it already has a working label, replace it only if the new working label is lower.

3. Select the minimum current working value in the network and make it the permanent label for that node.

4. If the destination node has a permanent label, go to step 5, otherwise repeat step 2.

5. Connect the destination to the start, working backwards. Select any edge for which the difference between the permanent labels at each end is equal to the weight of the edge.

PACKING CONSECUTIVE SQUARES — 1960

A fascinating gem of recreational mathematics is the puzzle involving consecutive integral squares starting from side 1 and up to a particular limit. Can a large square be found that can be completely covered without overlap by such a sequence of smaller squares?

Let's experiment: Squares of sides 1 and 2 cannot form a square: the best we can do is to fit them inside a square of side 3 leaving out empty space. Similarly squares of sides 1, 2, 3 cannot fill a square without leaving an empty space, nor can 1, 2, 3, 4.

The first requirement to solve this problem is to add up the areas of the consecutive squares until the result is a square number. But $1^2 + 2^2 = 5$
$$1^2 + 2^2 + 3^2 = 14$$
$$1^2 + 2^2 + 3^2 + 4^2 = 30$$
None of these are perfect squares.

If we keep continuing the series and go far enough, we eventually find that $1^2 + 2^2 + 3^2 + 4^2 + \ldots 24^2 = 4900 = 70^2$ is a perfect square. In fact astonishingly, this is not only the first, but the only way to add consecutive squares and obtain a square for the total. (The demonstration is a difficult exercise in number theory, and was itself an unsolved problem for a considerable time.)

Given that the areas of the first 24 consecutive squares equal the area of a 70-by-70 square, this raises the following beautiful geometrical puzzle: is it possible to pack 24 consecutive squares starting from square of unit side into the 70-by-70 square? Equality of areas is a necessary condition — but might not be sufficient. And, in fact, a complete packing has not yet been found and it has not yet been proven to be impossible. The problem might therefore be rephrased: How many of the first 24 squares can be packed into the 70-by-70 square?

The best answer known to date is "all but one," and in every known example so far, it is the 7×7 square that is left out, as our example shows. Can you do better?

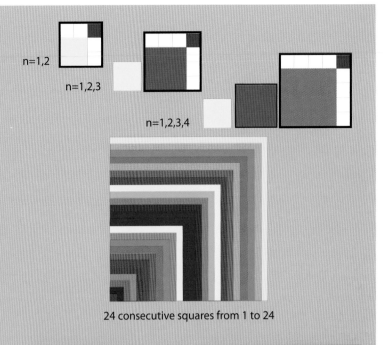

n=1,2

n=1,2,3

n=1,2,3,4

24 consecutive squares from 1 to 24

7

The first 1 to 24 consecutive squares packed in the 70 x 70 gameboard, with all but square 7 left out. Can you do better?

261

CHALLENGE

REQUIRES

COMPLETED

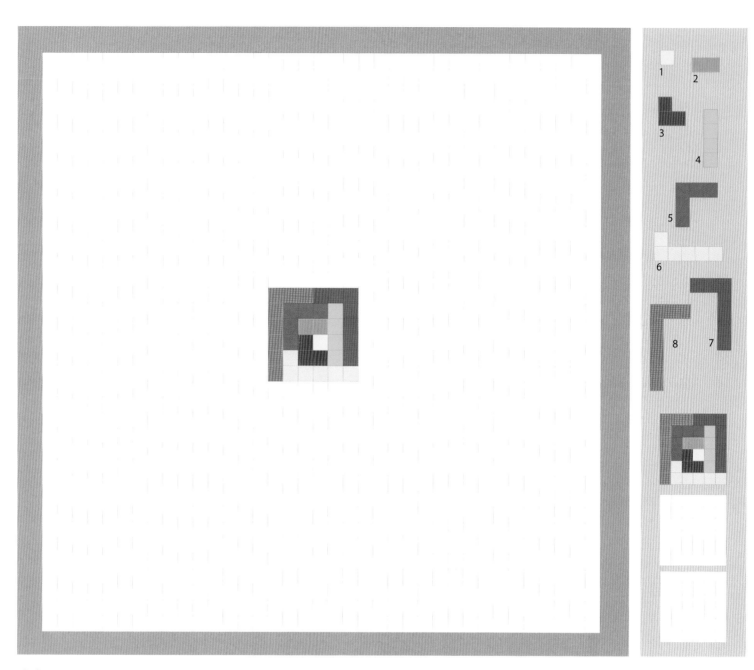

CONSECUTIVE POLYOMINO SPIRAL SQUARES — 1960

Can triangular numbers form squares? Obviously the first triangular number "1" is a square, and the second, the eighth triangular number "36" is also a square number as shown above. Which is the next square? We can see that by choosing a special selection of consecutive polyominoes starting with a monomino in the center, and adding to it a domino and further selected polyominoes (one of each group), a counterclockwise spiral configuration can be formed as shown.

The first eight consecutive polyominoes forms a polyomino spiral tessellating a six-by-six solid square, as shown.

Related to this construction principle, further interesting questions and problems can be stated:

1. Can you rearrange the pieces of the six-by-six square into other distinct square patterns?

2. Continuing the formation of the spiral selecting further consecutive polyominoes, at what stage will the next rectangle be formed, and what will be its proportion?

3. At what stage is the next square, and wwhat size is it?

262 CHALLENGE ●●●● ○
REQUIRES 🧠✏️
COMPLETED

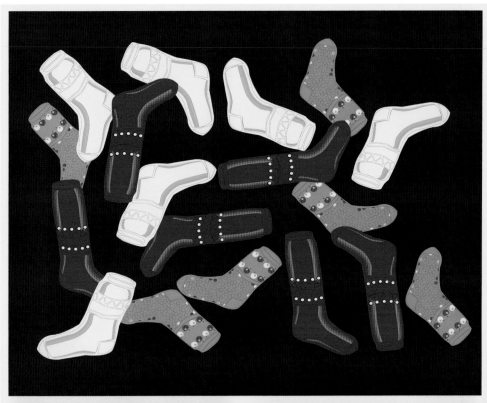

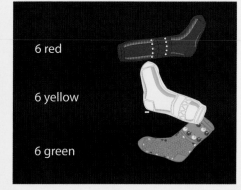

6 red

6 yellow

6 green

SOCKS IN THE DARK

In the drawer I have six red, six yellow and six green socks. In complete darkness, how many socks must I draw to get a pair of any color? And how many must I draw to have a pair of each color?

263	CHALLENGE ● ● ● ○ ○ ○
	REQUIRES 🧠✏️
	COMPLETED

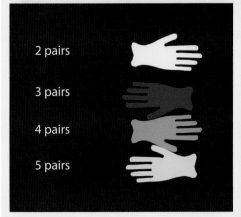

2 pairs

3 pairs

4 pairs

5 pairs

GLOVES IN THE DARK

In the drawer I have two pairs of yellow gloves, three pairs of red gloves, four pairs of green gloves and five pairs of blue gloves. In complete darkness, how many gloves do I have to draw to ensure having a complete pair of gloves in one of the colors and with proper handedness?

264	CHALLENGE ● ● ● ○ ○
	REQUIRES 🧠✏️
	COMPLETED

265	CHALLENGE ● ● ● ● ● ○
	REQUIRES 🧠✏️
	COMPLETED

LOST SOCKS VERSUS MURPHY'S LAW

Imagine that after washing five pairs of socks you discover that two socks are missing. Which scenario is more likely:

1. The two missing socks make a complete pair and you are left with four complete pairs.

2. You are left with three pairs of socks and two orphan socks.

Captain Edward A. Murphy stated: "Anything that can go wrong will, and at the worst possible time." Does Murphy's law rule the sock drawer?

8

PARADOXES, CELLULAR AUTOMATA, HOLLOW CUBE AND NIGHT CROSSING PUZZLE

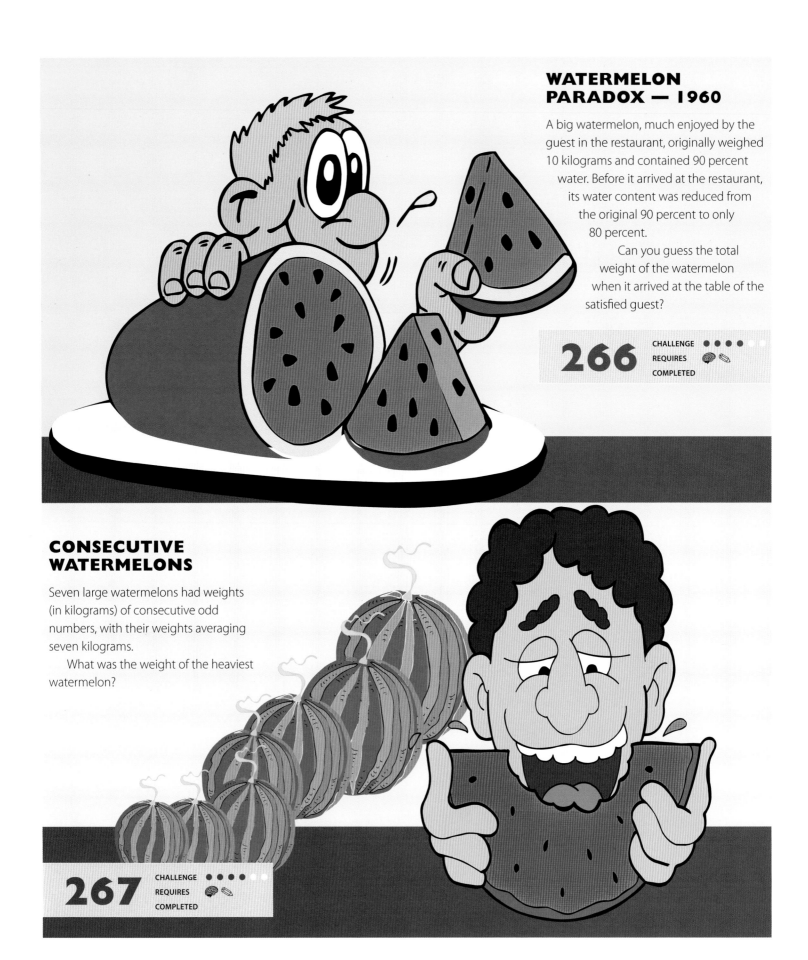

WATERMELON PARADOX — 1960

A big watermelon, much enjoyed by the guest in the restaurant, originally weighed 10 kilograms and contained 90 percent water. Before it arrived at the restaurant, its water content was reduced from the original 90 percent to only 80 percent.

Can you guess the total weight of the watermelon when it arrived at the table of the satisfied guest?

266 CHALLENGE ●●●● ○ ○
REQUIRES 🧠 ✏️
COMPLETED

CONSECUTIVE WATERMELONS

Seven large watermelons had weights (in kilograms) of consecutive odd numbers, with their weights averaging seven kilograms.

What was the weight of the heaviest watermelon?

267 CHALLENGE ●●●● ○ ○
REQUIRES 🧠 ✏️
COMPLETED

FREDKIN'S CELLULAR AUTOMATA — 1960

Edward Fredkin (born 1934) is a professor at Carnegie Mellon University, a visiting professor at Massachusetts Institute of Technology and an innovator in the field of digital physics. His primary contributions include his work on cellular automata, with his earliest and simplest self-replication system in 1960, a binary system in which each cell has two possible states: living or dead. Fredkin believed that a final grand theory of everything would be computational — that the universe is a computer.

The initial configuration of five living cells (red squares) with its adjacent neighbors, is transformed through successive generations according to the following simple rules:

1. If the number of neighbors is even, it will be dead (white) in the next generation;
2. If the number of neighbors is odd, it will be alive (red) in the next generation.

Take a look at the transformations below to the fifth generation and see the surprising outcome: the initial configuration of the five living cells in the first generation, is transformed through five generations into four identical copies of itself.

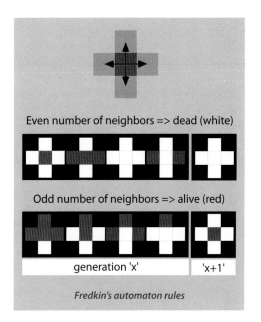

Even number of neighbors => dead (white)

Odd number of neighbors => alive (red)

| generation 'x' | 'x+1' |

Fredkin's automaton rules

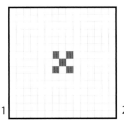

1

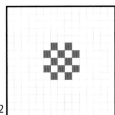

2

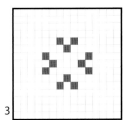

3

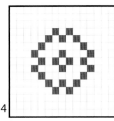

4

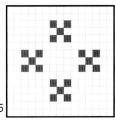

5

CONWAY'S AUTOMATON RULES

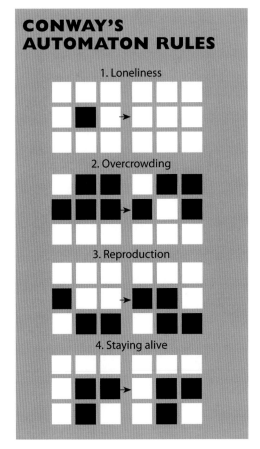

1. Loneliness

2. Overcrowding

3. Reproduction

4. Staying alive

CONWAY'S GAME OF LIFE

The Game of Life is a cellular automaton devised by the British mathematician John Horton Conway in 1970. The Game of Life is not a competition; you "play" alone, and you can't win or lose. The idea is simply to create an original configuration and see how it develops.

The Game of Life universe is an infinite two-dimensional orthogonal grid of square cells, each of which is either alive or dead. Each cell interacts with the eight cells that are horizontally, vertically, or diagonally adjacent to it. At each step in time, transitions occur according to the following rules:

1. Loneliness: a cell with less than two adjoining cells dies.
2. Overcrowding: a cell with more than three adjoining live cells dies.
3. Reproduction: an empty cell with exactly three adjoining cells comes to life.
4. Staying alive: a cell with two or three adjoining cells stays the same.

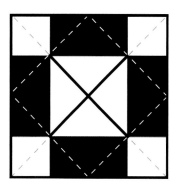

Front

Back

FLEXAGONS

Many interesting puzzles and topological discoveries can be made by just folding a square piece of paper. Such activities may be an excellent introduction to plane geometry for children and adults alike. The ancient Origami and flexagons are good examples.

A flexagon is a paper structure that can be folded along certain lines to form a series of loops or faces. Adam Walsh defined them as flat pieces of paper that have more faces to them than the obvious two.

In the early '50s, when Martin Gardner introduced flexagons to wide audiences, I was fascinated by them and invented two original folding puzzles. One is Flexi-Twist, which was the result of the self-imposed challenge to create novel and original paper folding structures, without the need for prefolding and taping, as the classic flexagons require, but still preserving an impressive number of faces and challenging folds.

The other puzzle was "Ivan's Hinge," the first patented puzzle in the new category of piano-hinge puzzles, as described by Greg Frederickson, the world authority on folding puzzles and structures, in his exciting book *Piano-Hinged Dissections* (see also Ivan's Hinge).

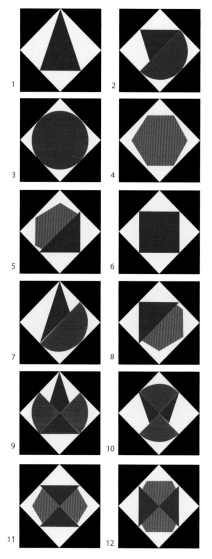

FLEXI-TWIST FOLDING PUZZLE

The Flexi-Twist is an original patented folding puzzle. Copy the square with the patterns printed on both sides. Crease and fold both ways along the interrupted lines, and cut through along the two diagonals of the middle yellow square. In succession, fold the square along the interrupted lines to create a half size square with the patterns shown on the right.

268 CHALLENGE
REQUIRES
COMPLETED

NIGHT CROSSING

The bridge will collapse in exactly 17 minutes. The four hikers have to cross the bridge in the dark. They only have one flashlight that must be used for each crossing. A maximum of two people can cross the bridge at one time carrying the flashlight, which must be returned after each crossing. Each hiker walks at a different speed and it takes the first one minute to cross the bridge, the second two minutes, the third five minutes, and the fourth 10 minutes. Consequently each pair is crossing the bridge at the rate of the slower hiker's pace (so for example, hiker one crossing with hiker three, will cross the bridge in five minutes). No tricks are allowed, the flashlight may not be thrown across, nobody can be carried, etc. There are two possible solutions, can you find both?

269

CHALLENGE ● ● ● ● ○ ○

REQUIRES 🧠🥜

COMPLETED

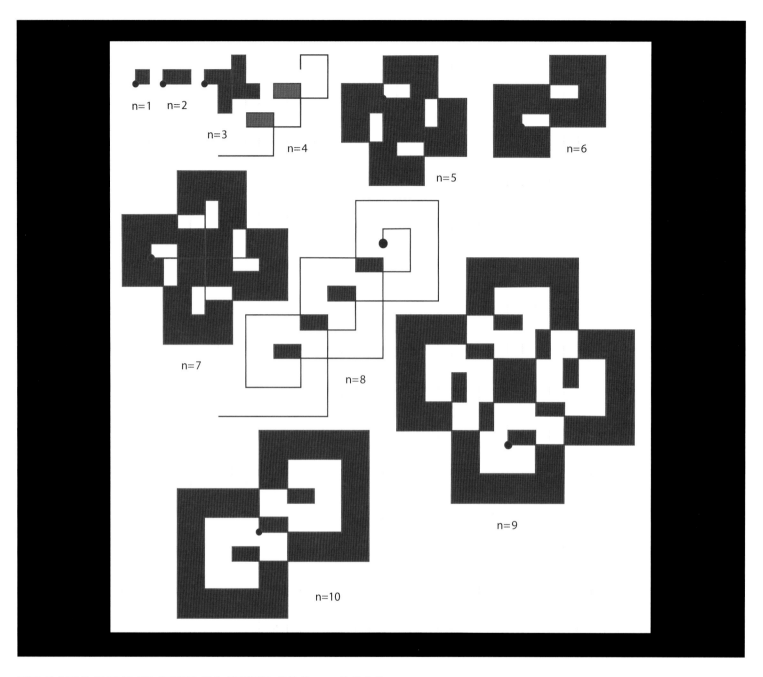

n=1 n=2 n=3 n=4 n=5 n=6 n=7 n=8 n=9 n=10

FRANK ODDS' SPIROLATERALS — 1962

Frank Odds, a microbiologist at the Institute of Medical Sciences at the University of Aberdeen, invented the concept of a spirolateral in 1962 while he was, as he describes it "doodling on graph paper during a not-very-interesting high school chemistry class." He proposed a simple set of rules for generating beautiful patterns full of surprises that he named spirolaterals.

From a very simple generative procedure, spirolaterals can be created based on the idea of defining geometric figures such as paths generated by a moving point. He thought about spirolaterals as paths traced by a worm according to the following rules: The worm starts moving a distance of one unit, turns, moves two units, turns, moves three units, and so on in consecutive order, always turning 90 degrees to the right, until reaching a specified limit n, which is the order of the spirolateral, and then repeating the pattern.

270 CHALLENGE ●●● REQUIRES 🧠✏️ COMPLETED

You can easily play spirolaterals as a paper-and-pencil game on squared grid paper.

The first ten spirolaterals are shown above. Can you continue the sequence for higher order spirolaterals: n=11 and n=13?

PRIME SPIRALS — 1963

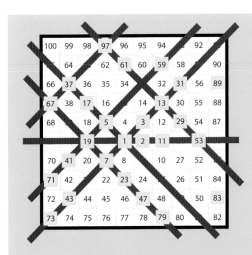

In 1963, Stanislaw Ulam, the famous Polish mathematician, was doodling numbers on a piece of paper during a boring lecture. He scribbled consecutive numbers in a square matrix, starting with 1 in the middle and spiraling outward as shown in our grid.

To his utter surprise, prime numbers tended to fall on diagonal and straight lines.

In his matrix, the first 26 prime numbers all fall on straight lines containing at least three primes, while some diagonal lines contain even more primes. The same mysterious line patterns appear in larger matrices as well, charting millions of primes in a spiral pattern, all forming similar configurations.

A law of nature or just a coincidence? No one knows yet.

Ulam also investigated spirals that started the matrix with whole numbers other than 1, like the one shown to the left, which starts with 17 in the middle. He was astonished to observe the strange patterns in the distribution of prime numbers in this spiral as well. Try your own spiral!

213	212	211	210	209	208	207	206	205	204	203	202	201	200	199
214	161	160	159	158	157	156	155	154	153	152	151	150	149	198
215	162	117	116	115	114	113	112	111	110	109	108	107	148	197
216	163	118	81	80	79	78	77	76	75	74	73	106	147	196
217	164	119	82	53	52	51	50	49	48	47	72	105	146	195
218	165	120	83	54	33	32	31	30	29	46	71	104	145	194
219	166	121	84	55	34	21	20	19	28	45	70	103	144	193
220	167	122	85	56	35	22	17	18	27	44	69	102	143	192
221	168	123	86	57	36	23	24	25	26	43	68	101	142	191
222	169	124	87	58	37	38	39	40	41	42	67	100	141	190
223	170	125	88	59	60	61	62	63	64	65	66	99	140	189
224	171	126	89	90	91	92	93	94	95	96	97	98	139	188
225	172	127	128	129	130	131	132	133	134	135	136	137	138	187
226	173	174	175	176	177	178	179	180	181	182	183	184	185	186
227	228	229	230	231	232	233	234	235	236	237	238	239	240	241

MAP AND POSTAGE STAMPS FOLDING — 1963

Folding postage stamps is a special case of the general map folding problem. You may have experienced this difficulty after you unfolded a big map and tried to refold it back to its original folded state. The Polish mathematician Stanislaw Ulam was the first to pose the question: In how many different ways can a map be folded?

The problem has frustrated researchers in the field of modern combinatorial theory ever since. Indeed, the general problem of map folding is still unsolved.

There is an old saying that is appropriate here: "The easiest way to refold a road map is differently!"

FOLDING A STRIP OF THREE STAMPS

Can you work out in how many different ways can you fold a strip of three stamps? You may only fold on the perforations and the end result must be of three stamps on top of each other. It does not matter whether the stamps are face up or face down. As we saw earlier, there are six possible permutations of three colors as shown = colors (3x2x1). How many can you achieve by folding? It should be noted that the problem of folding postage stamps has several possible variants:

1. Unlabeled (U) – For unlabeled stamps only the positions of the hinges (perforations) are considered, without regard to orientations of the stamps.
2. Labeled (N) – If the stamps are labeled and their orientations are taken into regard.
3. Symmetric (S) – Symmetric foldings

How many different folds can you achieve with a strip of three stamps?

FOLDING A STRIP OF FOUR STAMPS

Can you tell in how many different ways you can fold a strip of four stamps? You may only fold on the perforations and the end result must be of four stamps on top of each other. It does not matter whether the stamps are face up or face down.

FOLDING A SQUARE OF FOUR STAMPS

In how many different ways can you fold a square of four stamps? You may only fold on the perforations and the end result must be a stack of four stamps on top of each other. It does not matter whether the stamps are face up or face down. As we saw previously, there are 24 permutations of four colors (4x3x2x1). How many can you achieve by folding?

1   2   3   4

FOLDING A RECTANGLE OF SIX STAMPS

Six stamps are joined along a two-by-three rectangle, which can be folded in many ways along their perforated sides to create a stack of stamps. Four stacks are shown in their color sequences. Can you work out which stack is impossible to fold? It does not matter which side of a stamp is up in the final folded stack.

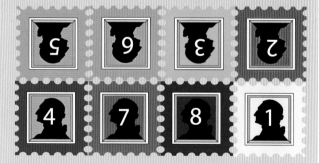

FOLDING A RECTANGLE OF EIGHT STAMPS

Can you fold the block of eight stamps along their perforations so that the stamps are stacked in order from 1 to 8?

ROLLING DICE ON A CHESSBOARD — 1963

In 1963, Martin Gardner introduced problems with dice rolling on chessboards of different sizes.

The die dimensions are the same as the unit cells of the boards and the die moves by rolling to an adjacent square, bringing a new face to the top of the die with each move.

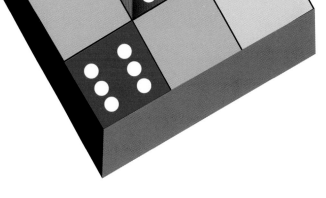

ROLLING DICE (1)

From the position shown, can you roll the die one face at a time six times so that it ends up on the bottom left square with its 6 facing up?

ROLLING DICE (2)

Can you roll the die in succession from the top square, one face at a time, so to end on the bottom left square with numbers from 1 to 6 facing up?

272

CHALLENGE ● ● ● ● ○ ○

REQUIRES 🧠 ✏️ ✂️

COMPLETED

BUCKMINSTER FULLER (1895–1983)

Buckminster Fuller, known to his friends as "Bucky," was one of the major innovators of the 20th century, with a staggering number of inventions to his credit. International recognition came with the enormous success of his Geodesic Domes, Dymaxion Map, Dymaxion House, Fullerenes or "Bucky Balls," and many others.

He believed synergy to be the fundamental principle of interactive systems and developed a major subject called *Synergetics: Explorations in the Geometry of Thinking*. He referred to himself as "Guinea Pig B" to show that his life was itself an experiment. He demonstrated his designs and ideas in models and prototypes built by himself, symbolizing his design philosophy of "more for less," to explain the basic principles of synergy.

I met Bucky twice. The first time in the 1960s attending his lecture in Tel Aviv. During the questions period after his lecture, I presented him with the first prototype of my new Mirrorkal invention, a novel modular mirror kaleidoscope puzzle inspired by his work which, when solved, created the portrait of Bucky. I was glad that he was very pleased to accept it, exchanging a few words with me to tell me that he loves good puzzles.

Around 20 years later in New York I was waiting for the elevator in the Edison Hotel. It arrived, its doors opened and I bumped head to head with the gentleman stepping out of the elevator. It was quite a blow to both of us, and when we recovered I recognized Bucky. The moment he saw me, he shouted: "Ouch, the mirror guy," recognizing me.

Bucky invited me for coffee and I had one of the most enjoyable experiences of my life. I will never forget listening spellbound to Bucky for more than two hours while drinking coffee.

> **"To me no experience of childhood so reinforced self-confidence in one's own exploratory faculties as did geometry. Its inspiring effectiveness in winnowing out and evaluating a plurality of previously unknowns from a few given knowns, and its elegance of proof lead to the further discovery and comprehension of a grand strategy for all problem solving."**
>
> — *Buckminster "Bucky" Fuller*

BUCKY'S SYNERGY — 1964

Synergy is two or more things functioning together to produce a result not independently obtainable. Buckminster Fuller was the main reason that Synergy became a common term. A large portion of his work involved exploring and creating synergy.

Bucky was adept at creating models to demonstrate his ideas. His skeleton tetrahedron beautifully visualizes his ideas of synergy. Two bent wire triangles may be combined in such a way as to create a perfect skeleton tetrahedron, a three-dimensional figure embraced by four triangles. Therefore, one plus one seemingly equals four.

Mel Stover, a friend of Bucky and a great magician, created a close-up magic act with Bucky's wire triangles. Mel showed his audience a perfect wire skeleton of the tetrahedron counting its four triangular faces. Then he lifted the tetrahedron and gave the two wire triangles (unbent) to the nearest person to recreate the tetrahedron, which of course, nobody could succeed in doing. Mel's cunning, well reversed trick was that while lifting the two bent wire triangles he unbent and straightened them into two flat triangles again.

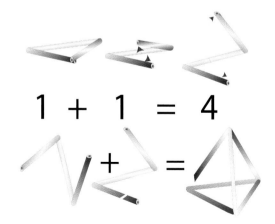

THE SYNERGETICS OF FULLER — 1964

According to Wim Zeiler in his *Synergetics between sun & energy*, "synergetics is the empirical study of systems in transformation, with an emphasis on total system behavior unpredicted by the behavior of any isolated components, including humanity's role as both participant and observer."

The fact that systems can be identified at every level from the quantum to the cosmic, when combined with humanity being composed of them as well as articulating their behavior, makes synergetics a discipline of immense scope, embracing a wide variety of scientific and philosophical studies including tetrahedral and close-packed-sphere geometries.

Buckminster Fuller invented the term synergetics and wrote a two volume work by that name in an attempt to define its scope. However, synergetics remains an unconventional and even radical subject with little mainstream support, and it is paid minimal attention by most traditional academic departments.

Fuller's work inspired a number of researchers to explore branches of synergetics. Hermann Haken studied self-organizing structures of open systems; Amy Edmondson tackled tetrahedral and icosahedral geometry and Stafford Beer explored geodesics in the context of social dynamics. A number of other researchers are now working on different aspects of synergetics, though many purposefully distance themselves from Fuller's comprehensive and all-embracing definition.

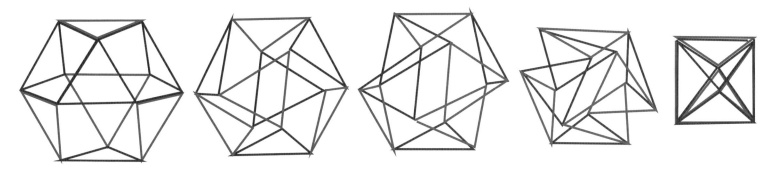

BUCKMINSTER FULLER'S JITTERBUG SYSTEM

One of the most beautiful polyhedral transformations is Buckminster Fuller's Jitterbug system. We have met the cuboctahedron among the family of Semi-regular Archimedean polyhedra, which Fuller called the "Vector Equilibrium." The jitterbug is Fuller's name for the continuous transformation that has four remarkable positions: from cuboctahedron, to octahedron, icosahedron, and dodecahedron (and vice versa).

Fuller achieved the transformation using sticks and fully flexible rubber vertices. The jitterbug motion allowed Fuller to shrink the cuboctahedron into an octahedron, and the opposite to expand the octahedron back into the cuboctahedron.

It is difficult to conceptualize the beautiful mechanism of the jitterbug motion without seeing it in slow motion, which you can experience when creating a model. The diagrams above illustrate a giant jitterbug at the Eureka exhibition in Zürich, Switzerland, in 1991. Jitterbug has eight triangular faces. As the faces rotate, they also shrink radially inward or outward from the center along four rotation axes.

There are many jitterbug-based toys and puzzles, made out of paper, metals or plastic, and lately out of ingenious magnetic rods.

The giant jitterbug at the Eureka exhibition in Zürich, Switzerland

ROLLING PORTRAIT CUBE — 1964

Rolling cube puzzles were first mentioned by Dudeney in his *Amusements in Mathematics,* published in 1917 and later widely popularized by Martin Gardner and John Harris. In our puzzle, a cube with faces of six famous people is rolled on a chessboard with two puzzle objectives:

Puzzle 1:
Starting with Einstein facing up on the bottom left square of the game board, the object is to visit each square of the board once, rolling the cube from square to square ending at the bottom right square of the board with Einstein on top once again (not necessarily facing the same way). This may sound simple, but Einstein is not allowed to be on top during the whole trip, except at the start and the end as stated.

Puzzle 2:
Can you complete the same trip starting with Einstein facing up on the fourth square of the second row from the top and make a closed tour, coming back to the same square without allowing Stalin to appear on top at any stage of the tour?

273

CHALLENGE ●●●●●
REQUIRES 🧠 ✏️ ✂️ 🔨
COMPLETED

| Einstein | Beethoven | Stalin | Newton | Queen Elizabeth | Shakespeare |

The rolling portrait and its net

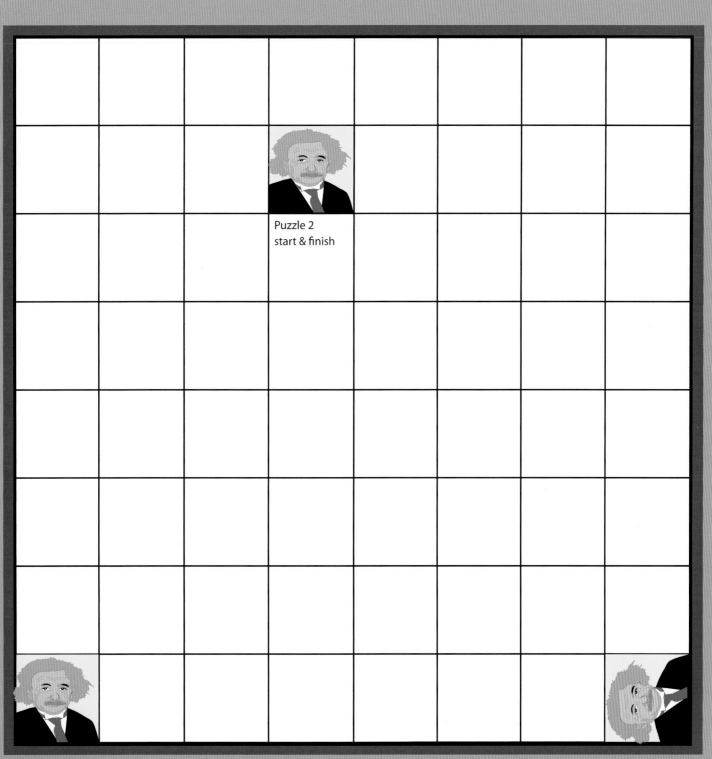

Puzzle 2
start & finish

Puzzle 1
start

Puzzle 1
finish

ROLLING PORTRAIT CUBE

Game board for Puzzle 1 and Puzzle 2

FIXED-POINT THEOREM — 1964

Two pictures have been placed on top of the other. They are identical, but one is larger than the other. The fixed-point theorem says that there is a point on the smaller picture that is directly above the same point on the larger picture. There can be only one such point. How can you find it?

There are hundreds of fixed-point theorems. This one is named the Brouwer theorem, after Luitzen Brouwer (1881–1966), a Dutch mathematician famous for his mathematical philosophy of intuitionism, which views mathematics as the formation of mental constructions that are governed by self-evident laws.

The right picture demonstrates how this point can be found. Overlapping a third picture that bears the same relationship to the smaller picture that the smaller picture bears to the larger, and repeating this sequence of adding smaller pictures, will ultimately result in the point we are trying to find, shown in red. Go ahead and check!

It is interesting to note: the theorem works even when the smaller picture is crumpled!

THE MONK AND THE MOUNTAIN — 1966

The monk climbs the mountain along a narrow track. He starts at 7 in the morning and reaches the top at 7 in the evening. His route was at varying rates of speed and with many stops. The next morning he starts going down at the same time and reaches the bottom again at the same time in the evening. Is there a spot along his path that the climber will occupy on both trips at precisely the same place at the same time on both days?

Finish:
Sunset 7 p.m.

274   CHALLENGE ● ● ● ● ○
REQUIRES 🧠 ✏️
COMPLETED

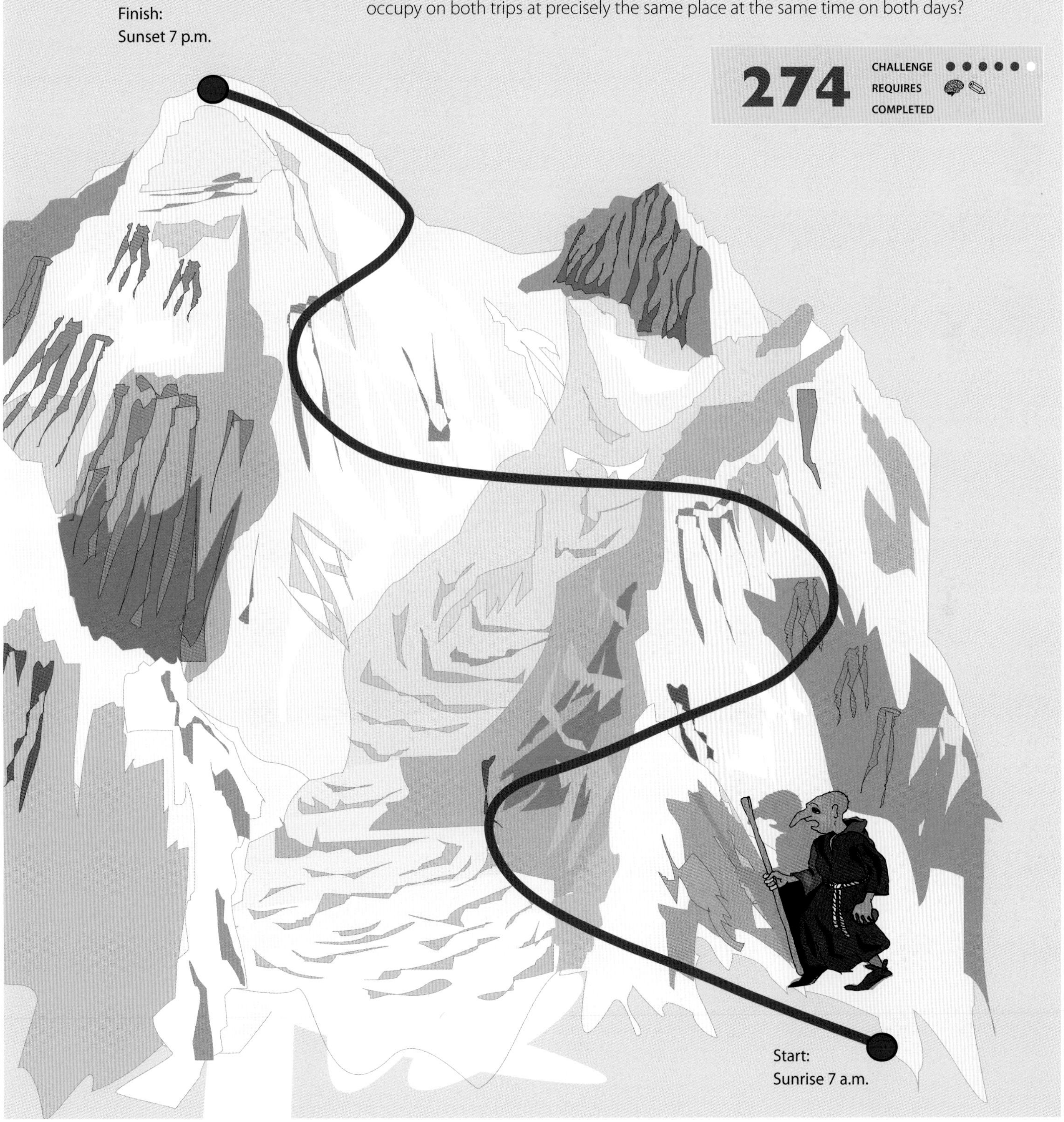

Start:
Sunrise 7 a.m.

SQUAREWHERE — 1974

Another challenging geometrical vanish puzzle of Russian origin appeared in my *Brain Drain puzzle* series, launched by Mattel in 1967.

The 18 colored shapes fit completely into the square above as shown. However, 17 of the 18 shapes can also cover a square, as the small square below can be left out. It sounds impossible, but it isn't. It can be done. Can you solve and explain the mystery?

275 CHALLENGE ●●●●○ REQUIRES COMPLETED

RATTLEBACK TOY — 1969

A mysterious object that spins in one direction and then reverses, the rattleback is a prehistoric object discovered by archeologists studying prehistoric Stone Age axes and hatchets. Rattlebacks seem mysterious because people expect something moving in one direction to continue moving in the same direction until some force intervenes to stop it; a law that physics defines as angular-momentum conservation.

A rattleback, or celt, has a very specific shape that causes it to spin in one direction only. When spun in the non-preferred direction, the rattleback's spin will slow down and it will begin to rock from end to end. This rocking motion is then transferred into spin

going in the preferred direction. In short, spin it the wrong way, and it will stop and spin the other way!

Originally made from wood with decorative carvings and patterns, rattlebacks are now commonly available as plastic toys.

To understand how a rattleback works, we need to take a look at its shape. The rattleback is flat on top and has an asymmetrical ellypsoidal bottom. The specific alignment of the long axis of the ellipsoid in relation to the long axis of the flat top causes the directional preference. In other words, they're not parallel so the rattleback is predisposed to spinning in one direction.

Today rattlebacks are produced in many different variations as popular science toys. A

giant version, big enough for small children to ride on, was created as a science exhibit at the Lasky Museum of Science and Technology in the late 1960s.

There are many explanations for the strange behavior of rattlebacks, though we still may expect a better physical description of how the rattleback behaves. After a century spent examining the toy, scientists are not likely to give up now. "The thing about scientists is that they really like toys," says Brian Pippard of the University of Cambridge. "They have an interest in anything that looks odd. And they're not happy until they can describe how it happens."

UNISTABLE POLYHEDRA — 1969

A unistable polyhedron (or monostatic polytope) is an n-dimensional solid that can stand on only one face and is of uniform density. In 1969, John Conway, Richard Guy and M. Goldberg constructed a unistable polyhedron, a prism of 17 sides and 19 faces shown below with its symmetrical cross-section, which holds the record. No such solid with fewer faces has yet been found.

Like a Weeble toy, when tilted, Guy's prism will right itself on one side. Some turtles, like the Indian star tortoise, have unistable shapes. No convex polygon in the plane is monostatic. This was shown by V. Arnold via reduction to the four-vertex theorem.

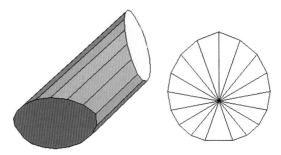

EQUILIBRIUM STATES

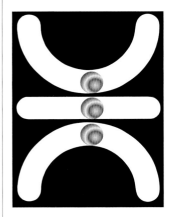

Three colored beads can freely move in the vertical channels. The behavior of the beads when disturbed demonstrate the three different types of equilibrium:
Top: stable;
Middle: neutral;
Bottom: unstable.

WEEBLE TOYS

A tilting doll or roly-poly toy is a toy that rights itself after being pushed over. It has a round base, approximately a hemisphere. Its center of mass is below the center of the hemisphere, so when it tilts, the center rises. If you push a tilting toy over, it wobbles briefly before trying to return to its upright position, of which the equilibrium is at the minimum gravitational potential energy.

INDIAN STAR TORTOISE

The shape of this tortoise enables it to return to an upright position has after being turned on its back. Mathematicians Gábor Domokos of the Budapest University of Technology and Economics and Péter Várkonyi of Princeton University designed Gömböc, a homogenous object with just one unstable and one stable balance point. They were able to construct a shape that always returns to an upright position in the same way as a bottom-weighted sphere; i.e. one with non-homogenous weight distribution. They observed its similarity to the star tortoise and tested 30 turtles by turning them upside down, finding that many of them were self-righting. For more on Gömböc, go to page 340.

NON-TRANSITIVITY PARADOXES — 1970

Most relationships are transitive, which is a binary relationship which states that if A is bigger than B, and B is bigger than C, then A must also be bigger than C. On the other hand, some relationships may not be transitive (if A is the father of B, and B is the father of C, it is never true that A is the father of C).

The well-known rock-paper-scissors children's game is non-transitive. In this game, the arbitrary relationships of rock breaking scissors, scissors cutting paper and paper covering rock create the winning rules. Ancient Chinese philosophers divided matter into five categories forming a nontransitive cycle: wood gives birth to fire, fire to earth, earth to metal, metal to water, water to wood.

In probability theory there are relations that are seemingly transitive when they are actually not. If this non-transitivity is so counterintuitive that it boggles the mind, these relationships are called non-transitive paradoxes or games.

A lot of ingenuity went into creating such paradoxes and games, which are perfect "sucker bets." One of the simplest and most astonishing of such games are non-transitive dice sets, like the set pictured on the right. Such dice were first designed in 1970 by Bradley Efron, a statistician at Stanford, and opened to a wide audience by Martin Gardner in his *Scientific American* column.

NON-TRANSITIVE DICE GAME

If you are playing a game using dice, you expect the numbers you throw to be random. The object of this game is to find out what is special about the four dice in this game.

Make a set of four dice as shown.

1. Ask a partner to choose one of the four dice, then you choose one out of the remaining three.
2. Taking turns, throw each die, and the higher number wins.

Can you always choose the die that will make you win in the long run?

276 CHALLENGE ● ● ● ● ● ○
REQUIRES
COMPLETED

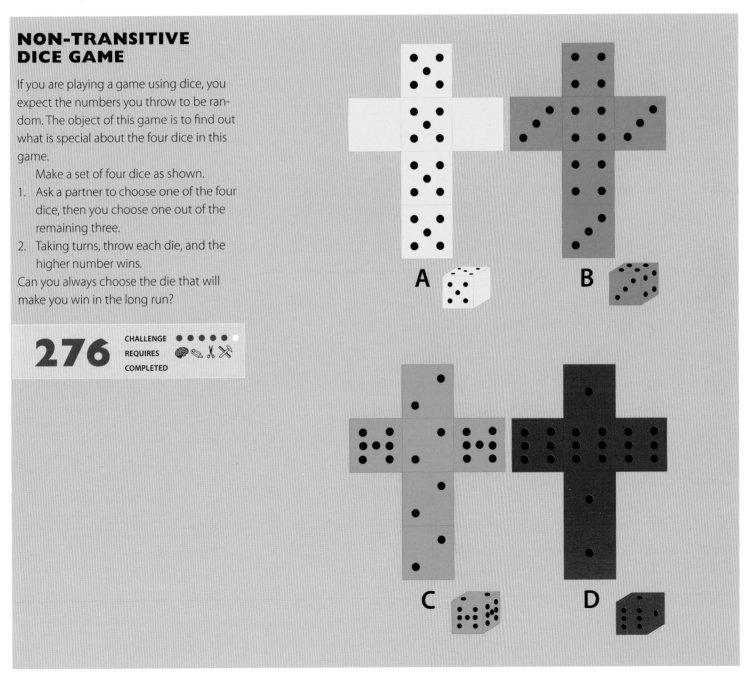

PENTAGORAS — 1970

A beautiful dissection puzzle based on golden ratio (see Chapter 2) was invented by Jean Bauer of Trigam, Switzerland. It consists of only three shapes, two isosceles golden triangles and a regular pentagon, as shown.

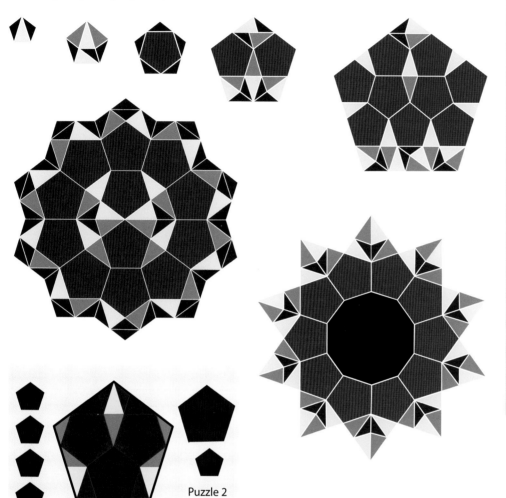

PENTAGORAS PENTAGRAM

The pentagram, a five-pointed pentagonal star, is the ultimate expression of the Divine Proportion. It was the secret symbol of Pythagoras and his followers, hiding the secret from which the golden ratio and golden triangles can be created. The pentagram is created using 23 triangles and five pentagons of the pentagoras puzzle.

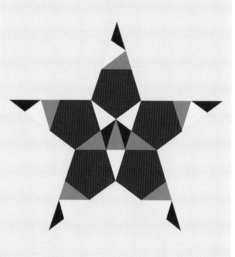

Puzzle 2

Puzzle 1

PENTAGOR

One of the most challenging puzzles in the Pentagoras series is a large dissected regular pentagon composed of three sets of different shapes: pentagons and two types of isosceles triangles, 17 pieces all together. Can you make puzzles 1 and 2, each time by using all 17 pieces?

277 CHALLENGE
REQUIRES
COMPLETED

PENTAGONS AND GOLDEN TRIANGLES

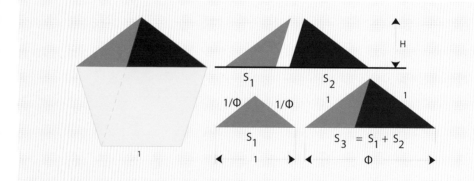

The pentagon, as with all regular polygons, has a number of internal correlations. Each side is opposite a point, and every diagonal connecting two sides internally has a golden ratio to one of the sides it connects.

HOLLOW CUBE — 1970

Imagine you are peering into a hollow cube from different angles and orientations. At the bottom of the cube there is a colored eight-by-eight square grid forming a picture. Each time you can see only a portion of the pattern you are looking at. But from the six different views, there is enough information to reconstruct the entire picture in the empty grid on the right.

278 CHALLENGE ●●●●○○
REQUIRES 🧠✏️
COMPLETED

SLOTHOUBER-GRAATSMA CUBE-PACKING PUZZLE

Can you assemble six 1x2x2 blocks and three 1x1x1 cubes into a 3x3x3 cube?

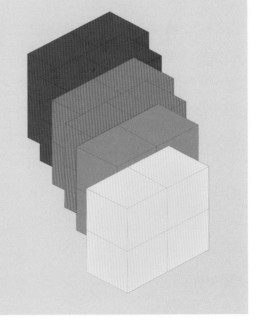

279 CHALLENGE ●●●●●○
REQUIRES 🧠✏️✂️
COMPLETED

CUBE-PACKING PUZZLES

One of the earliest cube-packing puzzles appeared in 1970 in a book by Dutch architects Jan Slothouber and William Graatsma, shown above. You can easily create the nine blocks from cardboard, but the solution may be elusive for this seemingly very easy puzzle. Conway later created many more difficult variations of the puzzle like the one shown on the right. When the secret to the solution of both puzzles is discovered, solving them becomes fairly easy.

CONWAY'S 5X5X5 CUBE-PACKING PUZZLE

Like in the 3x3x3 cube, the object is to pack 13 1x2x4 blocks, three 1x1x3 blocks, one 1x2x2 block, and one 2x2x2 cube into a 5x5x5 cube.

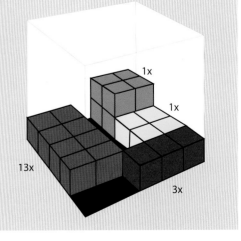

1x

1x

13x

3x

280 CHALLENGE ●●●●○○
REQUIRES 🧠✏️
COMPLETED

MASTERMIND — 1970

Mastermind, invented by telecommunications expert Mordecai Meirowitz in 1970, is a board game with an interesting history. After many rejections by leading toy companies, Meirowitz came to me for help with his early cardboard prototype of Mastermind. At the time, I was actively associated with Invicta Plastics, Leicester, UK, and succeeded to place his game with them. Together with Ronnie Sampson of Invicta, I was involved in the final design of Mastermind. Over 50 million copies later, the game is still marketed today. It was the most successful game of the 1970s. Sadly, Meirowitz died young in 1995 in Paris.

The basic idea of the game is similar to an early paper-and-pencil game called Bulls & Cows dating back a century or so. The aim is for one player, the code-breaker, to guess the secret code chosen by the other player, the code-maker.

The code is a sequence of four colored pegs chosen from six available colors. The code-breaker makes a series of pattern guesses, and after each guess the code-maker gives feedback in the form of two numbers, the number of pegs that are the right color and in the correct position, and the number of pegs that are the right color but in the wrong position. These numbers are normally represented by small colored pegs. If the code-breaker guesses the correct pattern in 10 or fewer turns, he wins, otherwise the code-maker wins.

CHVÁTAL ART GALLERY THEOREM — 1973

The Chvátal's Art Gallery theorem was the result of a request for an interesting geometric problem to work on by a young mathematician at the University of Montreal called Václav Chvátal in 1973. Victor Klee sent him the Art Gallery question: How many guards are at least necessary to guard a polygonal art gallery with n walls, i.e. the minimum number of vertices from which it is possible to see every point in the interior of the polygon?

A polygon is simple if its sides only meet at their endpoints, and never more than two at a time. In other words, the polygon should not have any self-intersections.

In the case of a convex polygon, the whole interior can be seen from any one point, but this is generally not the case. What is the minimum number of vertices that would answer the question for every admissible polygon?

Chvátal's solution was conceptually simple and involved a few special cases. A much simpler proof of the theorem was subsequently discovered by Steve Fisk of Bowdoin College. He learned of Klee's question from Chvátal's article, but found the proof unappealing. He gave the problem further thought and came up with the solution while traveling on a bus somewhere in Afghanistan.

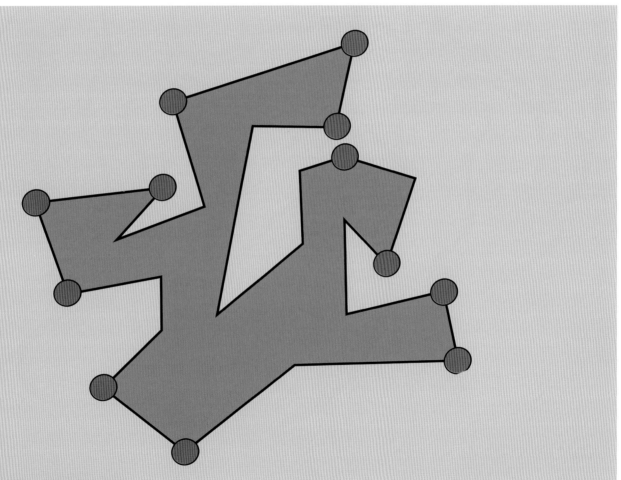

ART GALLERY PROBLEM

In this strangely shaped art gallery consisting of 24 walls, revolving security cameras are mounted in some corners. In the example shown, 12 cameras (red dots) have been installed.

However, cameras are expensive to install and maintain. What is the minimum number of cameras required so that every square centimeter of the gallery is covered by at least one camera?

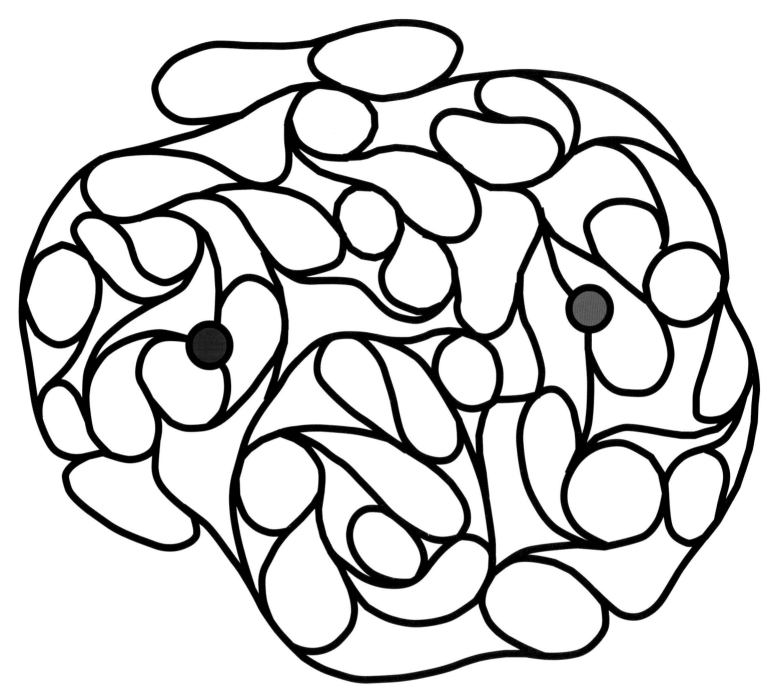

RAILWAY MAZES — 1974

In *A Lifetime of Puzzles : honoring Martin Gardner,* edited by Tom Rodgers and Erik and Martin Demaine, Roger Penrose describes railway mazes, an old, simple but ingenious paper-and-pencil game concept, the basic idea of which came from his father.

Railway mazes are connected networks of smooth curves, like the one above. Even for such simple railway mazes, the solutions may be quite elusive. The object is to find a smooth path along the route, from the starting point (red) to the end point (blue), with no reverse moves allowed.

You can design them to be railway mazes to solve. Many routes may return to the starting point, and there can even be "whirlpool" traps that, once entered, can never be left again. Can you find the solution to the railway maze above?

283 CHALLENGE ● ● ●
REQUIRES 🧠✏️✂️
COMPLETED

NON-PERIODIC TILING — PENROSE TILES — 1974

As we saw earlier, a periodic tiling is one on which you can outline a region that tiles the plane by translation, like the tessellations in Chapter 4. Aperiodic or nonperiodic tiling is a tiling obtained from an aperiodic set of tiles. Such a set would admit only nonperiodic tiling. For a long time, experts believed nonperiodic tiling did not exist. However, in 1964 Robert Berger constructed a set of more than 20,000 tiles that he later reduced to 104 tiles.

The most widely recognized examples of an aperiodic set of tiles are the assorted Penrose tiles. The aperiodicity of the Penrose prototiles suggests that a shifted copy of such a tiling will always be unable to match the original.

In 1974, Sir Roger Penrose proposed three sets of tiles that could be used to force aperiodic tilings only. His first set (P1) was composed of six tiles based on pentagons, inspired by Kepler. His third set (P3) uses only two tiles, a pair of rhombuses ("rhombs") as shown. But the most exciting was his second set (P2), using only two tiles that force nonperiodicity. John Horton Conway named the two tiles "kite" and "dart," which can create an endless variety of beautiful patterns, called Penrose Universes (see next page). Astonishingly, such patterns were later discovered in the arrangements of atoms in quasi-crystals.

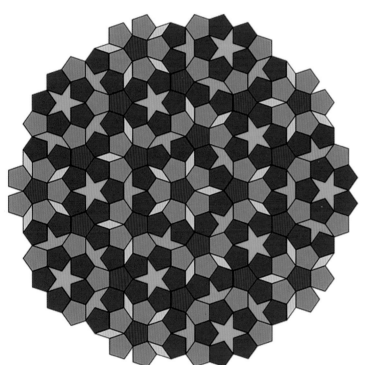

PENROSE P1 TILING

The Penrose tiling on the left was created using a set of four prototiles: pentagons, five-pointed stars or pentagrams, "boats" and a "diamond."

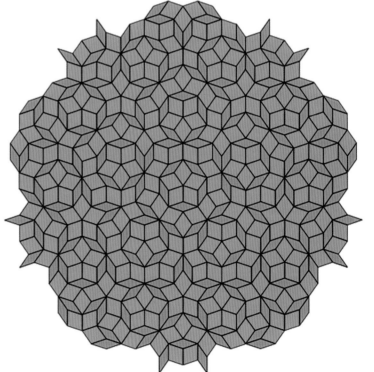

PENROSE P3 TILING

The Penrose tiling on the right was created using only two types of tiles: a pair of rhombuses ("fat" and "skinny").

PENROSE P2 TILING: THE CARTWHEEL

In 1974, Penrose discovered an aperiodic tiling that uses only two shapes, nicknamed kites and darts.

How can Penrose kites and darts be used to tile the plane to avoid periodicity and form only nonperiodic tilings?

Penrose solved the problem by marking the corners of the two tiles with H and T as shown below. To provide nonperiodic tilings it is sufficient to fit the tiles so that only corners of the same letter may touch. It is interesting to note that Penrose proof that the tiling is nonperiodic is based on the fact that the ratio of the number of pieces of the two shapes is the golden ratio Phi=1.618, an irrational number.

In some ways the pattern here, the cartwheel, is the most important Penrose tiling. The purple region at the center is outlined by a decagon consisting of a kite and dart edge. The outer portion of the pattern consists of two parts. There are ten yellow sectors and ten blue spokes. The spokes consist of "bowtie" units and they can be flipped 180 degrees and still fit their adjacent sectors.

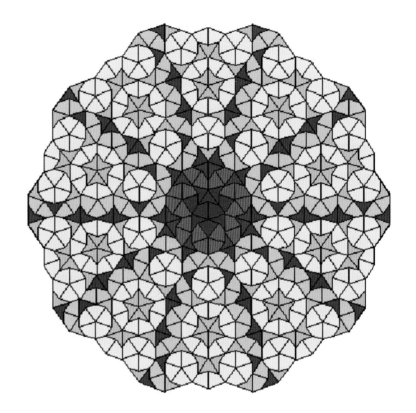

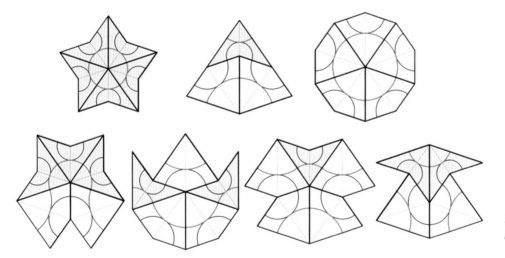

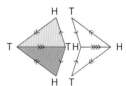

Kites and dart tiles and the matching rules, producing seven vertex figures in a P2 tiling

RUBIK'S CUBE — 1974

In 1974, a Hungarian Professor of Architecture named Erno Rubik invented a three-dimensional mechanical puzzle now known as Rubik's Cube. It was licensed by Rubik to be sold by Ideal Toy Corp. in 1980.

Each of the six faces of a classic Rubik's Cube is covered by nine stickers, each of one of six solid colors. Each face can be turned independently by means of an ingenious mechanism, thus mixing up the colors.

For the puzzle to be solved, each face must be returned to consisting of one color. The total number of possible cube permutations is: 43,252,003,274,489,856,000, a staggering number.

The Rubik's Cube was released worldwide after a deal was signed with Ideal in September 1979, to release, and the puzzle made its first international appearance at the toy fairs of London, Paris, Nuremberg and New York in January and February 1980.

Larry Nichols invented a 2×2×2 "Puzzle with Pieces Rotatable in Groups" in 1970 and filed a Canadian patent application for it. Nichols's cube was held together with magnets. U.S. Patent 3,655,201 was granted to Nichols on April 11, 1972, two years before Rubik invented his cube. Nichols assigned his patent to his employer Moleculon Research Corp., which sued Ideal in 1982.

I was invited by Rubik, Prof. David Singmaster and Tom Kramer to be a witness in the law suit and give testimony. In 1984, Ideal lost the patent infringement suit and appealed. In 1986, the appeals court affirmed the judgment that Rubik's 2×2×2 Pocket Cube infringed Nichols's patent, but overturned the judgment on Rubik's 3×3×3 Cube.

After its international debut, sales of the cube were briefly interrupted in 1983 (the year of the cube crash) in order to ensure that production was brought in line with Western manufacturing and packaging safety standards.

Taking advantage of an initial shortage of cubes, many imitations appeared. A Greek inventor Panagiotis Verdes patented a method of creating cubes beyond the 5×5×5, up to 11x11x11 in 2003. However, the world record is held by Oskar van Deventer, who built a 17x17x17 Rubik's Cube in 2012.

By January 2009 Rubik's cube had become the top-selling puzzle game of all time, with 350 million cubes sold worldwide.

four-unit fence

three-unit fence

OPAQUE FENCES — 1978

Opaque fences are minimal barriers that block any straight line of sight passing through a given figure. In 1978, R. Honsberger introduced the "opaque square" or "opaque fences" problem, generalized to opaque regular polygons and opaque cube problems by Martin Gardner and Ian Stewart.

What will be the shortest length of a fence that will block any straight line passing through a square of unit side?

The fence can be composed of any shape, straight or curved, and can consist of more than one piece. The most obvious solution would be to build a fence along the perimeter of the square as shown above, which would have a length of four units, but a better solution would be to build the fence along three sides only, reducing the length to three units. What do you think the shortest possible length is?

284 CHALLENGE ●● ○○○○
REQUIRES 🧠✏️✂️⚒️
COMPLETED

SPIDRONS OF ERDÉLY — 1979

Dániel Erdély, a Hungarian industrial designer and artist, created an amazing three-dimensional world of great mathematical beauty.

He discovered a new geometrical object, which he called "Spidron." Besides its aesthetics, the Spidron found a place in surprisingly diverse mathematical and art categories, such as plane geometry, tessellations, fractals, dissections, polygons, polyhedra and many other three-dimensional space-filling structures.

The main feature of the Spidron, essentially a plane structure, is its fantastic power of folding into amazing complex three-dimensional forms.

On the right is just a small kaleidoscopic sampling of the varieties of Spidron structures, showing some of the astonishing work of Erdély and his collaborators, Marc Pelletier, Amira Buhler Allen, Walt van Ballegooijen, Craig S.Kaplan, Rinus Roelofs and many others.

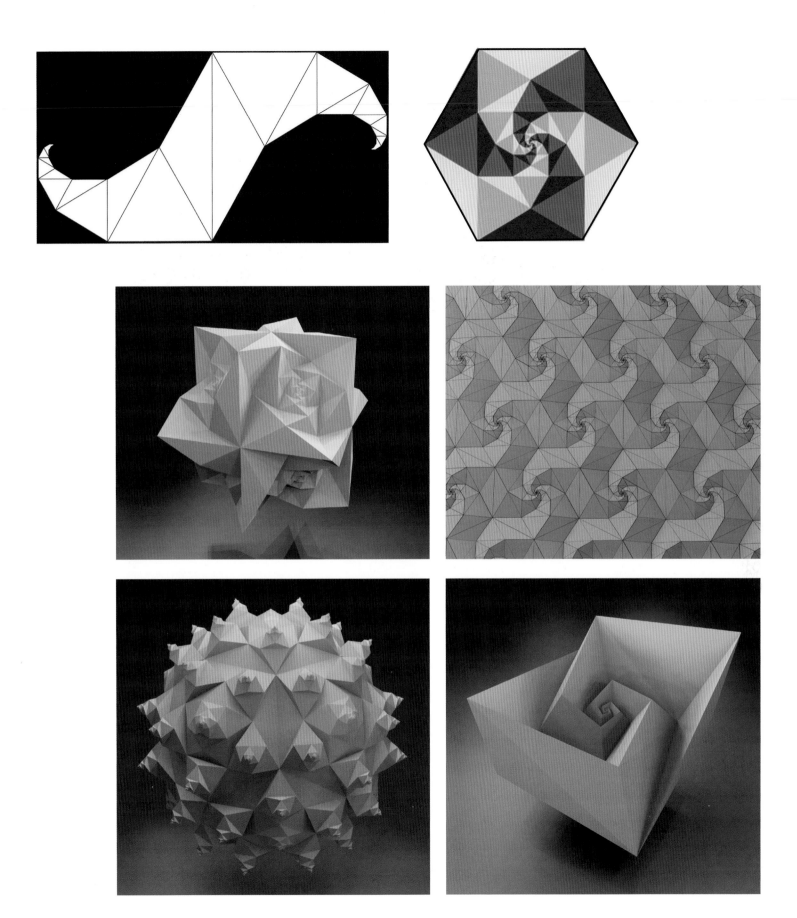

A small sampling of the varieties of spidron structures — © Rinus Roelofs

INTERLOCKING CYCLE PUZZLES — 1979

Conventional sliding block or sliding disk puzzles have an empty space into which the pieces can move. Knowing how to move pieces into the empty spaces is usually the key to solving such puzzles.

Churchill's puzzle, or the Hungarian Rings puzzle, and the Rally-Moeraki puzzle series incorporate the novel feature of having no empty spaces. The pieces in channels move like a chain, allowing the configurations and patterns to change by allowing disks to be transferred at the intersection points of the channels, from one channel to another. Moving any disk in a channel moves all the disks in that channel.

MOERAKI

In 1893, William Churchill patented a puzzle, an early example of a new category of mechanical puzzles. It was never commercially produced until 1982, when Endre Pap, a Hungarian engineer, patented and launched Churchill's puzzle under the name of "Hungarian Rings."

In 1979, I invented the Rally puzzle series, patented in 1981, and licensed to Meffert Novelties in 1982.

I was not aware of the existence of Churchill's patent until 1985 when my patent was granted with the Churchill patent listed in it as a reference.

Historically, the Rally puzzle series patent was the earliest produced puzzle in the subcategory of two-dimensional "interlocking cycle" puzzles, in the category of "Rubik's Twisty Puzzles," inspired by the Rubik's Cube and ever since growing in numbers to over 800 puzzles at present.

In 2011, the Rally puzzles were launched by Kasimir Landowski, Casland Games, Germany, under the name of Moeraki puzzle games series.

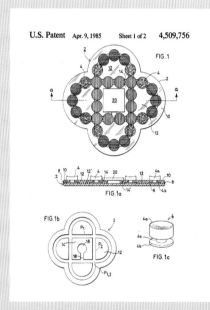

Moscovich patent, 1981

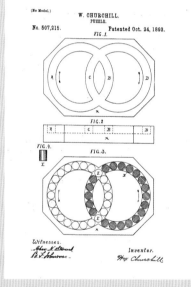

Churchill patent, 1893

Hungarian rings, 1982

RALLY-MOERAKI PUZZLE

The Rally-Moeraki puzzle series consists of sliding disk puzzles in which there are no empty spaces. In this sample, the 32 disks can move in a chain-like fashion in two elliptical channels as shown on the right. Each elliptic channel is composed of 18 disks, with four disks common to both channels.

Moving a disk in one of the channels will make all disks in that channel move clockwise or counterclockwise as you wish. Changing channels in succession will transfer disks from one channel to the other.

The disks are colored as shown. The basic object of the puzzles is to change the red square in the middle to a blue square in a minimal number of moves. A move is changing the configuration in a channel in any desired way, before changing it to a move in the other channel.

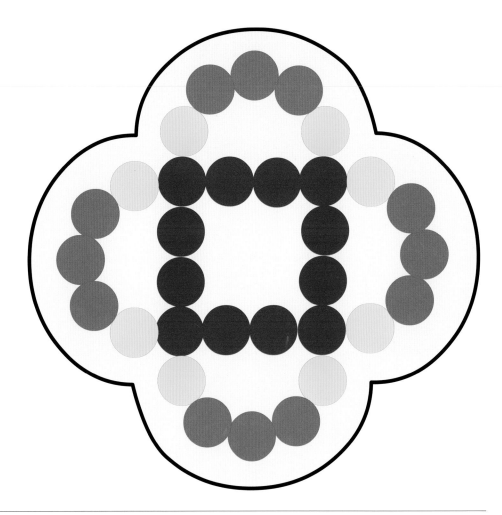

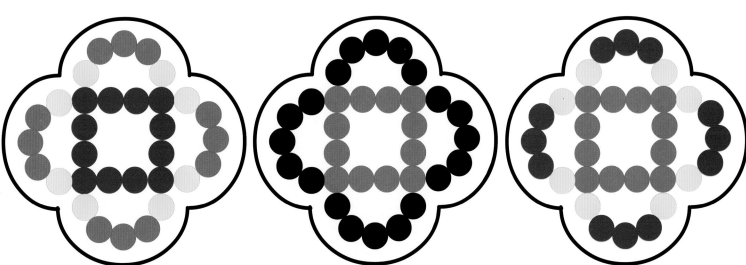

RALLY-MOERAKI PUZZLE PROBLEMS

What is the smallest number of moves to change the initial configuration into one of the other two?

1. Into a blue square in the middle (black = any color)
2. With the yellow disks in their initial positions

285 CHALLENGE ● ● ● ● ○
REQUIRES
COMPLETED

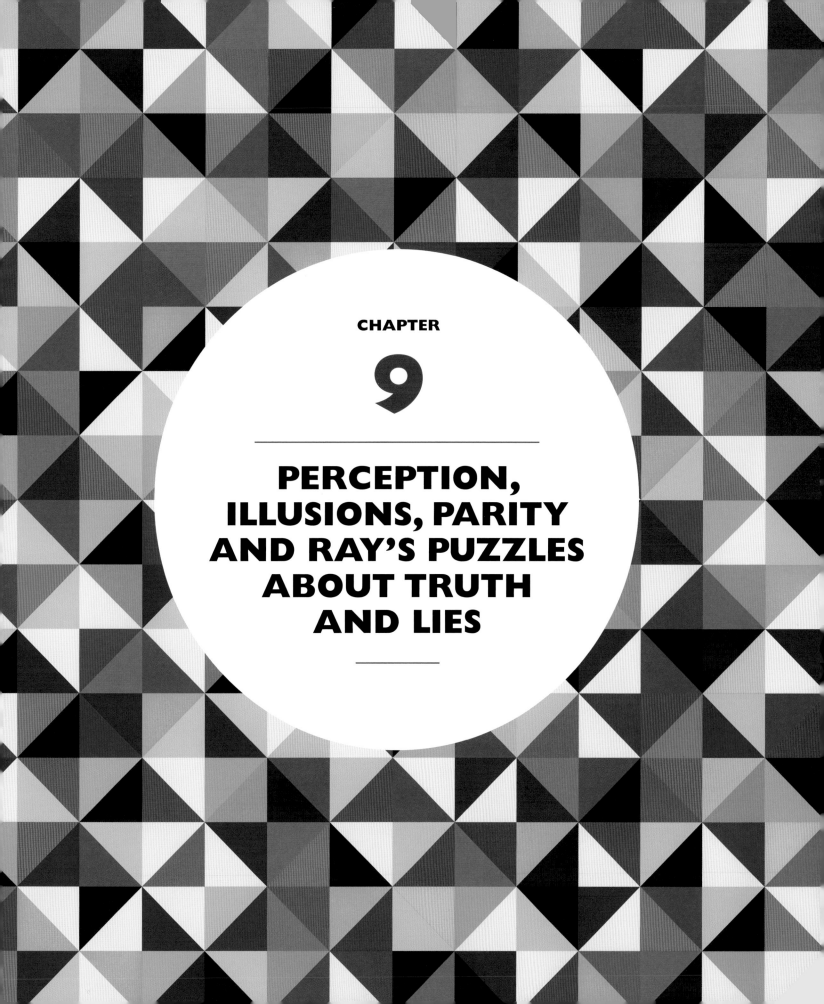

CHAPTER

9

PERCEPTION, ILLUSIONS, PARITY AND RAY'S PUZZLES ABOUT TRUTH AND LIES

BRONZE MERMAID — 1981

Angelo is in his rowboat, about to lower a solid bronze mermaid into the giant aquarium. Once the sculpture is successfully placed on her throne at the bottom, will the water level in the aquarium have risen, fallen, or stayed the same?

286 CHALLENGE ● ● ● ●
REQUIRES
COMPLETED

SCOTT KIM (1955)

Scott Kim is an American puzzle and computer game designer, artist, and author. He has created hundreds of puzzles for magazines such as *Scientific American* and *Games*. He is one of the world's most creative and original inventors of puzzles. Kim was born in 1955 in Washington D.C. He attended Stanford University, receiving a BA in music and a PhD in Computers and Graphic Design under Donald Knuth.

He is famous as one of the best-known masters of the art of ambigrams. In 1981, he created a book called *Inversions,* which concerned words that can be read in more than one way. It is a masterpiece in its category.

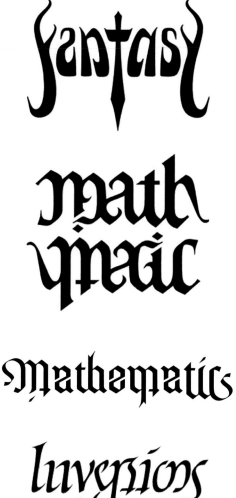

"Scott Kim is the Escher of the alphabet."

— *Isaac Asimov, science writer*

"Scott Kim has perfected a personal art form — one with grace, elegance, subtlety, and surprises. He draws on a deep understanding of letter forms and visual perception, and the resulting designs are highly original and gratifying. Many people will be delighted by what they see; some — I hope a good number — will go on to explore their own corners of the enchanting artistic space that Scott has revealed, for *Inversions* is an inspiring work."

— *Douglas Hofstadter, Pulitzer Prize winning author of* Gödel, Escher, Bach: an Eternal Golden Braid

"Scott Kim's *Inversions* is one of the most astonishing and delightful books ever printed. His book is interspersed with provocative observations on the nature of symmetry, its philosophical aspects and its embodiment in art and music as well as in wordplay. Over the years Kim has developed the magical ability to take just about any word or short phrase and letter it in a such a way that it exhibits some kind of striking geometrical symmetry."

— *Martin Gardner*, Scientific American

3 lines
1 triangle

4 lines
2 triangles

5 lines
5 triangles

6 lines
7 triangles

7 lines
11 triangles

8 lines
15 triangles

9 lines
21 triangles

KOBON TRIANGLE PROBLEM — 1983

The Kobon triangle problem is a problem in combinatorial geometry first stated in 1983 by Kobon Fujimura, a Japanese teacher and puzzle inventor.

This is the problem: What is the largest number of non-overlapping triangles that can be created by n straight line segments in the plane?

For n = 3, 4, 5 and 6 lines, the maximum number of triangles is 1, 2, 5 and 7. The maximum number of non-overlapping triangles for 7, 8, and 9 lines are 11, 15, and 21 respectively, as shown.

Saburo Tamura proved that the largest integer not exceeding k(k-2)/3 provides an upper bound on the maximal number of non-overlapping triangles realizable by "k" lines. For instance, for k=4, this means that 4(4-2)/3 is the largest integer, thus number of non-overlapping triangles, is 2.

In 2007, a tighter upper bound was found by Johannes Bader and Gilles Clement, by proving that Tamura's upper bound couldn't be reached for any k mod.6 congruent to 0 or 2. The maximum number of triangles is therefore one less than Tamura's bound in these cases.

Perfect solutions (Kobon triangle solutions yielding the maximum number of triangles) are known for k = 3, 4, 5, 6, 7, 8, 9, 13, 15 and 17.

For k = 10, 11 and 12, the best solutions known reach a number of triangles one less than the upper bound.

On his MathPuzzle site, Edd Pegg Jr. reported great progress on the problem, including Toshitaka Suzuki's beautiful 15 line and 65 triangle solution, as you can see on the next page.

What would these Kobon triangles look like if we introduce the restriction that the lines must form a single continuous broken line?

287

CHALLENGE ● ● ● ●

REQUIRES

COMPLETED

"Art, like morality, consists in drawing the line somewhere."

— *G.K. Chesterton*

SUZUKI'S SOLUTION

Toshitaka Suzuki's beautiful 15 lines with 65 triangles solution is the second best solution. At present, the best solution has 17 lines with 85 triangles.

TETRIS — 1984

Tetris was originally created by Alexey Pajitnov in the Soviet Union in the 1980s. It was the first entertainment software to be exported from the USSR to the U.S., where it was published for Commodore 64 and IBM PC. The Tetris game involves a popular use of tetrominoes, the special type of polyomino composed of four squares. Polyominoes were named by the mathematician Solomon W. Golomb in 1953, although they have been used in popular puzzles since as early as 1907.

Electronic Gaming Monthly's 100th issue named Tetris the "Greatest Computer Game of All Time." It was announced In January 2010 that Tetris has sold more than 100 million copies for cell phones alone.

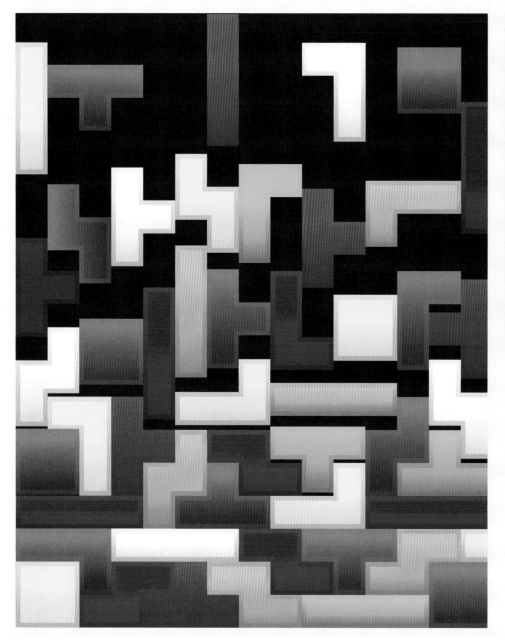

GIANT TETRIS

In 2012, hackers managed to transform MIT's Green Building into a giant Tetris game, operated from a wireless control podium at a comfortable viewing distance in front of the building.

HANDSHAKES (1)

At the board meeting, there were 17 members. And each of them was supposed to shake hands with every other person, but four members did not shake hands with each other. How many handshakes were there in total?

HANDSHAKES (2)

How many combinations of non-crossing simultaneous handshakes are possible with six people sitting at a round table?

HANDSHAKE PARTY

My wife and I invited four married couples to our housewarming party. Nobody shook hands with his or her spouse, and no couple shook hands more than once.

Before the guests left, I asked everybody how many people they had shaken hands with and I received the following replies: 8, 7, 6, 5, 3, 2, 1, and 0. How many people did my wife shake hands with?

NORTH POLE EXPLORER — 1986

An old classic about an explorer who starts his journey in a randomly chosen place, and walks one kilometer south, turns and walks one kilometer east, turns again and walks one kilometer north, and finds himself at his starting point, confronted by a bear. What color is the bear? The usual answer is "white," but is the North Pole the only possible starting point for his journey?

291

CHALLENGE ● ● ● ● ●
REQUIRES 🧠 ✏️
COMPLETED

HARRY ENG'S MAGIC — 1990

Harry Eng was born in 1932 and died in 1996. He was a school teacher, educational consultant, inventor, magician and a dear friend. Everything he did was intended to teach you to think. He did that to me for many years.

Harry became world famous through his bottles. During his lifetime he supposedly made around 600 of these "impossible bottles." There was no trick, glass was not blown around the objects. Everything inside the bottle was passed through the neck. We do know that Harry Eng invented a special device that could be taken apart, put inside a bottle, and then assembled inside the bottle. This would allow metal objects inside, to be bent and/or straightened. Once the job was done, the device could be taken apart and removed from the bottle.

Harry is a legend among magicians. He did very little magic, per se. No, his magic came from a different source altogether. His art is not the art of stage illusion, sleight of hand, prop magic or mentalism. Harry's magic was purely the art of thinking and creativity." The power of our lives, the very force by which we live, is in our minds," he would say. Impossibility was a way of life for Harry: "The impossible takes just a little longer," he said. And he was right!

> ## "The impossible takes just a bit longer."
>
> — *Harry Eng*

IMPOSSIBLE FOLDING PUZZLE

The object of the puzzle is to fold the sheet along the folding lines to make the big octagon pass through the square loop as shown. How would you do this?

This puzzle was invented by Harry Eng and received as a souvenir of the International Puzzle Party on June 11, 1994.

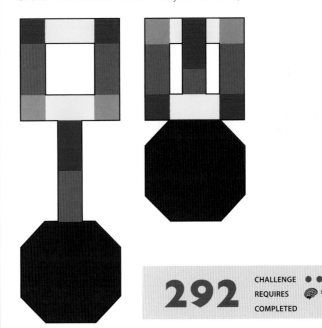

292 CHALLENGE ●●●●○○
REQUIRES
COMPLETED

MONTY HALL PROBLEM — 1990

This famous counterintuitive problem is often called the Monty Hall problem, named after Monty Hall, an American game show host who hosted "Let's Make a Deal."

Hall was known for tempting people to give up their prizes to gamble for a mystery gift behind a door. Martin Gardner introduced the problem in his October 1959 column.

The *Parade Magazine* columnist Marilyn vos Savant is most famously associated with the problem involving three doors and a luxury car behind one of the doors. She also provided the answer to the problem. Her answer resulted in thousands of letters of disbelief and accusations, among them about 1000 from people with PhDs, many of whom were mathematicians. Don't be surprised; the problem is really paradoxical and counterintuitive.

Even Paul Erdős, one of the most brilliant mathematicians of the 20th century, initially reacted with disbelief. It took his friends quite a while to change his initial reaction. It was only after watching hundreds of trial games of a computer simulation arranged by a colleague that Erdős concluded he was wrong.

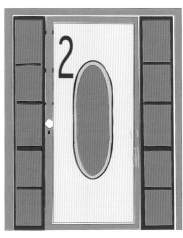

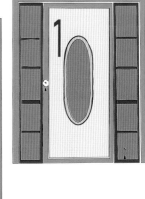

"Our brains are just not wired to do probability problems very well."

— *Richard Feynman*

MONTY HALL PROBLEM: THE RULES OF THE GAME

You have been selected to participate in the game show that offers you the chance to win a luxury car. The car is behind one of the three doors. A goat is behind each of the other two. You choose a door at random (move 1) that stays closed. At this point, the probability of success, i.e. choosing the car, is 1/3, or about 33 percent.

The host (who knows where the car is) does what he must do: he opens and eliminates an unpicked door to show a goat behind it (there's always at least one unpicked door with a goat behind it to open).

Now comes the critical moment when your host offers you the choice to switch your initial choice to the remaining door. To switch or not, that is the question (and your dilemma as well). Will switching make a difference in the initial probability of success? Marilyn's answer was always to switch.

The accompanying graphic on the next page shows all of the possibilities. The first column shows the three possible choices. The second column (move 2) shows the outcome if you decide not to switch and the third column the outcome if you switch. You will see that if you switch (move 3) —, your chances of success will double — from the original 1/3, to 2/3. Many may still not be convinced, and it's no wonder. Common sense suggests that a switch would make no difference. Two doors, one prize, you may as well just flip a coin. But Marilyn was right.

To convince you that she was, go to Monty Hall Problem 2, involving 10 doors, and play a number of game trials as Paul Erdős did.

Finally, remember that this is a case of conditional probabilities — the probability of something happening given that something else has already happened.

MONTY HALL PROBLEM WITH 10 DOORS

For those who are still skeptical, we are providing a different version of the problem with 10 doors, which should eliminate some of the mental blocks tens of thousands of people had when they first encountered the original problem.

As before, there is a luxury car behind one of the doors, and goats behind the other nine. You are allowed to choose which door stays closed. The host opens eight doors behind all of which are goats.

He leaves one door closed, and you once again have the option to switch. Will you? What are your chances of winning the car if you don't switch and stick to your initial choice? What are your chances of winning the car if you switch?

293 CHALLENGE ●●●●●○
REQUIRES 🧠✏️✂️
COMPLETED

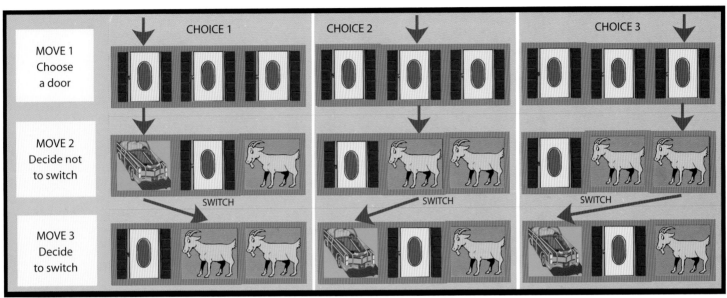

Note that in the first column the host must leave one unchosen door, even though he has a choice of two doors to open and eliminate. In the second and third columns the host has only one door to open and eliminate. In move 3, after the switch is made, an initial winning choice of doors (column one) becomes a losing door. On the other hand, an initial losing choice (columns two and three) converts to a winning choice.

DRACULA'S COFFIN — 1991

Can you find the proper lids to close the two coffins?

294 CHALLENGE REQUIRES COMPLETED

TREE PLANTING — 1991

Problems involving arrangements of n points on straight lines, so that exactly points will be on each line, are often known as "tree planting" or "orchard" problems, a group of difficult problems. Usually the object is to maximize the number of lines r. Curiously enough, a general solution for the problem hasn't been found, even for cases of k=3 and k=4, and their solutions are still waiting for a breakthrough.

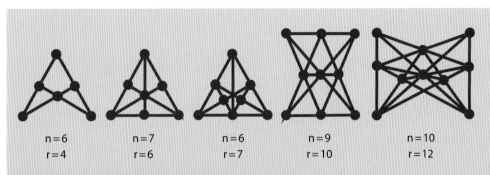

n=6
r=4

n=7
r=6

n=6
r=7

n=9
r=10

n=10
r=12

TREE PLANTING PROBLEM (1)

Maximum solutions for k=3 (three-in-a-row) from n=6 to n=10. Can you find the maximum solution for n =11 with r=16?

295 CHALLENGE ●●●●○○
REQUIRES 🧠✏️✂️
COMPLETED

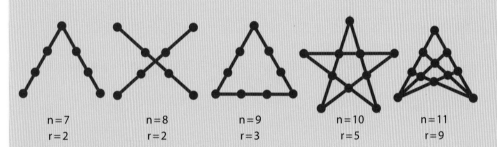

n=7
r=2

n=8
r=2

n=9
r=3

n=10
r=5

n=11
r=9

TREE PLANTING PROBLEM (2)

When k= 4, the problem becomes even more difficult. Maximum solutions for k=4 (four-in-a-row) from n=7 to n=11. Can you find the maximum solution for n=12 with r=7?

296 CHALLENGE ●●●●○○
REQUIRES 🧠✏️✂️
COMPLETED

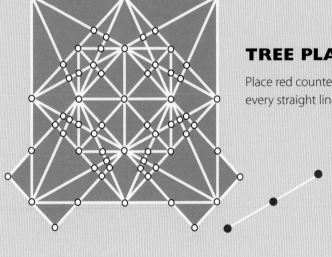

TREE PLANTING PROBLEM (3)

Place red counters on the white circles, so that you have exactly three red counters on every straight line. How many red counters will you need?

297 CHALLENGE ●●●○○○
REQUIRES 🧠✏️✂️
COMPLETED

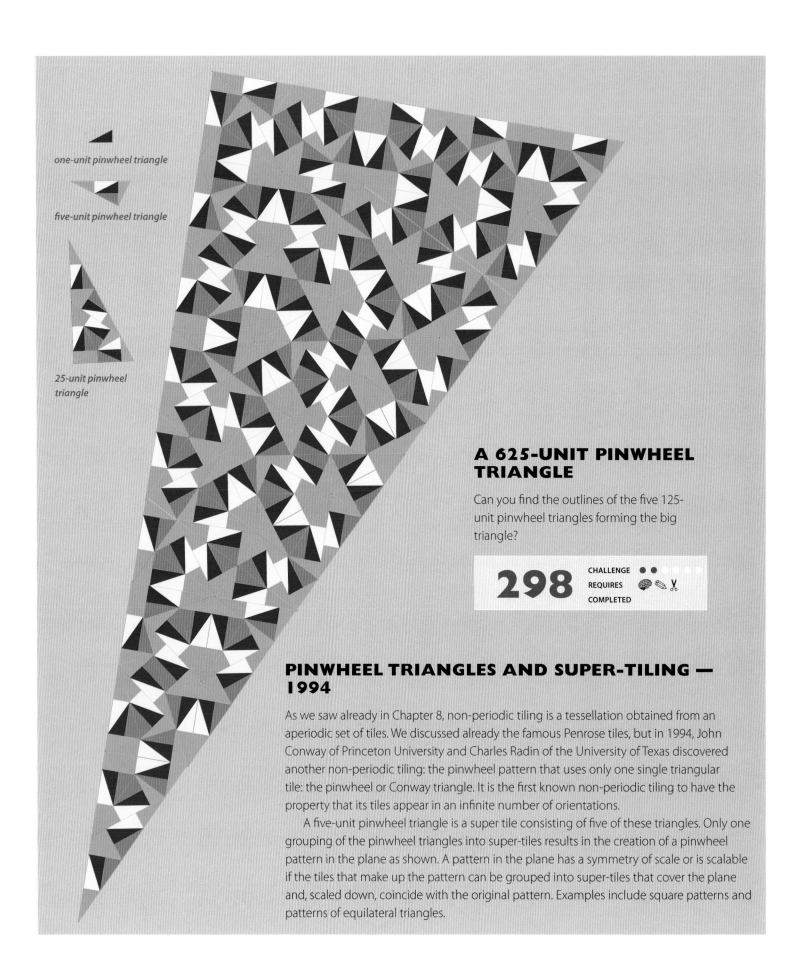

one-unit pinwheel triangle

five-unit pinwheel triangle

25-unit pinwheel triangle

A 625-UNIT PINWHEEL TRIANGLE

Can you find the outlines of the five 125-unit pinwheel triangles forming the big triangle?

298 CHALLENGE ●● ○ ○ ○ ○
REQUIRES 🧠 ✎ ✂
COMPLETED

PINWHEEL TRIANGLES AND SUPER-TILING — 1994

As we saw already in Chapter 8, non-periodic tiling is a tessellation obtained from an aperiodic set of tiles. We discussed already the famous Penrose tiles, but in 1994, John Conway of Princeton University and Charles Radin of the University of Texas discovered another non-periodic tiling: the pinwheel pattern that uses only one single triangular tile: the pinwheel or Conway triangle. It is the first known non-periodic tiling to have the property that its tiles appear in an infinite number of orientations.

A five-unit pinwheel triangle is a super tile consisting of five of these triangles. Only one grouping of the pinwheel triangles into super-tiles results in the creation of a pinwheel pattern in the plane as shown. A pattern in the plane has a symmetry of scale or is scalable if the tiles that make up the pattern can be grouped into super-tiles that cover the plane and, scaled down, coincide with the original pattern. Examples include square patterns and patterns of equilateral triangles.

PARITY — 1994

The term "parity" was first used in math to distinguish between even and odd numbers. If two numbers are both even or both odd then they have the same parity, otherwise they have opposite parity.

The parity of a pattern is conserved by an even number of moves. Many tricks with cards, coins and puzzles exploit the principles of parity, using a simple method called a parity check.

Parity also plays an important role in the physics of subatomic particles and wave functions.

You've undoubtedly heard of the daisy petal game, where someone pulls the petals off a daisy one by one reciting "He loves me, he loves me not." Parity can be used to quickly determine the outcome. If the total is even, the answer is negative.

This would be what mathematicians call a parity check, which is one of the most powerful tools in mathematics. It can often provide quick, elegant proof of a problem that might otherwise be difficult to obtain.

THE SIX GLASSES PROBLEM

Place six glasses as shown. Take any pair and reverse both. Continue reversing them as long as you please. Can you end up with all glasses upright, or with all glasses upside-down?

299   CHALLENGE ● ● ● ● ● ○
REQUIRES 🧠 ✏️ ✂️
COMPLETED

THE THREE GLASSES TRICK

Place three glasses as shown in the first arrangement. By turning over two glasses simultaneously, the object is to bring all of the glasses upright in exactly three moves.

You can easily accomplish this. After having done it, the trick is to invert the middle glass and challenge somebody to do it. This will be impossible. The first setup has an odd parity, the second an even parity. Whenever an even number of glasses (zero, two or even) is upright, the system has an even parity. When an odd number is upright, the system has an odd parity. Turning any two glasses in the second arrangement three times cannot change the system's parity.

THE SEVEN GLASSES PROBLEM

The object is to turn all of the seven glasses upright, inverting three glasses in each move. In how many moves can this be done?

300   CHALLENGE ● ● ● ● ● ○
REQUIRES 🧠 ✏️ ✂️
COMPLETED

THE TEN GLASSES PROBLEM

Ten glasses are placed as shown, five facing up, five facing down. Take any two glasses and reverse both. Continue reversing the pairs of glasses as long as you please. Can you end up with all of the glasses upright?

301   CHALLENGE ● ● ● ● ● ○
REQUIRES 🧠 ✏️ ✂️
COMPLETED

SELF-DESCRIPTIVE TEN-DIGIT NUMBER — 1994

There is a collection of puzzles based on the first 10 digits including zero. The most beautiful in this category is the Self-Descriptive Ten-Digit problem, as it was called by Martin Gardner. At the Ontario Science Center in Toronto, this intriguing puzzle is found in the mathematics exhibit, as shown above.

The object is to find a 10-digit number that has to be written in the 10 empty boxes of the second row. This number is determined by the 10 consecutive digits in row one, according to the following rule.

The first digit in row two indicates the total number of zeros in the 10-digit number; the second digit the total number of ones; the third digit the total number of twos, and so on, until the last digit, which indicates the total number of nines.

It's like a 10-digit number inventing itself. No wonder Martin Gardner called it the self-descriptive number. How can you start solving such a seemingly impossible, challenging problem?

Is there a solution at all, and if so, how many? Can you discover some insights to start with?

Daniel Shoham from MIT discovered some interesting facts related to this problem. He concluded that, because there are 10 different digits in row one, the sum of the digits in row two must be 10, and he determined the maximum number of possible values for each digit of the number in the second row. Can you follow his logic and find the unique solution to the puzzle?

HOW MANY NUMBERS?

Among my great collection of logic puzzles, there is a special selection of puzzles based only on the first 10 digits, involving zero to nine or, excluding zero, the digits from one to 10.

One of the early puzzles in this category is the Ten-Digit Number puzzle. How many different 10-digit numbers can you form, using only the digits from zero to nine? Numbers starting with zero don't count!

UNILLUMINABLE ROOM PROBLEM — 1995

In the 1950s, Erns Strauss posed the problem of whether there can be a polygonal room with its walls covered with mirrors, in which you can light a match somewhere leaving part of the room in darkness because the reflections cannot reach it.

There was no answer to the problem until 1995, when George Tokarsky of the University of Alberta in Canada answered it by saying that there is such a room, and the smallest possible one had a floor plan of 26 sides as shown below. If a match is held in just the right place, then at least one other point in the room is dark. Tokarsky

called it the minimal unilluminable room. In Tokarsky's room there is a particular point for the match to be held leaving part of the room dark, but if you move the match a bit the whole room is lit again. It should be noted that if a ray of light strikes a corner of the room exactly, then it is assumed not to reflect at all; it just gets absorbed at the join between the mirrors on the adjoining walls.

An improved solution was put forward by D. Castro in 1997, with a 24-sided room with the same properties.

The following problem still remains: Is there a room that is so complex that wher-

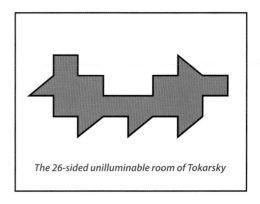

The 26-sided unilluminable room of Tokarsky

ever you hold the match there are some dark spots in the room? Nobody has the answer to this yet.

ILLUMINABLE ROOM

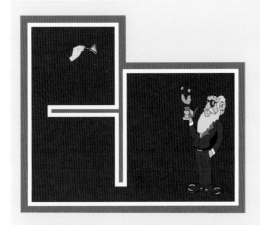

Imagine the walls of the L-shaped room in the layout on the left are all covered in mirrors from floor to ceiling. The room is in in complete darkness. The man at the top left lights a match. Will the man smoking a pipe in the bottom right of the room be able to see a reflection of the match?

304 CHALLENGE ●●●●○○
REQUIRES 🧠✏️✂️🔨
COMPLETED

PENROSE UNILLUMINABLE ROOM

In 1958, Roger Penrose used the properties of the ellipse to make a room that would always have dark areas, regardless of the position of the candle (yellow points). The red points are the foci of the half ellipses at the top and bottom of the room.

Can you draw in the areas left in the dark in each case?

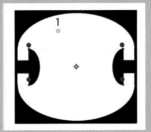
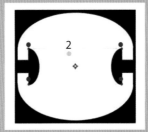
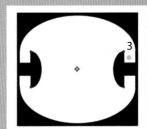

305 CHALLENGE ●●●●○○
REQUIRES 🧠✏️✂️🔨
COMPLETED

GÖMBÖC, THE WORLD'S FIRST SELF-RIGHTING OBJECT — 1995

Gömböc is the first known convex three-dimensional homogeneous body that, when resting on a flat surface, has just one stable and one unstable point of equilibrium.

A sphere weighted so that its center of mass is off center is an example, but it is inhomogeneous. The question of whether it is possible to construct a three-dimensional solid that is mono-monostatic and also homogeneous and convex was raised by Vladimir Arnold, a Russian mathematician, at a conference in 1995 in a conversation with Gábor Domokos.

The reason why many thought that Gömböc shapes don't exist was because in two-dimensions, there is no shape with only two points of equilibrium, the best one can achieve is two stable and two unstable equilibria.

Such a shape called Gömböc was developed by Gábor Domokos (head of Mechanics, Materials and Structures at Budapest University of Technology and Economics in Hungary) and a former student of his, Péter Várkonyi (at Princeton University). The Gömböc made the front page of mathematical journals, after its discovery and Domokos' appearance on British television on December 7, 2007, to explain how the Gömböc works. The Gömböc is an exciting object as seen above, the first perfect self-righting mechanism, and one of the most beautiful creative ideas of recent years. When placed on a flat surface in a random position, the Gomboc returns to the stable equilibrium point, in a similar way to "Weeble" toys. The Weebles, however, depend on a weight at the base, while the Gömböc is made of homogenous material; therefore, the shape itself is responsible for the self-righting action.

The single unstable equilibrium point of the Gömböc is on the opposite side. While it can be balanced in this position, the slightest disturbance will make it fall, rather like a pencil balanced on its tip. The Gömböc shape is not unique. It can have a wide range of possibilities, though most of them are quite similar to a sphere, with a very strict shape tolerance (about 0.1 mm per 100 mm). The shape of the Gömböc helped to explain the uncanny ability to return to their equilibrium position after being turned on their backs.

THE ETERNITY PUZZLE — 1996

Eternity is a tiling puzzle created by Christopher Monckton and launched by the Ertl Company in June 1999.

The puzzle consists of filling a large, almost regular dodecagon with 209 irregularly shaped smaller polygon pieces of the same color.

The puzzle was marketed as being practically unsolvable and a £1 million ($1.5 million) prize was offered for anyone who could solve it within four years. The prize was paid out in October 2000. A second puzzle, Eternity II, was launched in the summer of 2007 with a prize of $2 million, a complete solution of which is still missing.

The game quickly became an obsession, with 500,000 copies being sold worldwide. Eternity was the best-selling puzzle or game in the UK at a price-point of £35 ($55) in its launch month.

When Monckton was preparing for marketing, he believed that it would be at least three years before anyone could solve the puzzle. It was estimated at the time that the puzzle had 10,500 possible attempts at a solution, and that even if you had a million computers, it would take longer than the lifetime of the universe to calculate them all.

However, two Cambridge mathematicians, Alex Selby and Oliver Riordan solved the puzzle on May 15, 2000, before the initial deadline. Key to their success was the mathematical rigor with which they approached the problem of determining the tileability of individual pieces and of empty regions within the board. A complete solution was obtained within seven months of brute force search on two computers. It is shown below.

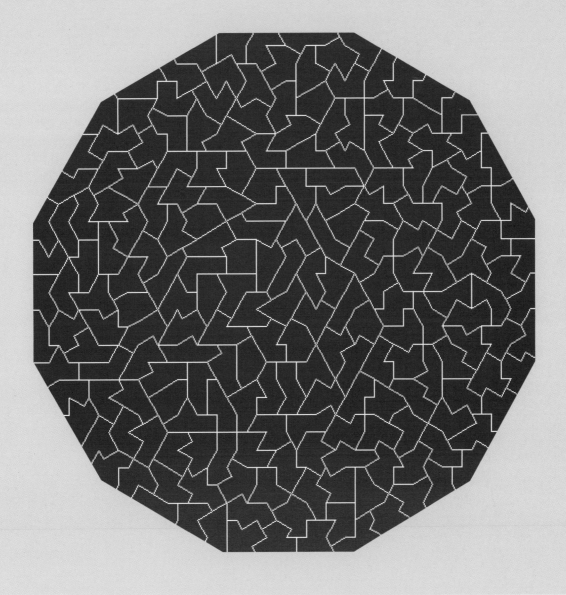

RAYMOND SMULLYAN (1919)

Ray Smullyan is, as Martin Gardner described him, a "unique set of perso-nalities that include a philosopher, logi-cian, mathematician, musician, writer and maker of marvelous puzzles." His first career was stage magic and his first loves were music and mathematics. He then earned a BSc from the University of Chicago in 1955 and his PhD from Princeton University in 1959. Raymond is internationally known as a mathematical logician. He is equally famous as a writer, having authored over 20 books, many of which have been translated into 17 lan-guages.

As Martin Gardner summarized: "Raymond truly has the wisdom of a Zen Master and sage; the artistry and finesse of a musician and magician, the heart, creativity and eloquence of a poet; the insight and analytical skill of a logician and mathematician, and the wonder of a wizard."

TRUTH CITY — 1996

The inhabitants of Truth City always speak the truth, while those of Liars City of course, always lie. You are on your way to visit Truth City, arriving at the crossroads of the two cities. As you can see, the signs are con-fusing and you are forced to ask the man standing at the crossroads for the right direction. Unfortunately, you don't know

306 CHALLENGE ●●● ○ ○
REQUIRES 🧠 ✎ ✂ ⚒
COMPLETED

whether the man is a liar or a truthteller. You are allowed to ask the man only one question. What is the question you must ask him, in order to be able to determine from his answer which is the right way for you to go to the truthtellers' city?

TRUTH AND MARRIAGE

The king has two daughters, Amelia and Leila. One of them is married, the other is not. Amelia always tells the truth, while Leila always lies. As in many fairy tales, the young man is allowed to ask just one single question to one of the daughters, to find out which daughter is the married one. His reward would be, of course, marrying the unmarried daughter.

The catch was that his question is not allowed to contain more than three words.

What is his question?

307 CHALLENGE ●●●●● ○
REQUIRES 🧠 ✎ ✂
COMPLETED

TRUTHS, LIES AND IN BETWEEN

In the cosmopolitan city of Nooneknows-truth there are those who always speak the truth, those who always lie, and those who alternately lie and tell the truth. Again you meet one of the residents. This time you are allowed to ask two questions. His answer must be sufficient for you to determine to which of the three groups the man belongs. What are the two questions you will ask him?

308

CHALLENGE ● ● ● ● ●

REQUIRES 🧠🥜

COMPLETED

RAYMOND'S SPEECH

"I am intensely interested in mysticism and religion, though I belong to no creed. I am interested in comparative religion — I would like to know what possible truth lies behind the various religions of the world. I believe that all religions are approaching the truth, but none have yet completely found it. The book that has had the most influence on me is *Cosmic Consciousness* by Richard Bucke, whose main theme is that a new type of consciousness is slowly coming to the human race through the process of evolution, and that the mystic and religious leaders of the past, as well as many artists and poets, possessed an advanced cosmic consciousness. Bucke cites the writings of many of these people, and they are really marvelous! I cannot recommend this book highly enough.

Politically, I am an extreme liberal, but not about all issues — for example, I absolutely refuse to be "politically correct"! Indeed, I am a bit of a maverick, and my epitaph will be:

IN LIFE HE WAS INCORRIGIBLE.
IN DEATH, HE'S EVEN WORSE!

In high school, I fell in love with mathematics and was torn between it and music.

Another interesting and wonderful thing happened to me whilst I was at Princeton: I frequently visited New York City in those days, and on one of my visits, I met a very charming lady musician. At one point I did a very clever trick which caused her to owe me a kiss! Instead of collecting the kiss I suggested we play for double or nothing. She, being a good sport, agreed. So she soon owed me two kisses, then with another trick four, then eight, then sixteen — then things kept doubling and escalating and doubling and escalating, and before I knew it, I was married! And I was married to Blanche, the charming lady musician, for 48 wonderful years. Unfortunately Blanche passed away in 2006 at the ripe old age of 100."

JERRY — THE MATHEMAGICIAN — FARRELL AND HIS FAMOUS ELECTION PUZZLE — 1996

Jeremiah (Jerry) Farrell, born in 1937, is an American Professor Emeritus of Mathematics at Butler University in Indiana. He is celebrated for creating the 1996 "Election Day" crossword in The New York Times. In addition, he has written puzzles for a number books and publications, including Scott Kim's puzzle column for Discover Magazine.

He attended the University of Nebraska, graduating in 1963 with degrees in mathematics, chemistry and physics. He later obtained a Master's degree in mathematics. In 1966 he was appointed to a position at Butler University in Indiana, where he was to work for the next 40 years, teaching a variety of different subjects within the mathematics department. Despite officially retiring in 1994, Farrell continues to teach every semester.

Farrell is best known for creating numerous crossword puzzles for The New York Times. In 1996, he designed the "Election Day" crossword, his most famous puzzle. One of the clues was "lead story tomorrow," with a 14-letter solution. However, in this puzzle there were two alternative answers, both correct: one was "Bob Dole elected," and the other "Clinton elected," while all of the "crossing" words were designed so that they could be either of two different words, to make either answer as needed. Will Shortz called it an "amazing" feat, and said it was his favorite puzzle.

In 2006 Farrell and his wife succeeded A. Ross Eckler Jr. as editors and publishers of the quarterly magazine Word Ways: the Journal of Recreational Linguistics.

Farrell is a member of the Flat Earth Society, as well as being awarded the title of "Omniheurist, First-Class" by Dennis E. Shasha, a New York University computer science professor, for being the first person to solve the puzzle embedded in Shasha's book, Puzzling Adventures. The solution took Farrell to a specific location in Greenwich Village for a meeting with the author.

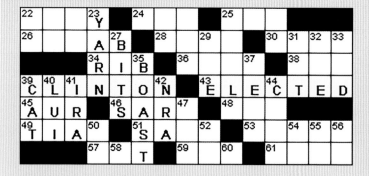
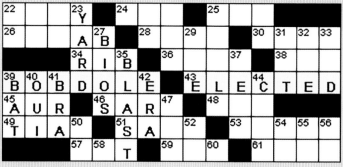

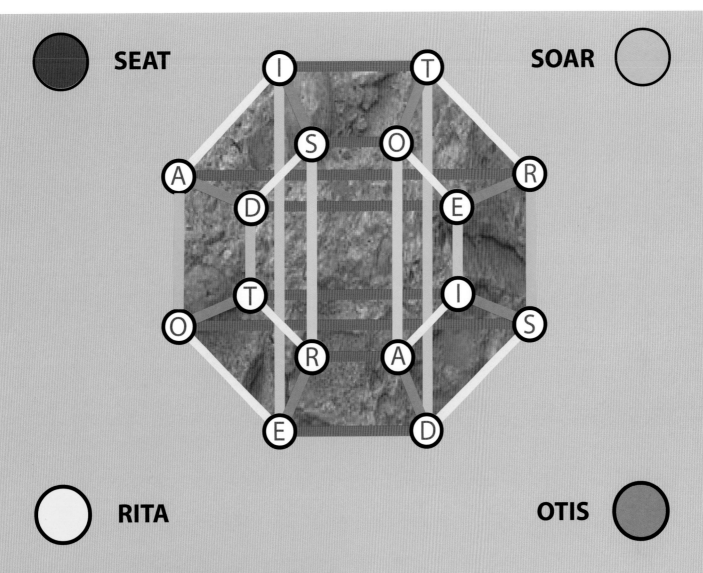

SEAT

SOAR

RITA

OTIS

ASTEROID, THE HYPERCUBE GAME

Asteroid is a four-dimensional magic feat, invented by Jerry Farrell, that you can easily perform to astound your audience. It is performed on a four-dimensional tesseract game board of 16 nodes labeled with the letters of the word "ASTEROID" in red and blue.

Ask a friend to choose a single letter from the tesseract. You will then ask him four questions:

Is your letter in the word SEAT? SOAR? RITA? or OTIS?

To make your magic more intriguing you can tell your friend that he or she can decide at will to answer the four questions truthfully or lying.

309 CHALLENGE ● ● ● ● ● ●
REQUIRES 🧠 ✎ ✂
COMPLETED

Suppose your friend answered the four questions with:

YES - NO - YES - YES

You can immediately tell him his chosen letter is T, complimenting him for telling the truth.

If he would have decided to lie, his answers would be:

NO - YES - NO - NO

You would again say that it is the letter T, but also that he lied! Can you explain how the magic works?

STEWART COFFIN'S POLYHEDRAL INTERLOCKING PUZZLES — 2000

Stewart Coffin is one of the world's most accomplished designers of polyhedral interlocking puzzles. There were very few examples of these until he began to explore beyond the world of orthogonal puzzles, in which all the pieces are at right angles to each other. To date, he has hundreds to his credit. A few have been commercially mass-produced in plastic. The most notable one is Hectix. His book, *The Puzzling World Of Polyhedral Dissections*, is a definitive work on interlocking puzzles with a geometric theme.

In some cases, Stewart has discovered a simple, but beautiful design and carried it to the extreme with amazing results.

He has been designing intriguing geometric puzzles and making them in his workshop since the early 1970s, creating more than 200 original designs. The ingenuity and originality of his designs have made him a much admired figure among puzzle enthusiasts and collectors worldwide.

Stewart was awarded the Sam Loyd Award in 2000, and then the Nob Yoshigahara Award in 2006 for his lifetime contribution to mechanical puzzles.

Nick Baxter's collection of Stewart Coffin originals

THE GRASSHOPPING GAME — 2002

On the first day of the International Puzzle Party (IPP) in Antwerp in 2002, there was a long and boring lecture. I was doodling on a squared sheet of paper, when out of the blue, an idea for a paper-and-pencil puzzle game surfaced. It was not the first time my subconscious provided me with original ideas.

The idea is as follows: Imagine a grasshopper jumping along a given line of integral length, according to the following rules.

Our grasshopper has to start jumping along the line from point 0, in successive jumps of consecutive lengths: 1-2-3-4-...-n, so as to make as many jumps as possible and finish the n-th jump at the end point of the line, at point n. If we have a line on which this can be achieved, our game ends and we have a solution. If not, the line has no solution. The jumps on the line are allowed in both directions.

Looks interesting, I decided, and I went on doodling systematically to find solutions,

starting from line one. By this time, I realized there are lines that have a solution, and other lines that don't, but it also became clear to me that the innocent grasshopping idea is more than just simple doodling. It looks like its solutions are generating an infinite number sequence, each solution a puzzle to solve, but what is the mathematics behind this sequence?

For the first eight lines I found two solutions, the trivial one for n=1, and the first real solution for n=4, as shown above. At this moment the boring lecture ended.

I went to my room and by the evening I had found 16 solutions up to n=40. It was hard work. The Grasshopping Game showed me that just moving a point along a line in consecutive unit distances can produce a challenging game with subtle mathematical principles and surprises.

Later in the evening I met Dick Hess and enlisted his help to find the general solu-

tion of the Grasshopping Number Sequence. The next day at breakfast I met Dick again. He politely thanked me for a sleepless night, but assured me that he would not give up on Grasshopping. Dick joined forces with Benji Fisher, and the next day the mathematics behind the Grasshopping Sequence was solved, providing the key to the Grasshopping Infinite Number Sequence.

Finally, there is the story of the Grasshopping Game. At the next G4G9 Gathering for Gardner in Atlanta 2010, I met Neil Sloane. I showed him the Grasshoping Game, incorporating my integral number sequence. Neil was enthusiastic about Grasshopping.

Today, the Grasshopping Number Sequence occupies a respectable place on the internet, among the giants of integer sequences, π, the primes, the Fibonacci and others in the exciting *Online Encyclopedia of Integer Sequences of Neil Sloane* under ref. A141000 — and I am quite proud of it.

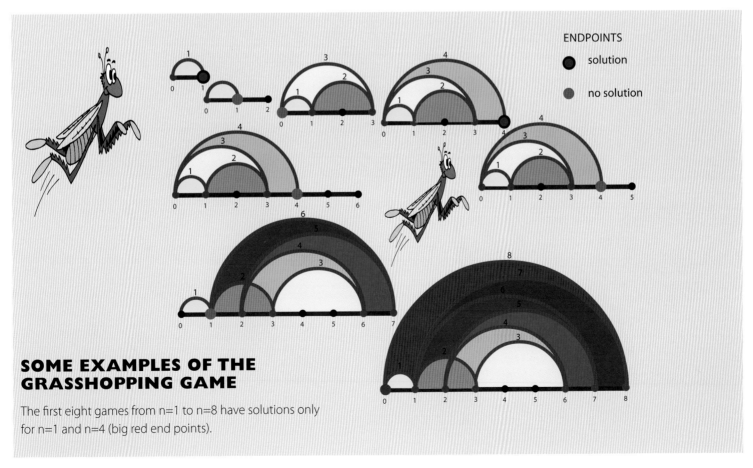

ENDPOINTS

● solution

● no solution

SOME EXAMPLES OF THE GRASSHOPPING GAME

The first eight games from n=1 to n=8 have solutions only for n=1 and n=4 (big red end points).

GRASSHOPPING PROBLEM: THE FIRST 40 LENGTHS

Given a line of integral length n, the object is to start jumping along the line from point 0, in successive jumps of consecutive lengths: 1-2-3-.........-n, so as to make as many jumps as possible and finish the n-th jump at the end point of the line.

On a line of length n, if this can be achieved the line has a solution, if not, there is no solution. The moves along the line can be in any direction but they are not allowed to leave the line. For greater lengths n, more than just one sequence of moves may be

possible. Can you find out how many solutions exist for the first 40 lengths? The first two solutions are marked.

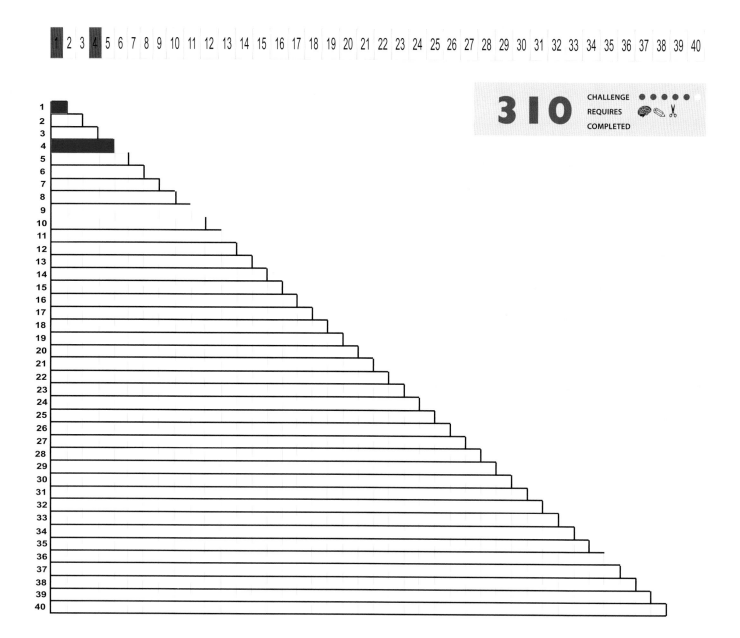

310 CHALLENGE ● ● ● ● ●
     REQUIRES
     COMPLETED

SID SACKSON (1920–2002)

Sid Sackson's *A Gamut of Games*, first published in 1969, contains the rules for a wide variety of paper-and-pencil, card, and board games.

Many of the games in the book had never before been published. It is considered by many to be an essential basic text for anyone interested in abstract strategy games.

Many people believe Sid Sackson to have been one of the most important and influential game designers ever. He was also an avid collector, and his collection is thought to have been possibly the biggest in the world at one time, containing around 18,000 pieces, including a number of prototypes and unique items that he kept in his New Jersey home until his death in 2002 at the age of 82.

If it were still complete today, the Sackson collection would be an incomparable record of the history of modern board gaming.

Sackson dreamed of one day curating a games museum, presumably using his vast collection as a foundation. Sid turned to me for my help and we spent endless hours discussing how his museum could be realized. Sid was disappointed that all of our endeavors were not successful. Sadly, after his death, his collection was scattered and sold off at auctions.

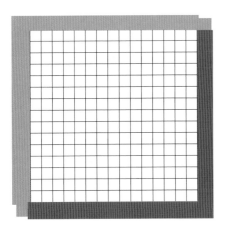

CUTTING CORNERS

Two players each play one of the colors (red or blue). They alternate moves by drawing a corner along the square grid. At least one of the connected edges must be of the opponent's color. At the end of the game, a section is won by a player if he has more sides of his color around it than there are sides of the opponent's color. If the sides are equally divided, they belong to neither player (*).

The sample game on the right was won by the blue player.

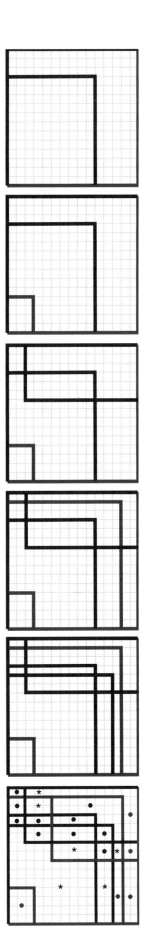

NOBUYUKI YOSHIGAHARA (1936–2004)

Nobuyuki Yoshigahara, nicknamed "Nob," was Japan's most renowned inventor, collector, solver and communicator of puzzles.

After graduating in applied chemistry from the Tokyo Institute of Technology, Nob embarked on a career in engineering, before switching paths to become a high school teacher of chemistry and mathematics.

Nob was a puzzle columnist who contributed to numerous journals, with columns in a variety of popular magazines such as *Quark*. He also wrote more than 80 puzzle books.

With a growing reputation as a puzzle inventor, he licensed his creations, which included the Rush Hour puzzle game, on a commercial basis to such companies as Binary Arts (now called ThinkFun), Ishi Press and Hanayama Toys. He was also a keen computer programmer, using computers to help solve mathematical puzzles.

Nob also enjoyed traveling the globe to take part in the International Puzzle Party, an annual forum for serious puzzle collectors. In 2005, a year after his death, the International Puzzle Party's puzzle design competition was renamed the Nob Yoshigahara Puzzle Design Competition in his honor.

In 2003, the Association of Game & Puzzle Collectors awarded Nob with the Loyd Award, for individuals who have made a significant contribution to the world of mechanical puzzles. Nob Yoshigahara was a celebrated inventor, collector and popularizer of puzzles, but above all a dear friend of mine.

COINS IN A GLASS — A TRIBUTE TO NOB

One morning, during breakfast at the Gathering for Gardner meeting in Atlanta, Nob Yoshigahara started to improvise. He took a glass, filled it with water up to the rim, took a bunch of pennies out of his pocket, and challenged me to tell him how many coins he would be able to drop into the glass before the water overflowed. I knew something about surface tension and so I did not fall into Nob's trap. I am sure he expected me to say just a few coins, three or four perhaps, but I boldly predicted 12 pennies.

With great patience Nob devoted the next 10 minutes to dropping 59 pennies into the glass until he ran out of pennies. I gave him a few more and after the 63rd penny the water overflowed. Nob got his usual applause and won the bet. How was this possible?

Water molecules have a strong attraction for one another. At the surface the molecules are strongly attracted downward, and surface tension causes the water surface to behave like an elastic membrane. When the coins are dropped in the glass this membrane stretches above the rim forming a stretched curved surface.

LILAC CHASER ILLUSIONS — 2005

The amazing afterimage illusions below were created by Jeremy Hinton sometime before 2005. He was designing stimuli for visual motion experiments when he discovered the configuration by accident.

In a version of a program moving a disc around a central point, Hinton neglected to erase the previous disc, thus creating the illusion of a moving gap. When he observed the afterimage of a moving green disc, he optimized the effect by fine-tuning the number of discs and the foreground and background colors.

In 2005 Hinton blurred the discs, resulting in them seeming to disappear when a viewer focuses on the central cross. He tried to enter the illusion in the ECVP Visual Illusion Contest, but was disqualified as he was not registered for that year's conference. Hinton then contacted Michael Bach, who placed an animated GIF of the illusion on his illusions web page, calling it the "Lilac Chaser," and later presented a configurable Java version.

The illusion became popular on the Internet in 2005, as one of the best afterimage illusions ever.

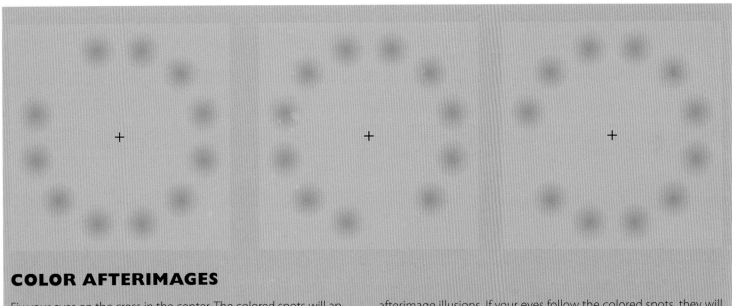

COLOR AFTERIMAGES

Fix your eyes on the cross in the center. The colored spots will appear to vanish in a few moments as the result of an effect called retinal fatigue, which occurs when the afterimage of an object cancels the stimulus of the object on the retina. But after a while you will also see slowly moving green afterimages appearing.

Variations of the "Lilac Chaser" color spots illusions created by Jeremy Hinton are among the most striking examples of color afterimage illusions. If your eyes follow the colored spots, they will remain of the same color: pink. However, if you concentrate on the black dot in the center, after a few moments the spots will all gradually disappear, to be replaced by moving green spots. The way in which our brains work is quite fascinating. There are, of course, no green spots, and the pink dots don't actually disappear

MECHANICAL PUZZLES — 2006

Jerry Slocum (1931-)

Jerry Slocum (1931) is an American historian, collector and author specializing in the field of mechanical puzzles. Prior to retiring and dedicating his life to puzzles, he worked as an engineer at Hughes Aircraft.

Containing more than 40,000 mechanical puzzles and 4,500 books, his personal puzzle collection is believed to be the world's largest.

Jerry Slocum has done more than anyone to advance the cause of mechanical puzzles. His numerous wonderful puzzle books started with *Puzzles Old & New* in 1986, the first comprehensive puzzle book to include all types of mechanical puzzles with hundreds of color illustrations of antique puzzles. Martin Gardner predicted in the introduction that the book would

"remain a classic for decades." In 1993, Slocum established the Slocum Puzzle Foundation, a non-profit organization dedicated to informing the general public on the subject of puzzles.

The first eight International Puzzle Parties took place in Slocum's home in Beverly Hills. They subsequently evolved into an annual invitation-only event circulating between North America, Europe and Asia.

Jerry Slocum has appeared on Johnny Carson's *Tonight Show*, *Martha Stewart Living*, and eight other nationwide TV shows. In 2006, he donated over 30,000 puzzles to Indiana University's Lilly Library, thereby establishing the first significant puzzle collection to appear in an academic setting.

POLYHEDRON 32 — 2002 PUZZLES

One of the more than 30,000 puzzles Jerry Slocum donated to the Indiana University, is the famous Polyhedron 32, made by Yashirou Kywayama in Japan in 2002.

Approximately 400 of the puzzles are on display in a refurbished exhibition room at the library named in Slocum's honor. Visitors to the Lilly Library can try replicas of puzzles that have absorbed people through the ages.

(Photo by Michael Taylor. Image courtesy of the Lilly Library, Indiana University.)

FAMILIES WITH CHILDREN — 2010

The following puzzle is part of Martin Gardner's *Families with Children* series of probability puzzles.

It seems likely that the probability of having a boy is the same as having a girl, but is this truly always the case?

See how these types of questions can be applied to a number of challenging probability puzzles by Martin Gardner involving conditional probabilities — that is, the probability of one event, given the occurrence of some other event. As you will see, the results can often be counterintuitive and sometimes very surprising.

TWO-CHILDREN FAMILIES

311

CHALLENGE ● ● ● ● ○ ○
REQUIRES
COMPLETED

A woman and a man each have two children. At least one of the woman's children is a girl. The man's older child is a girl. Is the probability that the woman's children are both girls equal to the probablity that the man's children are both girls?

TWO-DAUGHTER PROBLEM

312

CHALLENGE ● ● ● ● ○ ○
REQUIRES
COMPLETED

Suppose a mother is carrying two fraternal twins and wants to know the odds of having two girls.
1. What are her chances of giving birth to two girls?
2. What are the chances that at least one of the two babies will be a girl?
3. Given that one of the babies will be a girl, what are the chances that both of the twins will be girls?

THREE-CHILD FAMILIES

313

CHALLENGE ● ● ● ● ○ ○
REQUIRES
COMPLETED

What is the probability of a family with three children having at least one girl?

TWO FAMILIES WITH EIGHT CHILDREN

314

CHALLENGE ● ● ● ● ○ ○
REQUIRES
COMPLETED

There are two families, with eight boys in one and eight girls in the other. Since the probability of a child being a boy or girl are about the same, would you think in families of this size that four girls and four boys should be much more likely? Exactly how do the probabilities of a family having eight girls and a family having four girls and four boys compare?

THE BOY BORN ON TUESDAY — 2010

At the Gathering for Gardner conference in Atlanta 2010, Gary Foshee, a very creative puzzle designer, gave a lecture consisting of the following three sentences:

"I have two children.
One is a boy born on a Tuesday.
What is the probability that I have two boys?"
Gary added, deadpan as always:
"The first thing you may think is, what has Tuesday got to do with it?
Well, it has everything to do with it."
And then he stepped down from the stage.

After the Gathering, Gary's "Tuesday Boy" problem became a widely discussed controversial topic on blogs around the world.
The puzzle is a variant of Martin Gardner's series of "Boy or Girl Paradox" puzzles, summarized in this book (see earlier in this chapter).
Gary's addition of Tuesday created endless polemics and many interpretations. What is your solution to Gary's problem?
The main issue became how to properly interpret Gary's presentation. Let's try to clear some of the issues involved.
The basic problem, disregarding Tuesday, can be interpreted as:

Of all the families with one boy and exactly one other child, what is the probability of these families to having two boys? The possible outcomes of two children:

| Boy – Girl | Girl – Boy | Boy – Boy | Girl – Girl |

One of the four possible outcomes is two boys, the probability of which is 1/3 (the two girl outcome can be eliminated).

So what does Gary say after all the polemics?
"There is definitely an argument to be made based on choice.
My solution was based on set theory.
Look at the entire set of all families with two children.
Then look at a subset: those with two boys.
Then look at another subset: those with a boy born on Tuesday.
If you look at it that way, then 13/27 is the correct answer.
But if you start putting in other factors about how the children were chosen, and from which set, then yes, there is an argument the answer could be different.
Indeed, it's a very tricky and controversial puzzle."

I loved Gary's problem and the controversy it created. On the next page I have tried to follow Gary's interpretation and solution to his puzzle.

METHOD 1

Note: Gary didn't say that only one boy was born on a Tuesday. He obviously meant "at least one."

Summing up, there are 7+7+7+6 = 27 different equally likely combinations of children with specified gender and birthdays, and 13 of these combinations are two boys. The answer is 13/27, which is quite different from 1/3.

Child 1: BOY

Child 2: GIRL

	M	T	W	T	F	S	S
M		1					
T		2					
W		3					
T		4					
F		5					
S		6					
S		7					

Boy born on Tuesday.
Girls born on any day of the week.
There are seven
different possibilities.

Child 1: GIRL

Child 2: BOY

	M	T	W	T	F	S	S
M							
T	8	9	10	11	12	13	14
W							
T							
F							
S							
S							

Girls born on any day of the week.
Boy born on Tuesday.
There are seven
different possibilities.

Child 1: BOY

Child 2: BOY

	M	T	W	T	F	S	S
M		1					
T		2					
W		3					
T		4					
F		5					
S		6					
S		7					

Boy born on Tuesday.
Boys born on any day of the week.
There are seven
different possibilities.

Child 1: BOY

Child 2: BOY

	M	T	W	T	F	S	S
M							
T	8	2	9	10	11	12	13
W							
T							
F							
S							
S							

Boy born on any day of the week.
Boy born on Tuesday.
There are six possibilities,
one of them — when both boys
are born on an Tuesday —
was counted before.

METHOD 2

Another visual way of calculating the answer to the Tuesday Boy problem of Gary, is credited to Bill Casselman in *Science News Magazine*.

The total number of families is 27. The number of families in which both children are boys is 13 as shown. So the probability is 13/27.

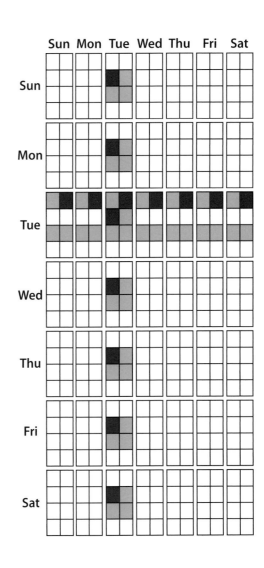

SLIDING BLOCK PUZZLES

Sliding block puzzles are puzzles in which a player has to slide pieces along specific paths (usually flat pieces and on a board) to form a particular final configuration.

As we saw already in Chapter 5, the 15 puzzle is the original form of the sliding block puzzle. It was invented by Noyes Chapman and created a puzzle craze in 1880. In the sliding block puzzle players are prohibited from lifting any piece off the board, as distinct from other tour puzzles. This characteristic distinguishes between sliding puzzles and rearrangement puzzles, and makes finding moves and the routes created by the moves within the two-dimensional confines of the board an important part of the solution to sliding block puzzles. Sliding puzzles are essentially two-dimensional in nature, despite the fact that the sliding is made possible by mechanically interlinked pieces. This kind of puzzle has now been computerized and can be played on line.

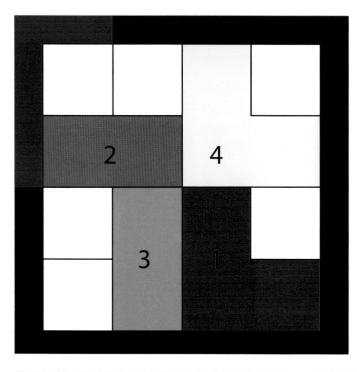

Simplicity sliding-block puzzle prototype by Oskar van Deventer

SIMPLICITY SLIDING PUZZLE — 2011

In the December 2011 issue of *The Economist,* a question was posed: What is the hardest possible simple sliding-block puzzle?

Two inventors are at the forefront of inventing the best and really hard sliding block puzzles — James Stephens and Oskar van Deventer.

According to Ed Pegg Jr, the Simplicity puzzle of James Stephens, proto-typed by Oskar van Deventer, deserves the label "hardest simple sliding-block puzzle." The object is to move the red block to the upper left corner. The puzzle can be solved in 18 moves.

OSKAR'S TWISTY PUZZLES

Starting from the '80s, inspired by the enormous success of the Rubik's Cube, a new category of mechanical puzzles appeared in an ever-increasing number and variety. Oskar van Deventer, a Dutch designer, is the genius behind this category, inventing and creating hundreds of twisty and other puzzles.

One of Oskar's latest creations is the "Over The Top" puzzle. This twisty 17x17x17 mechanical puzzle debuted at the New York Puzzle Party Symposium in 2011, which was three-dimensional printed at Shapeways.com. Consisting of 1,539 individual plastic pieces, it is a giant in every sense compared to the mass-market Rubik's Cube.

An electrical engineer by training, van Deventer designed the Over The Top puzzle in 2010, taking over 60 hours to develop it. After Shapeways three-dimensionally-printed the pieces, van Deventer had to sort them, color them individually, and finally assemble them, all by hand, adding an additional 15 hours of labor to the final puzzle. Van Deventer offers a peek inside the mechanics of the Over The Top puzzle and shows many of his other creations on his YouTube channel.

Oskar van Deventer with his "twisty puzzles"

"When I heard about the world records being set for Rubik's Cubes, like the 7x7x7, 9x9x9, and 11x11x11 created by Panagiotis Verdes from Greece, and Lie from China for his 12x12x12, I wanted to try to set a new record myself," Van Deventer said in an interview with Jane Doh for *Wired*. "With sponsorship and prototyping help from my good friend Claus Wenicker, I began designing and testing a number of prototypes, and my third attempt was printed successfully with Shapeways."

The 17x17x17 Over The Top puzzle is by no means the end of his impressive carreer. His Iconosaix puzzle for example, a face-turning regular solid with 20 triangles, was recently produced by MF8.

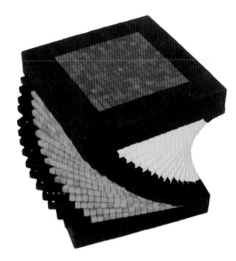

Over The Top puzzle

COMBINATORIAL PUZZLES

A combinatorial puzzle, also called a sequential move puzzle, comprises an assembly of parts that can be twisted and turned into a number of different combinations. Beginning with a random setting, the puzzle is successfully completed when a pre-defined combination is arrived at, which is usually something like groups of identical colors or sequential numbers.

The best known puzzle of this type is the Rubik's Cube, a cubic puzzle in which each of the six faces can be turned independently. Each face is a different color, with the nine parts on each face having the same color. They are then manipulated until the colored parts are distributed randomly, and the puzzle is solved when each of the six faces are again of the same color.

The rules governing how the pieces can be rearranged are generally defined by the way in which the puzzle is constructed, which results in certain restrictions with regard to possible combinations. With the Rubik's Cube, for example, a wide range of combinations can be achieved by randomly placing the colored stickers on the cube; more than can actually be accomplished by rotating the cube's faces.

ANTONIO PETICOV (1946)

Antonio Peticov is a Brazilian mathematical artist, born in 1946. With his varied and diverse creations of high technical and esthetic quality, he opens a colorful world, a hidden universe of subtle mathematical content and inspiration.

MATHEMATICS AND ART — 2012

Although modern opinion holds that mathematics and art are two unrelated fields, there are many visual artists for whom mathematics is the focal point of their work. Mathematical artists have used various elements, including polyhedra, tessellations, impossible figures, Mobius bands, distorted or unusual perspective systems, fractals, etc.

The world of mathematical art, however, is much bigger and more varied than most of us imagine. A considerable number of contemporary artists consider mathematics — from Fibonacci numbers and the digits of Pi to tetrahedra and Mobius strips — to be the inspiration for their work.

O Mestrado

The First Impressions

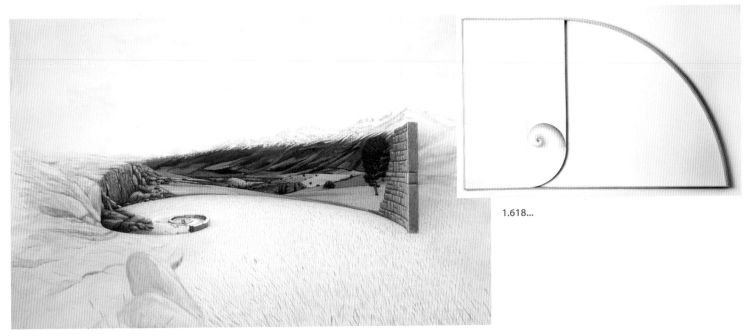

The Well

1.618...

Reincarnation

TEJA KRASEK

Holds a B.A. degree in painting from Arthouse — College for Visual Arts, Ljubljana, and is a freelance artist who lives and works in Slovenia. Her theoretical as well as practical work is especially focused on symmetry and mathematical concepts as a linking concept between art and science.

Krasek's work concentrates on melding art, science, mathematics and technology. She employs contemporary computer technology as well as classical painting techniques.

BIOGENESIS

Slovenian artist Teja Krasek explores the fractal boundaries between art, science, and mathematics.

"Art and mathematics blend and swirl. They conjure images and emotions in unexpected and profound ways — and allow us to transcend space and time."

Teja Krasek, **Earth 3000 A.D.**

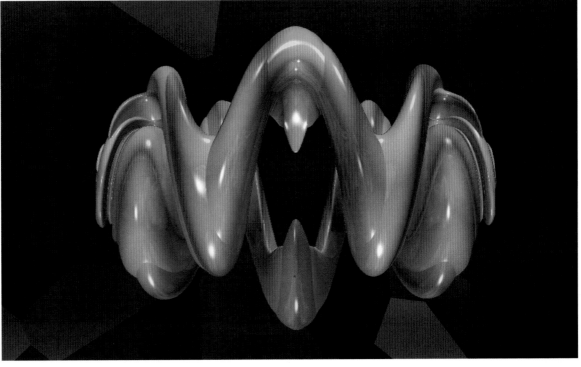

Teja Krasek, **Halloween Torus**

CHAPTER

10

ANSWERS

1 Eight different size squares as shown:

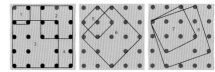

2 There is no problem on the contrary. Would you have any problems, if Julia Roberts were your blind date?

3-4 After you have performed the two tests, obviously, there must be some meaningful difference between the two seemingly random number patterns, since in the second test your performances were so much better. What can it be?

It was not something you were aware of while playing the game, but your subconscious was. It discovered the secret helping you to solve test 2 more efficiently while your conscious was not yet aware of it. It discovered the pattern unknown to your conscious while performing the tests. You may have discovered the secret only after you did the tests..

The numbers in the second grid are grouped in such a way that your eyes follow a repeating pattern. The number panel in the second test is divided into four quarters along the middle of the sides. Thus number 1 is in the upper right quarter, number 2 in the lower left quarter, number 3 in the upper left corner and number 4 in the lower right corner, and the same cycle is repeated by number 5 again from the upper right quarter, until you come to number 90. This procedure narrows the search for the next number to a quarter of the search area and makes it more efficient.

Your subconscious discovered the secret and made good use of it, while you were not yet consciously aware of it. This is quite an impressive demonstration of the power and creativity of your subconscious mind, solving a problem while your conscious is not aware of it, and a good indication of how your mind solves problems in general.

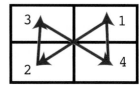

5 By the end of the 16th day the snail reached the height of 80 cm. Near the end of the 17th day it reached the top of the window at 90 cm.

6 1 + 2 + 3 - 4 + 5 + 6 + 78 + 9 = 100
12 + 3 - 4 + 5 + 67 + 8 + 9 = 100
123 + 4 - 5 + 67 - 89 = 100
A large number of variations of the problem have been devised, many of which allow operations other than addition and subtraction.

7

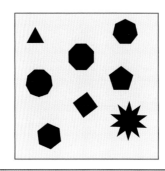

8 With an infinite number of birds randomly distributed along the wire, 50 percent of the birds will be watched by one of their neighbors, and another 25 percent will be watched by two of their neighbors, while 25 percent will be left unwatched.

The situation is similar to tossing a coin twice: 50 percent chance of one head, 25 percent chance of two heads, and 25 percent chance of two tails.

9 A-2, B-3, C-2

10

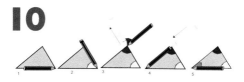

The sequence clearly shows that the rotation of the pencil described in sequence the three angles of the triangle, with the pencil ending on the line where it started, but pointing in the opposite direction. This is convincing evidence that the three angles add up to a straight angle.

11 It is possible to visit all floors of the building. The maintenance man makes 30 trips to visit each floor, pressing the "UP" button 18 times and the "DOWN" button 12 times as shown.

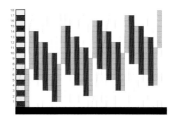

12 The circle is the best enclosure.

13 The highest point is shown. It is the only plank which is slanted and not parallel to the ground like all the rest.

14

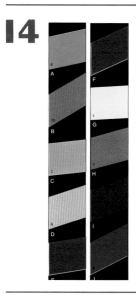

15 The rule was 1 block straight, turn right 2 units, turn right 3 units, and so on, until 9 units, when the sequence started all over again, until the point the rule changed.

16

1-17	8-5	15-2	22-23
2-14	9-4	16-20	23-22
3-3	10-8	17- 25	24-12
4-24	11-13	18-16	25-6
5-16	12-1	19-19	
6-18	13-10	20-21	
7-7	14-11	21– 9	

17

$(8 + \pi)$ meters:

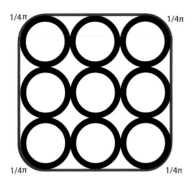

1/4 π   1/4π

1/4π   1/4π

18

Cake: $1.75
Ice-cream: $0.75

19

She was wrong. $1200 is 125 percent of $960 – she made a $240 profit. $1200 is 80 percent of $1500 – her loss was $300. The combined sale had a loss of $60.

20

There are eight different ways to seat men and women in a row without women having to sit next to each other:

W M M W, W M W M, M W M W, W M M M, M W M M,
M M W M, M M M W, M M M M

When n is 1, 2, 3, 4, 5… the answers are 2, 3, 5, 8, 13… and so on, interestingly enough, according to the Fibonacci number sequence.

21

Three identical squares inscribed in an equilateral triangle are dissecting it into 28 regions.

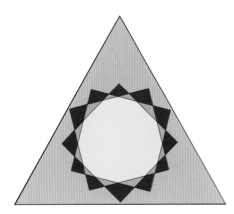

22

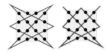

Four solutions with the smallest number of intersections

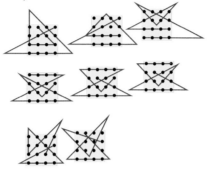

Two symmetrical solutions

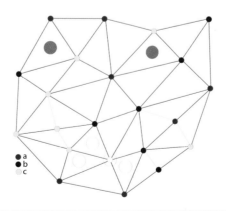

Among the 14 different possible solutions (not counting rotations or reflections as different) there are two symmetrical patterns and four solutions having the smallest number of intersections (2).

23

At the outset two complete triangles were created, but at the last dots a third complete triangle seems unavoidable, no matter how the colors are chosen. This will always happen no matter how the triangles are colored. This is the conclusion of Sperner's lemma for triangles: if there is an odd number of complete edges on the boundary, then there is an odd number of complete triangles. If there is an even number of complete edges on the boundary, then there is an even number of complete triangles.

An edge among 'a' and 'b' is a complete edge.

- a
- b
- c

24

The probability of rolling either of them on any given toss is 1/6 + 1/6 = 2/6 or 1/3.

25

The information is inconclusive. Seen from above the rooftop may be convex as rooftops usually are, but it also be concave.

26-28

Puzzle 1 – 23 squares (below)
Puzzle 2 – 47 squares
Puzzle 3 – 16 squares

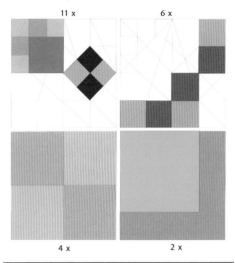

11 x   6 x

4 x   2 x

29

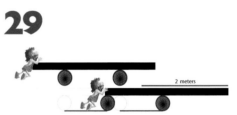

2 meters

Surprisingly enough the load always moves farther than the rollers. If the roller rolls through one revolution, it advances forward a distance equal to "pi" times its diameter, while the load it supports moves forward a distance twice that distance as shown above. This is because the load moves forward relative to the roller at the same time that the roller is moving forward on the ground. If the rollers have a circumference of one meter, then the slab will move forward two meters per revolution. This is called the "roller and slab theorem."

Evidence of wheeled vehicles appears from the mid-fourth millennium BC, near-simultaneously in Mesopotamia and Central Europe, so that the question of which culture originally invented the wheeled vehicle remains unresolved and under debate. The earliest well-dated depiction of a wheeled vehicle (a wagon—four wheels, two axles), is on the Bronocice pot, a ca. 3500–3350 BC clay pot excavated in a Funnelbeaker culture settlement in southern Poland.

30

Labyrinth is generally synonymous with maze, but many contemporary scholars observe a distinction between the two: maze refers to a complex branching (multicursal) puzzle with choices of path and direction; while a single-path (unicursal) labyrinth has only a single, non-branching path, which leads to the center. A labyrinth in this sense has an unambiguous route to the center and back and is not designed to be difficult to navigate.

31

1– Leibniz was wrong.
A total of 12 can occur in only one way (red die=6, blue die=6). On the other hand, a total of 11 can occur in two ways (red die=6, blue die=5, or red die=5, blue die 6).

Thus their probabilities are different 1/36 and 2/36 respectively, as you can easily read from the table.
2 – With a pair of dice you can't get a 1. There are 6 possible even numbers: 2 - 4 - 6 - 8 - 10 - 12. There are 5 odd numbers: 3 - 5 - 7 - 9 - 11. There are 18 ways of getting an even number and 18 ways of getting an odd one, as shown in the table. So the chances are even.

32

The sums from three through 18 can come up in 6 x 6 x 6 = 216 different ways when throwing three dice. Seven can come up in 15 different ways (7 percent), and 10 in 27 ways (12.5 percent).

33

The probability that both of you will throw the same number is one in six. Therefore the probability that one of you will throw higher than the other is five in six. This is halved to give the probability that one of you will get a higher number than the other: 5/12.

34

You may not throw a six in six throws of a single die, so obviously, the probability is not one or 100 percent. In fact, you have to calculate the probability of not throwing a six, six times in a row. The probability of not throwing a six in a throw is 5/6, and in six throws: 5/6 x 5/6 x 5/6 x 5/6 x 5/6 x 5/6 = 0.33. Therefore, the probability of throwing a six in six throws is high: 1-0.33 = 0.67 or 67 percent.

35

6/6 x 5/6 x 4/6 x 3/6 x 2/6 x 1/6 = 0.015

36

Two and a half revolutions counterclockwise.

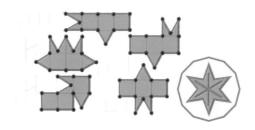

37

1 – up, 2 – down, 3 – up, 4 – down

38

1. There is a large number of different four-unit area polygons that can be made with the Egyptian rope. Elton M. Palmer, from Oakmont, Pennsylvania, ingeniously correlated this problem with polyominoes, specifically to tetrominoes. Each of the five tetrominoes can be the basis for a large number of solutions, simply by adding and subtracting triangles to accommodate the 12 equal lengths. Some solutions are shown using the five different tetrominoes.
2. Any area from 0 to 11.196 can be encompassed by the Egyptian rope. Eugene J. Putzer, Charles Shapiro and Hugh J. Metz suggested a star configuration solution as shown. By adjusting the width of the star points the largest area is that of a regular dodecagon.

39

It is sufficient to remove five triangles out of the nine to eliminate all the red triangles. The quickest descending sequence of the remaining red triangles is by removing triangles 1, 2, 3, 4, and 7, as shown. There are 120 triangles of all sizes altogether: 59 pointing up and 61 pointing down.

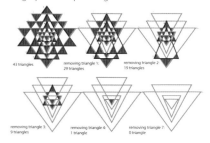

40

16,807 unit measures of flour: 7 x 7 x 7 x 7 x 7. The ancient Egyptians pushed mathematics to a very advanced level. Ahmes' puzzle may be the world's oldest mathematical puzzle, found in the ancient Egyptian "Rhind Papyrus", written by the scribe Ahmes (c.1650 BC). A great number of variations of the same puzzle can be found in recreational math literature. The solution is the sum of geometric progression of five terms, of which the first term is 7, and the multiplier is also 7.

41

Only one. All the others were coming from St. Ives. Ahmes' puzzle has inspired many variations, among them the St. Ives riddle. Leonardo of Pisa (Fibonacci) published the rhyme in his book *Liber Abaci in 1202*, though it is unclear how he got access to the Rhynd papyrus at that time.

42

Even if it is red's move, red cannot stop blue from winning.

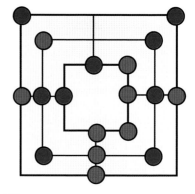

43

1. Pythagoras' proof:
The yellow square in the first diagram is equal to the sum of the two yellow squares in the second diagram, convincingly proving the theorem.
2. Leonardo's proof:

The interrupted lines divide Leonardo's diagram into four congruent quadrilaterals.

3. Baravalle's proof:

The fourth step can be explained by Cavalieri's theorem: if a parallelogram is transformed by transformation without changing its hight and base, its area does not change.

44

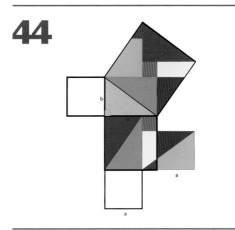

45

Oedipus solved the riddle by answering: Man - who crawls on all fours as a baby, then walks on two feet as an adult, and then walks with a cane in old age.

46

The numbers of the Tetraktys can be arranged in $(10!)/(2 \times 3)$, different ways, which is 604,800 ways.

47

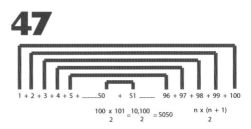

$$1 + 2 + 3 + 4 + 5 + \ldots\ldots 50 + 51 \ldots\ldots 96 + 97 + 98 + 99 + 100$$

$$\frac{100 \times 101}{2} = \frac{10,100}{2} = 5050 \qquad \frac{n \times (n+1)}{2}$$

It took Gauss just a few seconds to discover the pattern and conceptualize that there are 50 sums of 101 in the sequence of addition, which gives a total of 5050.

He did not need a calculator or a piece of paper to come to this result. Gauss' feat works for any number n, not just for 100, according to the general pattern: $1+2+3+ \ldots + n = n(n+1)/2$.

It is interesting to note that this general formula is also the formula for triangular numbers. Babylonian cuneiform tablets show that the formula for deriving triangular numbers has been known since antiquity. For any number n, its triangular number (or the sum of the first n integers) can be calculated as $n(n+1)/2$, which is exactly the formula which Gauss used for n=100, a formula that was visualized by ancients by figurate numbers.

48

The third stick is broken in Golden Ratio.

49

Zeno's conclusion was that it would take Achilles an infinite amount of time to catch the tortoise. Achilles gets closer and closer, but he never catches up with the tortoise; his journey is divided into an infinite number of pieces. Before a moving object can travel a certain distance, it must travel half that distance. Before it can travel half the distance it must travel a quarter of the distance, and so on forever. The original distance cannot be traveled, and therefore motion is impossible.

Since we know motion is possible, the first fault in Zeno's race is the assumption that the sum of infinite number of numbers is always infinite. This is wrong. The infinite sum of $1 + 1/2 + 1/4 + 1/8 + 1/16 + 1/32 + 1/64 \ldots$ is equal 2, which is known as a geometric series. (A geometric series is a sequence which begins with one and its successive terms are those multiplying the previous term by a fixed amount, say x, which in this case is 1/2. Infinite geometric series are converging to a finite number when x is less than one.)

The distance that Achilles travels and the time it takes him to reach the tortoise can both be expressed as an infinite geometric series with x less than one, and so the total distance Achilles traverses to catch up with the tortoise is not infinite. The same goes for the time required. Suppose Achilles gave a 10-meter head start to the tortoise, and he runs at one meter per second, ten times faster than the tortoise. It takes Achilles five seconds to cover half this distance. Half the remaining distance will take him 2.5 seconds, and so on, covering the total distance in a finite 10 seconds according to the aforementioned infinite geometric series. By that time the tortoise has moved to the point of 11 meters. We know that Achilles should pass the tortoise at a point of 11.11 meters from Achilles' starting point, taking him 11 seconds to reach that point, winning the elusive race. The usefullness of Zeno's paradoxes is that they gave birth to the idea of convergent infinite series, crystallizing a number of mathetical concepts, the main one among these, the notion of limits. Interest in paradoxes has been strongly revived during the Renaissannce when more than five hundred collections of paradoxes are known to have been published.

50

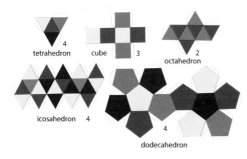

tetrahedron 4   cube 3   octahedron 2

icosahedron 4   dodecahedron 4

51

Solid	Vertices (V)	Edges (E)	Faces (F)	V - E + F
Tetrahedron	4	6	4	4-6+4
Cube	8	12	6	8-12+6
Octahedron	6	12	8	6-12+8
Icosahedron	12	30	20	12-30+20
Dodecahedron	20	30	12	20-30+12

52

Puzzle 1 – 60 different ways corresponding to the five positions of each of the 12 faces of the dodecahedron.

Puzzle 2 – The missing colors are 1 - 2 - 3 - 4, respectively.

Puzzle 3 – The cross-sections can be triangles, squares, rectangles, pentagons, hexagons and decagons.

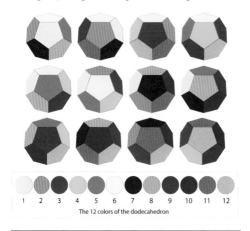

1 2 3 4 5 6 7 8 9 10 11 12
The 12 colors of the dodecahedron

53

The tile is dissected into 16 equilateral triangles and 32 isosceles triangles with angles 15, 15, and 150 degrees.

A quarter of these equal the area outside the dodecagon.

54 To solve the thereom it is sufficient to prove it for one-quarter of the circle as shown: The crescent between the semi-circle with the center P, and the quarter circle with the center O, passing through A and B, has the same area as the right-angled triangle.

OA = r, then AB = r √2 Then the area of the one-quarter of the circle is: C= 1/4 r² π

The area of the semi-circle of diameter AB is:
D= 1/2 (r√2/2)² π = 1/4 r² π

Thus, C=D, and since the area of the crescent (i.e. the CIRCULAR SEGMENT) is part of C as well as part of D, it follows that the area of the crescent (i.e. the LUNE) is the same as that of the triangle OAB, which proves the theorem.

The sum of the areas of the two red lunes equals the area of the blue triangle.

The sum of the areas of the four red lunes equals the area of the blue square.

55

The area of the black hexagon equals the sum of the areas of the six red lunes and the areas of the two red semicircles.

The area of the hexagon + 3 areas of circle of diameter AB = large circle + areas of the six crescents.

The area of the hexagon equals the area of the circle of diameter AB plus the area of the six crescents, as show in the sequence illustrated above.

56 The abacus may be operated in one of two ways. The groove (line) may hold as many pebbles (counters) as the number base. We then empty the groove (line) when full, put one on the next groove (line) and reject the remainders.

Alternately, the groove (line) may accommodate one less than the base. We then put the next pebble (counter) added to a full groove (line) on the next one before emptying it. For example, what number is represented in the picture?

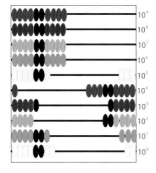

57 The Riemann hypothesis implies results about the distribution of prime numbers. Along with suitable generalizations, it is considered by some mathematicians to be the most important unresolved problem in pure mathematics .

Many other problems in number theory, such as those involving the distribution of primes, have been shown to be related to the Riemann hypothesis, so answering this would provide insight into a whole range of other problems! For instance, the Prime Number Theorem gives a good approximation of how many primes are less than a given number, but the Riemann hypothesis is related to a conjecture about how good that approximation is!

58 For centuries searching for patterns in primes was a daunting task. The last prime smaller than 1000 is 997. The next prime is 1009, at a distance between the two of 12 numbers, which should be visualised by the color violet.

59 The moment of a force about a point is equal to the magnitude of the force multiplied by its perpendicular distance from the point. On a uniform rod, unit weights can be balanced equidistant from the fulcrum.

The lever can change mechanical energy involving a small force into mechanical energy involving a large force. A heavy load is lifted a small distance by an effort five times less, which is the mechanical advantage of

the machine. As the effort is five times farther from the fulcrum than the load, it will have to move five times as much when the load is raised.

You push the handle a long way and make the blade move a short way, but with an increased force, so that you can lift a heavier load of soil. The spade turns, or pivots low down on the ground.

60 One can determine the density of an object (O), by comparing its weight to that of the water it displaces in a bathtub.The weight of the water, which has the same volume as "O' is called O's buoy-ancy, and the ratio of O's weight to that of the displaced water is called O's "specific gravity."

Step 1 – The block of gold weighs exactly as much as the disputed crown.

Step 2 – The same weighing is repeated with both objects immersed in water and the displaced water from both objects measured as shown. If the displaced water were the same in both cases the crown would be proven solid gold. But this was not the case. The crown displaced more water, proving that the crown was alloyed with a metal less dense than gold, the volume of which is greater than the volume of solid gold. The crown turned out to be a fake, and Archimedes'fame continued as he made many other discoveries. The discovery of the fact that a body in liquid gains a lift (i.e. it becomes lighter) due to the upward force called buoyancy, which is equal to the weight of the dis-placed liquid, established the science of hydrostatics. Ever since Archimedes'discovery, this method has been used to assay metals, identify jewels and measure the density of materials. We can compare the weight of a substance with the weight of an equal volume of water by means of Archimedes'principle. The ratio of these weights is called the specific gravity of the body:

$$\text{specific gravity} = \frac{\text{weight of object}}{\text{weight of equal volume of water}}$$

61 In a system of a single rope and pulleys, when friction is neglected, the mechanical advantage gained can be calculated by counting the number of rope lengths exerting force on the load.

In our example the mechanical advantage is 6, and the man can exert sufficient force to lift the big load. The force on the load is increased by the mechani-cal advantage; however the distance the load moves, compared to the length the free end of the rope moves, is decreased in the same proportion.

In equilibrium, the total force on the pulley must be zero. This means that the force on the axle of the pulley is shared equally by the two lines looping through the pulley.

The pulley simply allows trading force for distance: you pull with less force, but over a longer distance.

62 Surprisingly, already in the seventh group we can find the complete set of the first 10 digits: (0, 1, 2, 3, 4, 5, 6, 7, 8, 9)?

π = 3.1415926535 8979323846 2643383279 5028841971 6939937510 5820974944 5923078164

63 The astonishing and truly counterintuitive answer is that the circumscribed and inscribed circles will never become infinitely large (or infinitely small) but in both cases will have a finite limiting value. In case of the circumscribed circle and polygons the radius of the limiting size of the biggest circle is 8.7 units (and 1/8.7 in case of the inscribed circles and polygons). In both cases the limiting polygons will have an infinite number of sides, becoming circles. It is interesting to note that in 1940, when Kasner and Newman first gave the answer to this problem, their published result was 12 units, which was believed to be true until C.J. Bouwkamp in 1965 provided the right answer.

The specially designed beautiful graphics convincingly visualizes the concept of the problem.

The white areas show the finite growth domains with the size of the limiting circles and the limiting polygons of an infinite number of sides.

64

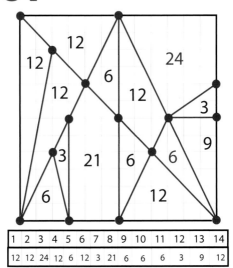

1	2	3	4	5	6	7	8	9	10	11	12	13	14
12	12	24	12	6	12	3	21	6	6	6	3	9	12

65 In all solutions:
1– pieces 6 and 7 are duplicated (in our color pattern they are also of the same color (red and black).
2 – and pieces 1-2; 9-10; and 11-12 form pairs always found together (blue-black; red-black).

66 Earth's circumference at the Equator is 40,075.16 km. It is believed that Eratosthenes' value is between 39,690 km and 46,620, not a bad result at all.

67 The four shapes on the left have the same area. The shapes on the right have equal perimeters. The two circles are identical and have the same area and perimeter.

The same perimeters in the other three shapes on the right all enclose a smaller area than the other three shapes on the left.

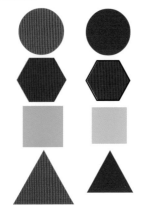

68 The earliest attempt to make practical use of steam was Hero's door-opening blueprint from 50 AD. Simple mechanical principles, utilizing chains, pulleys, levers and containers with air and water did the "magic": the priest lit a fire on the altar. The air in the two containers is heated, expanding and pushing the water from the lower spherical container over a siphon into the hanging basket over a pulley. The descending basket will start pulling the ropes or chains, thus activating the hinges and the doors will "magically" open. When the fire is extinguished and everything cools down, the doors will automatically close again, due to the action of the counterweight at the lower right. (From the Latin edition of *Hero's Spiritalium liber*, 1575)

69 Josephus and his accomplice occupy places 31 and 16 to survive.

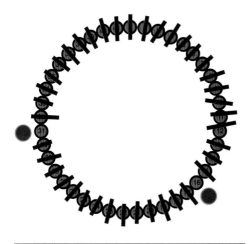

70 The three lowest rings in the necklace are Borromean Rings, three mutually interlinked rings, named after the Italian Renaissance family who used them on their coat of arms.

If you cut any of the two rings in the second row from bottom you will be able to separate the necklace into the maximum possible number of parts – three parts consisting of 1 – 1 – 9 rings.

71 In simpler English it says: Diophantus's youth lasted 1/6 of his life. He had the first beard in the next 1/12 of his life. At the end of the following 1/7 of his life Diophantus got married. Five years from then his son was born. His son lived exactly 1/2 of Diophantus's life. Diophantus died 4 years after the death of his son. How long did Diophantus live? Here is the equation to reflect the several ages of Diophantus:
$1/6x + 1/12x + 1/7x + 5 + 1/2x + 4 = x$
So the solution (x) is 84 years.

72 1 – infinity, 2 – approaches (in value), 3 – and so on, 4 – equal to or less than, 5 – equal to, 6 – therefore, 7 – summation, 8 – equal to or greater than, 9 – smaller than, 10 – square root, 11 – similar to (proportional), 12 – corresponds to, 13 – intersecting circles, 14 – plus or minus, 15 – approximately equal to, 16 – identically equal to, 17 – not equal to, 18 – diameter, 19 – perimeter, 20 – tangent, 21 – radius, 22 – sector, 23 – circular segment, 24 – scalene triangle, 25 – rhombus, 26 – parallelogram, 27 – trapezoid, 28 – diamond, 29 – equilateral triangle, 30 – right-angled triangle, 31 – circle area, 32 – isosceles triangle, 33 – acute angle 34 – right-angle, 35 – obtuse angle, 36 – congruent, 37 – tetrahedron, 38 – parallelopiped, 39 – cube, 40 – sphere, 41 – cone, 42 – octahedron, 43 – regular pentagon, 44 – regular hexagon, 45 – regular septagon, 46 – regular octagon, 47 – cylinder, 48 – pyramid,

49 – regular nonagon, 50- rectangular prism, 51 – semicircle, 52 – parallels, 53 – intersection, 54 – secant to a circle, 55 – arc, 56 – central angle, 57 – inscribed angle, 58 – circumsized circle, 59 – inscribed circle, 60 – perpendicular, 61 – factorial, 62 – symbol for pi, 63 – percent, 64 – vector, 65 – because, 66 – end of proof, 67 – natural numbers, 68 – integers, 69 – a is not an element of b, 70 – there exists, 71 – line segement AB, 72 – line AB

73 First, he takes the goat over and returns. He then takes the wolf over and returns with the goat. Next, he takes the cabbage over. He returns and finally takes the goat over.

74 This is how the trips can be organized. It takes only seven trips. All four of us begin at the spaceport: Rigellian, Denebian, Terrestrial and me.
1– I take the Denebian up to the liner.
2– I return alone.
3– I take the Rigellian up to the liner.
4– I return with the Denebian.
5– I take the Terrestrial up to the liner.
6– I return alone.
7– I take the Denebian up to the liner.
And we all go through the airlock, into the tender care of the beautiful hostesses.

75

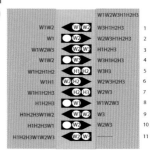

H HUSBANDS   W WIVES

76 S-Soldiers - B-Boys

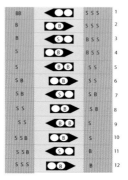

77

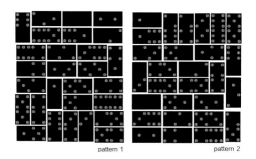

78

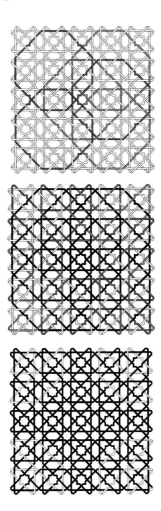

pattern 1   pattern 2

79 January: 1, February: 1, March: 2, April: 3, May: 5, June: 8, July: 13, August: 21, September: 34, October: 55, November: 89, December: 144.
The now familiar sequence of numbers shows the number of rabbit pairs in each month, starting from January (when the first pair was introduced) till December. The total number of pairs at the end of the year is 144.

80-81 Every natural number may be expressed as the sum of distinct non-consecutive Fibonacci numbers in more than just one way. For example, 232 is demonstrated below.

1 1 2 3 5 8 13 21 34 55 89 144 233
1   3   8   21   55   144 = 232

By definition, the first two Fibonacci numbers are 0 and 1, and each subsequent number is the sum of the previous two. Some sources omit the 0 and begin the sequence with two 1s. If we take a calculator and look at the list of decimals, we can see that the numbers are getting closer and closer to each other approaching a limit that is really astonishing. The truly amazing result is Phi, the Golden Ratio.

Who could have believed that this innocent looking line division, which Euclid defined for purely geometrical purposes, and the man-invented number series would have such important consequences in nature and the whole of mathematics and science? The Golden Ratio, or Phi, plays an important role as a fundamental building block in nature. We have mentioned that we can form similar recurrence sequences using different starting numbers. For example the Lucas sequence starts with 2 and 1. The resulting sequence is: 2, 1, 3, 4, 7, 11, 18, 29, 47, 76, 123 ... having little in common with Fibonacci

numbers. Apart from the first three numbers, none of the numbers of the Lucas sequence are Fibonacci numbers. Still, is there any relationship between Fibonacci and Lucas sequences or, for that matter, any other possible recurrence sequence? Yes, there is. If we create the procedure described above with the numbers in Lucas sequence or, for that matter in any recurrence sequence, they are all going to approach the Golden Ratio, the ultimate in the amazing coincidences involving Golden Ratio, Fibonacci numbers and Pythagorean theorem.

82 The next square in the sequence will be the 14th Fibonacci Number 377.

83 The pattern is composed of 25 interlocking closed loops of three shapes in different orientations as shown.
shape 1 – nine identical shapes
shape 2 – 12 identical shapes in four different orientations
shape 3 – four identical shapes in four different orientations

84 A perpetual motion machine is impossible, according to the Law of Conservation of Energy, which is fundamental for science.

Leonardo's design is one of the most ancient gravitational perpetual motion machine concepts. The idea is that once started, the balls rolling in compartments will produce a greater moment about the center on the descending side (farther from the center) of the wheel than the moment of those on the ascending side (nearer to the center than those on the descending side). This will cause the wheel to rotate in a clockwise direction. As successive weights are brought over the top, according to the theory they would fall down to the outer position and so keep the wheel turning. However, if we give the wheel a complete turn, so that each ball returns to its original position, the whole work done by the balls will, at the most, equal the input done on the wheel. The system can never gain energy during its motion. The wheel will never go on turning; it will only swing a bit and settle in a balanced position. Its motion can be explained by the theorem of moments.

85 Gamow's idea is based on the rotational symmetry of the numbers "6" and "9," attaching sixes to the spokes of a wheel. Once set in motion the weight of nines appearing on top would keep the wheel in motion eternally. Regretfully, mathematical ideas do not always translate into physical reality. "Impossible," you will say and, all attempts to build a perpetual motion machines have failed. One has never been constructed. But it was and still is in fashion to try to build perpetual motion machines. Often they are very complicated fakes ending up in shop windows to attract attention.

86

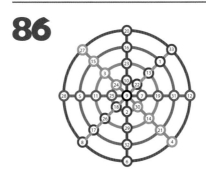

87 The final number of grains on the chessboard is the result of a geometrical progression.
$2^{64}-1 = 18,446,744,073,709,551,615$

A geometric progression is a sequence of numbers where each term after the first is found by multiplying the previous one by a non-zero number called the common ratio. The sum of terms of a geometric progression is known as a geometric series.

88 One of the many possible solutions

89 The ingenious inner structure of the egg is basically very simple: a small viscous liquid-filled cylinder is put in a slanted position inside the egg. A small heavy piston is moving slowly inside the cylinder. Its descent, from its highest to its lowest position, when the egg is in a vertical position, is timed for a period of about 70 seconds. In the middle 10 seconds of this descent, the egg will be balanced on its thin end.

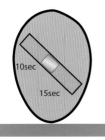

90

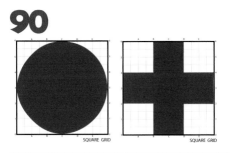

91 Hans Holbein (1497–1543), the great court painter to Henry VIII, created perhaps the most famous and striking example of a hidden anamorphosis (see Holbein), "The Ambassadors," a portrait of the two French ambassadors Jean de Dinteville and George de Selve. If you position yourself at an oblique angle relative to the right side of the picture plane, you can see the image of a skull.

The painting was originally hung on a staircase in Jean de Dinteville's chateau, so that the skull may have appeared from below left or down the stairs. Although numerous explanations have been offered about the symbolic presence of the skull, including that it is a play on the artist's name – Holbein, which means "hollow bone" in German, the reason for its inclusion is still unclear.

92

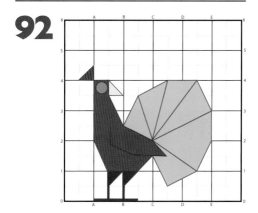

93

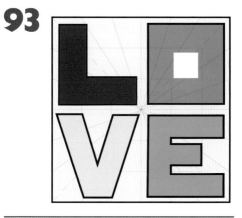

94 The word is "mastermind."

95 "Illusion is the first of all pleasures."

96

97

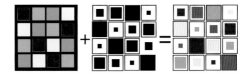

98

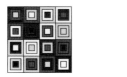

99

An order-4 magic square. Amazingly, there are 86 different ways the magic constant of 34 can be achieved. All of them are forming interesting patterns as shown.

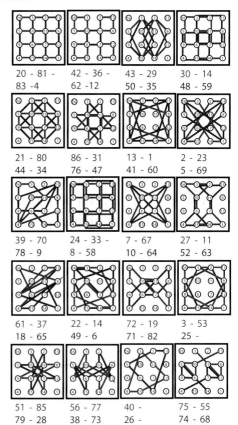

20 - 81 -
83 -4

42 - 36 -
62 -12

43 - 29
50 - 35

30 - 14
48 - 59

21 - 80
44 - 34

86 - 31
76 - 47

13 - 1
41 - 60

2 - 23
5 - 69

39 - 70
78 - 9

24 - 33 -
8 - 58

7 - 67
10 - 64

27 - 11
52 - 63

61 - 37
18 - 65

22 - 14
49 - 6

72 - 19
71 - 82

3 - 53
25 -

51 - 85
79 - 28

56 - 77
38 - 73

40 -
26 -

75 - 55
74 - 68

100

Sixteen moves is the minimum. Note that the knights have to move around in a cycle in the same direction.

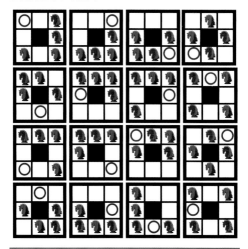

101

The solution requires 16 moves:1- 3-4 2- 4-9 3- 11-4 4- 4-3 5- 1-6 6- 6-11 7- 12-7 8- 7-6 9- 6-1 10- 2-7 11- 7-12 12- 9-4 13- 10-9 14- 9-2 15- 4-9 16- 9-10. Note that both puzzles can be solved by representing the problems by planar graphs. The squares of the chessboards as the nodes of a graph, and the possible moves between them as the connecting lines of the graphs. The graph of puzzle 1 is shown in two topological variations (the second graph topologically unfolded). The solutions are easily obtained. The solutions are not unique, and one solution is shown for each puzzle.

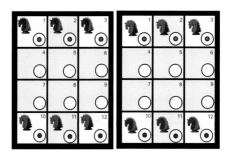

102

Your path will be a loxodrome, a spherical spiral also known as a rhumb line, which cuts a meridian at a constant angle (not a right angle).

103

Since such a division is impossible, the brothers divided the 17 horses by borrowing one horse from a neighbor. Now they had 18 horses that they could divide in the required relationships (after which they returned the borrowed horse) into nine, six and two horses.

Since 1/2 + 1/3 + 1/9 = 17/18 < 1, each of the brothers actually received more: 9> 17/2; 6> 17/3; 2 > 17/9 !

104

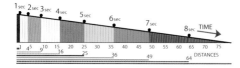

Of course, the time the ball travels to the end of the plane depends on the inclination, but the ball's speed at the end of the inclined plane did not change. No matter what its inclination, its speed is always the same at the end of the inclined plane. A ball that rolls a distance down an incline in one second rolls four times as far in two seconds, nine times as far in three seconds, and 16 times as far in four seconds. You can easily test this by rolling a ball on a ruler, provided the angle of incline is sufficiently small, so that the ball will remain rolling for as long as 4 seconds. 1,4,9,16,25, 36, 49, 64…

The striking number sequence visualizes a pattern that arises in motion. After n seconds' descent, the ball is exactly at the mark n2, i.e. the distance a falling body travels increases as the square of the time, and what is even more interesting, this is true no matter what the angle on the inclined plane. Galileo used the inclined plane for his famous experiments with falling bodies, since the motion of a body on it is similar to that of free fall, except that its velocity is slowed by the slope for easier observation and measurement.

105

Many numbers are not squares. So are there more numbers than squares? Whenever you encounter two infinite sets and try to decide which is the bigger of the two, you are liable to run into the kind of paradox pointed out by Galileo. You should know:
1- With infinite sets you can't compare their sizes.
2- Infinite sets are not like finite sets - in an infinite set you can have a part that is "equal" to the whole of it.
3- Counting by "pairing-off" or "point-to-point correspondence" does not work for infinite sets.

106

It's astonishing that if the angle is kept small, the time of a pendulum swing varies not with the size of the swing but only with the length of the pendulum itself, which is quite counterintuitive. Whether it makes a long swing or a short one, the period will be the same. The strange motion of a pendulum obeys certain laws:
1) The period of oscillation does not depend on the weight of the bobs.
2) The period does not depend on the distance traveled.
3) The period of oscillation is proportional to the square root of the length of the pendulum.
A pendulum's period, or the time (T) it takes to go through one cycle, can be expressed by the simple

formula: $T = 2\pi\sqrt{(L/g)}$, where L is the length and g is the rate of acceleration due to gravity, which is 9.80 m/sec². Since g is the only variable besides the length, a pendulum is a simple way to measure the gravity of a planet. A 1-meter- long pendulum will complete a swing in about 1 second on earth and 2.5 seconds on the moon.

107 A mechanical Antigravity Paradox. Can an object defy gravity and run uphill?

Leybourn uphill cones are sometimes attributed to Galileo.

The double cone seemingly rolls uphill when it actually descends the inclined tracks, as can easily be seen when the motion is viewed from the side.

As the double cone moves "up" the track, the increasing width of the tracks lowers the cone so that its centre of gravity moves down, seemingly defying gravity.

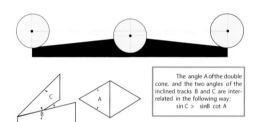

The angle A of the double cone, and the two angles of the inclined tracks B and C are interrelated in the following way:
sin C > sin B cot A

108 The sum of the inradii is a constant and independent of the triangulation chosen. This is convincingly visualized by comparing the sizes of the two sets of circles. The beautiful theorem was a Sangaku problem. According to an ancient custom of Japanese mathematicians, the theorem was inscribed on a tablet hung in a Japanese temple to honor the gods and the authors of the theorems in the 1800s.

109 Six spheres can be removed as shown, and the box won't rattle.

110 Yes! By combining the two packings, you can squeeze in another circle, altogether 106 circles, as shown. Again, the best solution is not necessarily the most ordered and regular one. Needless to say, how important such packing problems are for manufacturing, specifically when you want to cut as many circular parts out of a sheet as possible, etc.

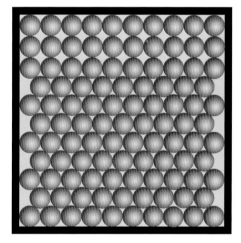

111 1- Weights allowed on one side of the scale only: For weighing consecutive objects from 1 to 40 kilograms, we shall need a binary set of six weights of:

1 - 2 - 4 - 8 - 16 - 32 = 40 kg

2- Weights allowed on both sides of the scale: For weighing consecutive objects from 1 to 40 kilograms, in this case, we shall need a ternary set of weights (the first 4 powers of 3): 1 - 3 - 9 - 27 kg.

112 There are six possible ways to arrange the three identical boxes. A single weighing can decide between two possibilities; two weighings among four possibilities, three weighings among eight, and so on.

In general, n weighings will determine 2^n possibilities at most.

In our case, say we have:
Weighing 1: 1 > 2
Weighing 2: 1 < 3
Conclusion: 3 > 1 > 2, and the problem is solved.
If weighing 2: 1 > 3 there may be two possibilities:
1 > 2 > 3 or 1 > 3 > 2 and a third weighing will be needed to compare 2 and 3.

113 Three weighings will be needed at the most. 1- Divide the 21 rods into three groups of seven. Put one group on each side of the scale. There are two possible outcomes:
a) Scales balance b) Scales tilt.

If the two scales balance, the group containing the heavier rod is the uinweighed set. If the scales tilt, obviously the heavier side of the scale holds the group with the heavier rod. Take the heavy group, divide it into two groups of three rods with one box left over, and put one group of three on each side of the scale.

2- Again, there are two possible outcomes:
a) Scales balance b) Scales tilt.

If the scales balance, the unweighed box is the heavier one, and no more weighings are needed. Otherwise, one more weighing will be needed, by putting one rod on each side of the scale, with one left over.

114 The eight coins are divided into two groups of three and a pair.

In both possibilities shown above two weighings are sufficient to find the false coin, as visualized above. In the second weighing in outcome two, if the scales are in equilibrium, the false coin is of course the one not weighed.

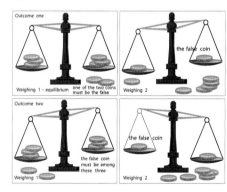

115 Volumes of spheres are to each other as the cubes of their corresponding radii, as shown above. The volumes of the two groups in equilibrium are 729 units each, as shown below:

1 1 27 64 125 216 512 512 = 1458

1 216 512 = 1 27 64 125 512 = 729

116 1- There are six different equilibrium situations (three pairs of reflections).

1

2- There are 17 distinct equilibrium solutions not counting the reflections as different.

The probability of achieving equilibrium by random distribution of the weights is about 4/100 = 1/25

2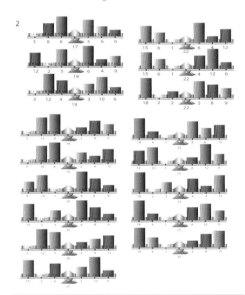

117

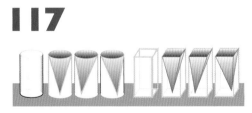

The water in the three cones and three pyramids will exactly fill the cylinder and prism of the same base and height as the cones and pyramids. This relationship can be expressed in the formula:

The volume of a cylinder or prism is the area of the base multiplied by the height, and the volume of a cone or pyramid is one-third of the volume of the corresponding cylinder or prism.

118 According to Desargues's theorem, all the intersections among the extensions of the four triangles fall along the side where the triangles come into contact with their shadows. The result of this constraint is that all the other points around it are "freed up": each of them is authorized to be the potential light source, or the projection center.

119 To construct a number in a following row, add the two numbers directly above it, left and right.

120 The equilateral triangle provides an elegant geometrical analogy to solving the problem. Each point in the triangle represents a unique way to break the stick. The sum of the three perpendiculars is constant and equal to the altitude of the triangle, which is the length of the stick. The three lines will form a triangle only when the point is inside the small middle triangle. In such a case not one of the three perpendiculars will be longer than the sum of the other two, which is the condition of forming a triangle. On the other hand, if the point is outside the middle triangle, one perpendicular is sure to be longer than the sum of the other two. Since the area of this triangle is 1/4 of the total area of the big triangle, the probability of randomly selecting a point is also 1/4.

121 Since calculus is outside the scope of this book, let's try to solve the problem by an intuitive explanation. If we imagine an infinitely small hole in a six-inch sphere, it leaves nearly all the volume of the sphere intact: any sphere with a diameter greater than six inches can have a six-inch hole drilled completely through it. The bigger the sphere, the larger the hole has to be for its length to be six inches. Calculus reveals that the volume of the remaining napkin-ring-shaped solid remains the same, no matter the hole's diameter or the size of the sphere. Quite amazingly the residual volumes of the drilled spheres and even drilled Earth, are exactly the same. Now that sounds quite counterintuitive. But even though the Earth is vastly larger than the sphere, the drill had to take out proportionally more in order to make the thickness of the hole the same. The volume left does not depend separately on the initial size of the sphere or of the hole, but only on their relation, which is forced by requiring the hole to be exactly six inches long. By now you may have discovered that this problem is the three-dimensional analog of the Circle Chord Rings problem shown on the next page. The accompanying graphics visually demonstrates the counterintuitive relationships of both problems.

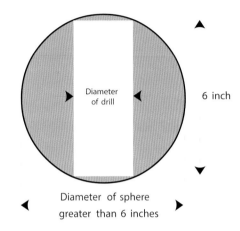

Diameter of drill

6 inch

Diameter of sphere greater than 6 inches

122 Surprisingly enough, there is enough information to work out the areas of the rings. And it will sound even more counterintuitive to discover that all the rings have the same area.

The areas of the rings depend only on the length of the chords which are all of the same length S for all the rings.

By Pythagoren theorem:
$R^2 = (R - h)^2 + (S/2)^2 = R^2 2Rh + h^2 + (S/2)^2$
Area of the larger circle = πR^2.
Area of the smaller circle = $\pi r^2 = \pi (R-h)^2 = \pi (S/2)^2$.
The difference between the area of the two circles is :
$\pi (R^2 - r^2) = \pi S/2^2 = \pi$

Thus the area of the rings depends only on the length of the chord and π.

Note that the diameters of the circles are not given. If we imagine that the diameter of the smaller becomes zero. the ring then becomes the area of the larger circle of which the diameter is then the length of the chord and its area the area of the circle.

123 The binary language is the language of computers. It is based on the base 2 number system, uses only 0s and 1s (represented here as white and black circles), which correspond to the possible states of an off-on switch. Each place to the left in a binary number represents the next highest power.

2^9	2^8	2^7	2^6	2^5	2^4	2^3	2^2	2^1	2^0	
512	256	128	64	32	16	8	4	2	1	
○	○	●	●	●	○	●	○	●	○	= 234
●	○	○	●	○	○	●	○	○	○	= 580
●	○	○	output	●	○	●	●	○	○	= 612
●	●	●	●	●	●	●	●	○	●	= 1021

124 -125

It is possible for a sphere to touch 12 other spheres of the same size, as shown by the first layer of spheres. Six spheres surround the middle sphere in a plane to which three spheres can be added on top and below.

This is the maximum number of spheres that can "kiss" at one time, fitting into a sphere that has a diameter three times larger than the middle sphere. Thus the kissing number of identical spheres is 13.

Problems involving kissing numbers are related to many important fields in mathematics, including error-correcting codes - codes employed to send messages over noisy electric channels, etc.

The number of spheres in the following layers is given by the simple equation: $10\,F^2 + 2$ where F = Frequency (the number of layers):

The number of spheres in the second layer = 42; And in the third layer, 92. Thus the number of spheres in a 3-frequency system as shown is 147.

126

The cycloid is the solution to the brachistochrone problem. The ball on the cycloidal track will be the first to arrive. This is only one of the remarkable properties of the cycloidal curve. A descending ball will roll on an inverted cycloid in shorter time than it would roll on any other path, straight or curved, and this in spite of being the longest. It is quite amazing that the shortest path (straight line) is not the quickest route. The cycloid is therefore called the curve of the quickest descent - or the brachistochrone. The ball descending along the cycloid reaches a higher speed in the early part of its descent and thus reaches the end first, and in some cases, may dip below the horizontal level before rising again. It is even more surprising that a descending ball will reach the bottom in the same time no matter which point on the inverted cycloid it started from. Galileo discovered that the period of a pendulum depends only on its length, which is true only for small oscillations. By making the pendulum wrap around a cycloid, it becomes true for oscillations of any amplitude.

127

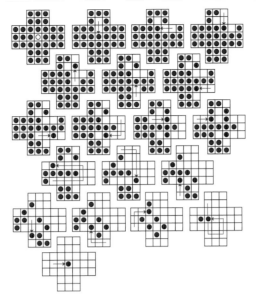

128

Four wheels:

Red: 3-bit binary numbers. A unique solution. Longer binary wheels are used to code messages in telephone transmission and radar mapping. University of California mathematician Sherman K. Stein called such binary structures memory wheels; they have also been called Ouroborean rings, a name derived from the mythological snake that ate its tail.

Green: 4-bit binary numbers One solution shown.

Yellow: 5-bit binary numbers

Blue: 6-bit binary numbers

129

The intuitive answer is that since one meter is inconsequential compared to the circumference of the Earth, the rope would hardly budge. But in this case intuition is wrong.

A little analysis shows why. $2\pi(r + x) - 2\pi r = 1$ meter
$2\pi r + 2\pi x - 2\pi r = 1$ m $2\pi x = 1$ $x = 1/2\pi$ $x = 1/6.28$
x = 16 centimeters.

It is even more astonishing that this result is independent of the radius of the Earth, or even of a ping-pong or tennis ball. This fact is visualized by the diagram.

130

As we can see, the results are always parallelograms, called parallelograms of Varignon. This beautiful principle is called Varignon's theorem, after Pierre Varignon (1654–1722). The area of Varignon's parallelogram is half that of the quadrilateral and its perimeter is equal to the sum of the two diagonals of the quadrilateral.

131

In general, the number of different ways regular polygons can be divided into triangles is: 1 - 2 - 5 - 14 - 42 - 132 - 429 - 1430 - 4862 - .... These number are called Catalan numbers, after Eugene Charles Catalan (1814–1894). They appear in many problems of combinatorics. A convex polygon of n sides needs n-3 diagonals to triangulate it, and to divide it up into n-2 triangles.

132

Number of intersection points (V): 9
Number of regions (F): 11
Number of edges (E): 18
Euler discovered the formula for any connected graph in the plane, also known as "Euler Characteristic:
V - E + F = 2 which in our example is: 9 - 18 + 11 = 2
The insight that the result of Euler's formula is always 2 is

one of the most beautiful and important expressions of mathematics.

133

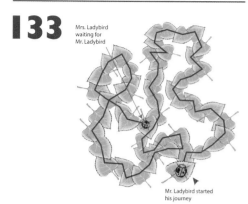

Mrs. Ladybird waiting for Mr. Ladybird

▶ Mr. Ladybird started his journey

The two following puzzles were created by the author as a tribute to the genius of Euler. They are Euler's Graphs in disguise, and can be solved by applying Euler's Theorem. The leaves can be considered as nodes (points) of a graph. If a leaf has an even number of boundary crossings (overlaps with other leaves), Mr. Ladybug can enter and leave it, whereas a leaf with an odd number of boundary crossings can be entered and left, but when Mr. Ladybug reenters, he cannot leave again. Observing the leaves, the only leaf with an odd number of crossings is the leaf on which Mrs. Ladybug will be waiting. Drawing a line through all the leaves that have only two crossings, and marking the multiple crossed leaves, you can easily complete a continuous line through all the flowers, never retracing the line. In general, according to Euler's theorem, a graph like this can be traversed if only 0 or 2 of the leaves have an odd number of adjacent leaves. If it's "0," you can start anywhere because it's a closed loop. If it's "2," those two leaves are the start and finish points. That's what we have in this case - the start point has one adjacent leaf, and the finish point has three adjacent leaves. All the other leaves have an even number of adjacent leaves.

134
The illegal alien spaceship entered from the top left planet and intended to leave from the lowest planet on the right where it was intercepted by the waiting defense forces. In the given graph there are only two points with an odd number of edges. It can only be traced without crossing a route more than once if one of these points is the beginning or the end. We know that the upper (North) point is the entry and so the other lower point is the only possible end of the route, or the potential exit point. One of the many possible crossings of all routes is shown.

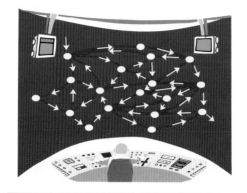

135
This is a version of the famous Buffon's Needle experiment, which you can easily perform yourself and which will allow you to calculate the number π with fair accuracy. Georges Louis Leclerc, a French mathematician, showed that if a needle is dropped from a random height onto a piece of paper covered with parallel lines, the length of the needle being equal to the distance between the lines, then the probability of the needle falling across a line is equal to 2/π.

If the needle is shorter than the distance between the lines then the probability that the needle will fall across a line is 2 c/π × a, where a is the distance between the lines and c is the length of the needle. Thus by throwing the needle at random a large number of times (n) and counting the number (m) of times the needle falls on a line, we can calculate an experimental value of π as:

π = 2 c × n/a × m or π = 2n/m if c = a.

At first it seems almost magical that the answer involves π. This beautiful experiment was long forgotten until 1812, when Simon Laplace (1749–1827) published a major work in probability popularizing the needle experiment. In 1901, Lazzarini, an Italian mathematician, was patient enough to make 3,408 throws in the course of such an experiment, obtaining a value for of 3.1415929, a result that contains an error of only 0.0000003. Compare your results with the results of the author's experiment.

136
To get 100 heads in 100 tosses of a coin:
1 head: 1/2 = 0.50
2 heads: 1/2 × 1/2 = 1/4 = 0.25
3 heads: 1/2 × 1/2 × 1/2 = 1/8 = 0.125
100 heads:
$(1/2)^{100}$ = 1/1,000,000,000,000,000,000,000,000,000,000

It is theoretically possible to get 100 heads in 100 tosses of a coin, but it is mind-bogglingly unlikely because there are so many different configurations of mixed heads and tails. Still, for the same reason, it is equally unlikely that you will get any other specific sequence. All the sequences shown have equally the same likelihood of occurring.

137

length 1   length 2   length 1.74   length 3   length 2.732

138-143
A knight's tour is impossible on a three-by-three and four-by-four board. The five-by-five and six-by-six boards have 128 and 320 different tours, some of them closed. The total number of tours on the seven-by-seven board is over 7,000, while the number of tours on the eight-by-eight board is in the millions.

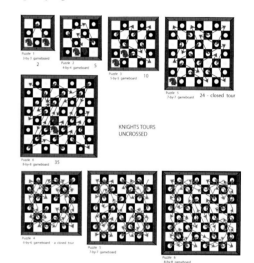

KNIGHTS TOURS UNCROSSED

144
The right figure is also an unknot.

145

There can be eight different configurations of the overlapping string, of which only two are knots. The other six are only loops, not forming knots when the two ends of the string are pulled apart straight. Therefore, the probability of a knot is only 1 in 4.

146 Max Delbruck of the California Institute of Technology, who received a Nobel prize in 1960, proposed this beautiful problem and provided a solution of 36 links. Yuanan Diao in 1993 provided a solution of 24 links, shown here.

Yuanan Diao showed (*Journal of Knot Theory and Its Ramifications* vol 2. #4 (1993) pp 413–425) that this polygon with 24 vertices is a minimal trefoil on the cubic lattice, and that no knots other than trefoils are possible using only 24 vertices.

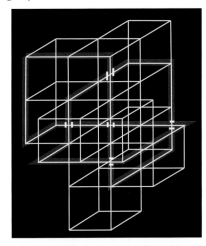

147

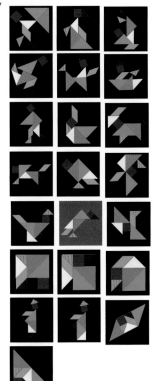

148

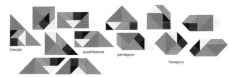

149

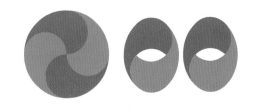

150 The astonishingly elegant four-piece solution of Sam Loyd

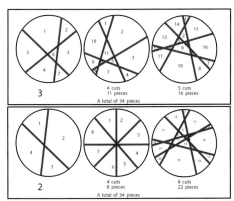

151 a) The maximum number of pieces the three cakes can be cut into by three, four and five, a total of 12 straight line cuts, is seven, 11, and 16 pieces respectively, a total of 34 pieces. This solution can be considered the minimal "best" solution, but there can be other solutions, if more than two cuts (lines) are allowed to meet at a point.

For example, such as cutting the cakes by two, four and six cuts into four (max), eight, and 22 (max) pieces respectively, etc., in which some cuts are not minimal solutions.

This problem is a simple example of a branch of mathematics called combinatorial geometry, in which there is a fascinating interplay between shapes and numbers.

b) With the requirement of cutting the cakes into identical pieces we have to cut each cake radially from the center into 12 pieces, altogether 36 pieces (in which case there will be a piece of cake for you and me as well).

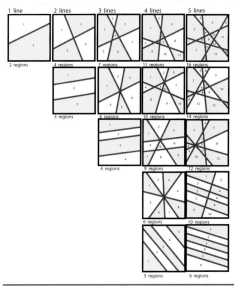

152 As a general rule, to obtain the maximum number of separate regions try to place each new cut across all the previous lines. In that way every nth cut creates n new pieces. For example, if two cuts can make four regions, a third cut cutting the two previous lines will create three new regions, etc. This rule is seen in the first row, which illustrates the maximum number of regions. Minimizing the number of regions is easy: make all the cutting lines parallel as shown in the last examples. Determining the maximum number of pieces in which it is possible to divide a circle (square) for a given number of cuts is called the circle cutting or pancake cutting problem. The minimum number is always n+1, where n is the number of cuts, and it is always possible to obtain any number of pieces between the minimum and maximum. Evaluating for 1, 2… gives the sequence of maximal number of regions into which a plane can be cut by straight lines: Sn = 2, 4, 7, 11, 16, 22, 29, 37…

153 Placing a tetrahedron inside the sphere will produce the maximum number of space regions: 15 parts.

This is a problem where a model might help. The parts are the following: four at the vertices, six at the edges, four at the faces of the tetrahedron (the original cuts) and the tetrahedron itself.

The total is 15 regions. This number is the maximum number of regions in general, into which three-dimensional space can be divided with four plane cuts. The "cake number series" goes on as: 1, 2, 4, 8, 15, 26, 42, 64, 93…

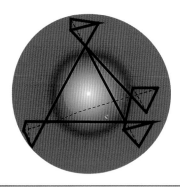

154

A pattern of a minimum 12 matchsticks meeting at eight vertices, three in each. The best known answer for the pattern in which four matchsticks meet in each vertex is Heiko Harborth's solution of 104 matchsticks meeting at 52 points. No better arrangement requiring a lesser number of sticks has been found. It is also interesting to note that no solution exists for five or more sticks.

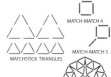

◀ 111 ▶

Optimal Stacking three - block solution

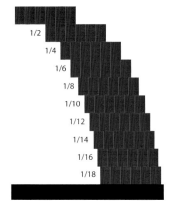

1/2
1/4
1/6
1/8
1/10
1/12
1/14
1/16
1/18

155

The surprising answer to this problem is that the offset can be as large as one wishes, which sounds quite unbelievable. When you move the top block over the rest so that it just balances, its center of gravity rests over the edge of the block below. Each time you move a block over, you are finding the center of gravity of a new stack of blocks - the block you move plus the blocks above it. The edge of each block acts as a fulcrum supporting all the blocks above it. By considering the positions of the centers of gravity of the blocks as the stack is built, it can be seen that the first block will be moved 1/2 a block length along the second block, the top two blocks will be moved 1/4 of a block length along the third block, the top three blocks will be moved 1/6 of a block length along the fourth block, and so on.

With an infinite supply of cards or blocks, the offset is the limit of the following number sequence as shown in our stack of 10 blocks: 1/2 + 1/4 + 1/6 + 1/8 + 1/10 + 1/12 + 1/14 + 1/16 + 1/18........which is called the Harmonic Number Series. This series diverges very slowly and it would take a great number of blocks to achieve even a small offset. For example, with a deck of 52 cards, the maximum overhang is about 2 and 1/4 card lengths.

156

1- You can see it outside on the vertical wall of the cube. 2- You can see it outside at the bottom of the cube. 3- You can see it inside on the floor of the cube

157

In the extreme, where the mail is evenly distributed, there would be three letters in each mailbox, except one with four letters, which is the least number for most letters in a mailbox.

This is a very simple example of the so called Pigeonhole principle, or Dirichlet's box principle, a principle of reasoning that can be applied to solve and prove a great variety of different problems.

158

Even though it is not so obvious at the outcome, the same principle can give us the answer to this problem as well, by going through a proof to show that there must be pairs of people who have exactly the same number of hairs.

We can estimate the number of hairs on one person's body and come up with a safe upper limit for the number of hairs in one square centimeter. We can also estimate the number of square centimeters on the human body. If we multiple the two numbers we get an estimate of the number of hairs on a human body. We might multiply this number by 10 to get an upper limit. This way we could get a result that no one human being has more than, say, 100 million hairs.

This fact guarantees that there must be at least two people on Earth who have the same number of hairs. According to the Pigeonhole principle, with 6.3 billion people on Earth, all having fewer than 100 million hairs, we must have pairs of people who have the exact same numbers of hairs. Imagine we have 100 million rooms and labeled them from 1 to 100 million, and we line up all 6.3 biilion people telling each of them to step into the room whose number is the number of hairs on their body. What happens after 100 million people go their appropriate rooms?

Even in the worst case scenario if all of them go into different rooms, there are still a lot of people left. So after 100 million, there must be a duplication.

159

This will always be the case.

160

No matter how you choose the circles and the triangle, the chain will always close - the sixth circle is always tangent to the first.

161

The size of the three green circles approaching infinity becomes a triangle and then the central circle becomes the triangle's incircle.

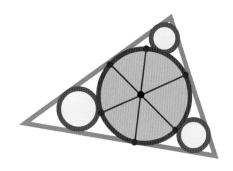

162

The area of a smaller circle is 1/9 of the area of the big circle. The gaps between the smaller circles and the big circle equal the area of two more smaller circles, for a total area of nine smaller circles. Therefore, the total of yellow areas equals the area of a small circle, and the same of the red areas.

163

From the 455 possible groups of triplets, 35 triplets must be selected to provide one of the seven solutions to the puzzle, properly distributing the 105 possible number pairs into the triplets. Three pairs are involved in each triple.

The number of ways to make such a selection is a large number and systematic procedure is needed to sort them out. Ingenious geometric methods were devised to create Steiner triplets. One such is demonstrated for the solution of our puzzle. Around the outer disk 15 points are distributed equally around the circle. The inner rotating wheel with a pattern of colored triangles is turned about point 15. It is rotated each time (counterclockwise) two units at a time to seven different positions to provide the seven sets of triplets, as shown.

The classic designs on Kirkman problems were described by W.W. Rouse Ball and H.S.M. Coxeter in *Mathematical Recreation & Essays*, and in a later edition completely revised by J.J.Seidel.

Steiner triple systems are not possible for every n. The number of pairs of n objects is: 1/2 n (n - 1), and the number of required triplets is 1/3 of the number of pairs or, 1/6 n (n - 1), since each object must be in a triplet.
This happens when n is divided by 6 and there is a remainder of 1 or 3 (in mathematical language (n is congruent to 1 or 3 modulo 6). The order of the triplets and the order of the numbers (or the altered numbers) don't change the basic pattern of a solution. Therefore, Steiner Triple Systems are possible only for n=3, 7, 9, 13, 15, 19, 21, and so on…

The Steiner triple system for n=3 is trivial. We have three pairs forming one triplet. The Seven Bird Family puzzle is the next Steiner triple system, and is not very difficult to solve. Each day three of the birds fly out. After seven days every pair of birds will have been in exactly one of the seven triplets. Number the birds from 1 to 7 and find the triplets.

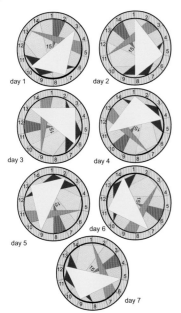

1	1 2 15	3 7 10	4 5 13	6 9 11	8 12 14
2	1 5 8	2 3 11	4 7 9	6 10 12	13 14 15
3	1 9 14	2 5 7	3 6 13	4 8 10	11 12 15
4	1 4 11	2 6 8	3 5 14	7 12 13	9 10 15
5	1 3 12	2 9 13	4 6 14	5 10 11	7 8 15
6	1 10 13	2 4 12	3 8 9	5 6 15	7 11 14
7	1 6 7	2 10 14	3 4 15	5 9 12	8 11 13

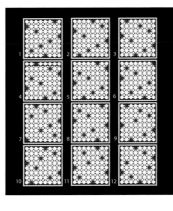

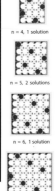

n = 4, 1 solution

n = 5, 2 solutions

n = 6, 1 solution

n = 7, 6 solutions

164 The problem in general deals with the question: given a square game board of order n (sides of n unit squares), how many chess queens can be placed on it so that no queen attacks another? Likewise, how many counters can you place on different gameboards so that no two are in the same row, column or diagonal? There are 12 basically different solutions for the full-size chessboard (not counting reflections and rotations as different).

The puzzle can be played as a competition game for two players.

165 Möbius strip bisected: the result will be a structure in one piece, with two edges, two twists and double length. Trisected: the result will be two linked bands, one of them a Möbius of the same lengt, the other a band of twice the length with two complete twists.

166 The result will be an interesting structure, consisting of one piece with two sides, three edges, no twists and two holes. Topologically speaking, the twist in the upper part gets cancelled out by the one in the lower part.

167 The outcome of bisecting Manning's surface is a plane square ring, with two edges, two sides and zero twists.

168 One of the fascinating properties of a pendulum is that once set in motion, it will continue to oscillate in the same plane if undisturbed. Its changing paths in the sand below can be explained only by the fact that the Earth turns beneath the pendulum. The apparent rotation of a pendulum varies with the latitude at which it is installed. Its rate at points between the poles and the equator is equal to 15 degrees per hour multiplied by the sine of the latitude. Thus at the poles the pendulum would take 24 hours to make a complete circle, but at the Equator it would remain on its original path.

169

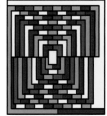

170 McGregor's map was, of course, an April Fool's joke.

The Four Color problem is solved; it is now the Four Color theorem, telling us that for any map in the plane, four colors are sufficient.

After its publication, Martin Gardner received hundreds of letters with the map colored with four colors; one such solution is shown.

171-172 The width of a closed convex curve is defined as the distance between the parallel lines bounding it. Curves of constant width have the same width regardless of their orientation between the parallel lines. There is an infinite number of them including the circle, which has the largest area, and the well known Reuleaux triangle, which has the smallest area. The two irregular wheels have outlines which are curves of constant width. One is a Reuleaux triangle, which is created by drawing a circular arc from one corner passing through the other two. The width of such a curve in every direction is equal to the side of the equilateral triangle.

Any regular polygon with an odd number of sides can be rounded up like the Reuleaux triangle to create a curve of constant width, like our other wheel based on a pentagon. Such curves are called Reuleaux polygons. Therefore like a revolving circle, when the three wheels are revolving, the platform with the cocktail glasses remains in the same position, without rising tilting.

The Reuleaux Triangle can be rotated inside a square, and it is the base of the fascinating Watt's square drill bit.

173 There are two distinct solutions, as shown.

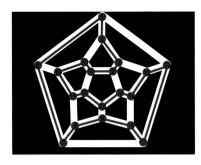

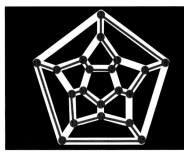

174 There are 21 pairs of points.
Each pair of points can be reached from a third point from which the arrows point towards each pair.

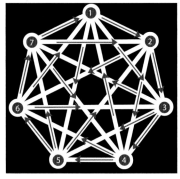

175 1- Berlin, 2- Cairo, 3- New York, 4- Paris, 5- Amsterdam, 6- Tokyo, 7- London

176 The simplest idea about how to solve the problem is to find the weight of every Hamiltonian circuit and to select the one with the minimum weight. The total number of Hamiltonian circuits starting at a particular vertex for a graph with n vertices is given by (n-1) x (n-2) x ... x 3 x 2 x 1 or (n - 1)!

From the table we can see the difficulty of solving such problems. For graphs with a very large number of vertices, no computer could cope with the problem. There is no known efficient algorithm that will solve the traveling salesman problem. There are only algorithms enabling us to find circuits whose weights are fairly

close to the minimum possible weight (approximate algorithms).
An approximate algorithm:
1- Start at the vertex on which the circuit is supposed to begin and end.
2- The next vertex to be chosen is the one connected by the edge of least weight.
Starting at vertex 2 we get a circuit of weight:
4 + 8 + 11 + 2 + 10 = 35.

By comparing this result to the minimum weight circuit obtained by checking all circuits starting at vertex 2, one of which is shown, we can see has weight 29.

Our result of 35 is still much better than the maximum weight circuit starting at vertex 2, which adds up to 54. There are other approximate algorithms which may give better results than the nearest neighbor algorithms.

Growth of (n-1)! table:

n	(n-1)!
3	2
4	6
5	24
8	5,040
12	39,916,800

177

HAMILTONIAN CIRCUIT 1-

HAMILTONIAN CIRCUIT- 2
(page 196)
No Hamiltonian circuit is possible without retracing at least one of the lines as shown.

178 Puzzle 1- the green line, puzzle 2- the red circle.

179 You can solve the problem by first working out how many pairs are possible among the nine dogs. There are 36 pairs, which you can easily list, as shown on the right, with the 12 triples you have to find. After you have worked out the number of pairs, it will be clear that in each triplet three pairs are involved, which means there can be 12 triplets altogether. Following a systematic procedure you can eliminate pairs by choosing the sequence of triplets: Each set is called a "Steiner triple system" of order nine. Steiner systems have a great importance in combinatorial theory. The next Steiner triple system is of order 13. In general the Steiner triplets problem is to arrange n objects into triplets so that each pair of n objects appears in a triplet just once.

The number of pairs is : 1/2 n (n - 1)

The number of the required triplets is 1/3 of the number of pairs: 1/6 n (n -1)

Such a system of triplets is possible only when both these numbers are integers.

The orders of Steiner triple systems are numbers that have a remainder of 1 or 3 when they are divided by 6 or, in mathematical language, when n is congruent to 1 or 3 modulo 6.
Therefore, the sequence of possible values for n is: 3 - 7 - 9 - 13 - 15 - 19 - … and the number of triplets 1 - 7 - 12 -26 - 35 - 57- …

Problems like these started in the 19th century as recreational mathematics problems pioneered by Jacob Steiner, a Swiss geometer, but today they are in the center of modern combinatorics block-design theory, with applications in many fields of modern science. For n greater than 15 the number of solutions is not known, but it has been proved that every value of n does have a solution, and also that for n=19 the number of solutions is in hundreds of thousands.

1 - 2	
1 - 3	Day 1 1 2 3
1 - 4	Day 2 1 4 7
1 - 5	Day 3 1 5 9
1 - 6	Day 4 1 6 8
1 - 7	Day 5 2 4 9
1 - 8	Day 6 2 5 8
1 - 9	Day 7 2 6 7
2 - 3	Day 8 3 4 8
2 - 4	Day 9 3 5 7
2 - 5	Day 10 3 6 9
2 - 6	Day 11 4 5 6
2 - 7	Day 12 7 8 9

Puzzle 1
Unique solution

Day 1	1 2 3	4 5 6	7 8 9
Day 2	1 4 7	2 5 8	3 6 9
Day 3	1 5 9	2 6 7	3 4 8
Day 4	1 6 8	2 4 9	3 5 7

Puzzle 2
Unique solution

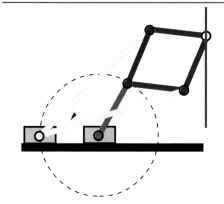

180 The first mechanical device to produce exact straight-line motion was Peaucellier's Linkage invented in 1864. Theoretically the point will describe an exact straight line.

181 Schematic representation of the famous Watt's Linkage. When the two rotating rods (blue and green) are of equal lengths, the midpoint of the red linkage (white point) traces a figure eight pattern over the full travel of the mechanism, the middle part of which is shown, is a good approximation of straight lines. The true curve of motion of Watt's linkage is a mathematical curve known as Bernouilli's Lemniscate, an elongated figure 8, a segment of which was close enough to a straight line for Watt's purposes. From strips of card with holes punched in them and joined by eyelets, you can easily create Watt's linkage, and many others, and perform experiments.

182 Initially, your acceleration would be the surface acceleration of gravity: $g = 9.8 \text{ m/s}^2$ It would decrease as you approach the center of the Earth. Passing through the center you would be weightless for a moment, falling at 27,000 kilometers per hour. You would arrive on the opposite side after about 42 minutes. That's a crucial moment of your journey. Unless you grab something to hold on to, you would fall back oscillating for a return trip of 84.5 minutes.

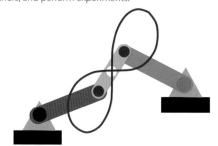

183 Confronted with this puzzle many have a conceptual "block," and are unable to place the riders properly. But the solution is quite simple as you can see.

The strip with the riders has to be moved left until it is properly aligned forming two galloping horses with their riders.

Loyd sold this puzzle to P.T. Barnum earning $10,000 in a few weeks, a fortune in those times.

184 There is no possible way.

185 No pencil really changed color. Nothing has really changed or dissappeared. The

ingenious design principle, as Martin Gardner properly called it the "concealed distribution," causes changes in the lengths of the pencils resulting in the perception of one pencil changing its color.

186 One red pencil changed color.

187 Puzzle 1 - Altogether seven bins are needed as shown. Note that the bins are fully packed with the 33 weights.

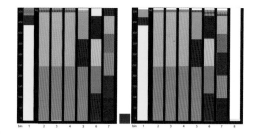

Puzzle 2 - Paradoxically enough, after throwing out the weight of 46 kilograms, one more bin is needed to pack the remaining weights. This result seems counterintuitive, but now the bins cannot be fully packed resulting in an additional bin as shown.

188 Puzzle 2 requires 15 moves to solve, puzzle 3: 24 moves, puzzle 4: 35 moves, puzzle 5: 48 moves.

189-190 Lucas' famous puzzle was manufactured as a children's toy. It is an ingenious model for the concept of geometric exponential series. Even today, a version of the puzzle can be found in toy stores worldwide.

Only seven moves are needed for the transfer of the three coins: for transferring four coins 15 moves will be necessary, for five coins 31, and for six coins 63. In general for transferring n coins $2^n - 1$ moves will be necessary. The solution to Babylon can be quite elusive because it is so easy to make an incorrect move.

One possible hint towards the solution:
1- Move the smallest disk from its present column to the next, always in the same cyclic order and
2- Thereafter, move any disk except the smallest. This rule seems arbitrary, but you will find there is always one legal move you can make within it – until the puzzle is suddenly and miraculously solved (not necessarily in the fewest number of moves).

191 If your guess was seven, you were wrong, since there were ships already on the move

before you started your journey. The ships leaving Le Havre will meet 15 ships, 13 ships at sea and one in each harbor. The meetings are daily, at noon and midnight.

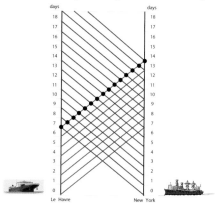

192 Flatlanders may not be able to sense the approach of the cube. Those Flatlanders who won't be removed by the collision into another dimension, will experience the first contact of the cube approaching with its vertex first, as a point and then as a triangle that will grow in diameter reaching a maximum and then becoming hexagons, and then again into a big triangle, which will diminish again to a point that will dissappear.

In case the cube changes its course and approaches Flatland edge first, then the first contact will be a line that changes into growing rectangles becoming a square, and again rectangles, which would diminish again into a disappearing line. Interesting Flatland paradoxes may be associated with this event:

The cube traveling through Flatland can remove an object from a sealed chest – without opening the chest and without breaking any of its walls. A chest in Flatland is just a closed two-dimensional figure. Flatlanders can't enter it without opening it or breaking its walls.

193 With a cube there can be three different slicing sequences, depending how the cube is oriented when entering the plane as shown. The most interesting is the vertex first entry, which provides a variety of polygons including hexagons as shown. The two pentagons among the shapes are impossible to obtain.

A square section of a tetrahedron and a hexagonal section of a cube are shown.

face entry -identical squares

edge entry - thin rectangles with 2 parallel sides equal to edge, growing to a maximum

vertex entry- point to small triangles which grow into bigger triangles, hexagons, which at the middle of the cube becomes a perfect regular hexagon, and than back again.

194 Two mice can be eaten by the cats as shown. A simple closed curve (a bent or curved line) is one that does not cross itself. If you imagined it as a loop of string you could always rearrange it as a circle. Such a line divides the plane into two regions: an inside and an outside.

How can you tell whether points in a simple closed curve are inside or outside? One time-consuming way would be to trace or shade everywhere a point can get to without crossing any line. But there is a much more elegant and shorter way of finding out whether a point is inside or outside a simple closed curve. Draw a straight line from a point in question to the outside of the curve, and then count the number of times the straight line crosses the curve. If it crosses the curve an even number of times, the point is outside; if an odd number of times, it is inside. This rule is the famous "Jordan Curve Theorem" of mathematics. This rule works in our problem as well, even if some parts of the closed curve are hidden. All inside regions are separated from each other by an even number of lines and any inside region is separated from any outside region by an odd number of lines. In our puzzle the cats are outside of the fence and they can catch only two unfortunate mice who also happen to be outside the fence.

195 Puzzle 1- 16 - 32 - 24 - .8
Puzzle 2- Shown left
Puzzle 3- There are 16 skeleton cubes, as shown right.

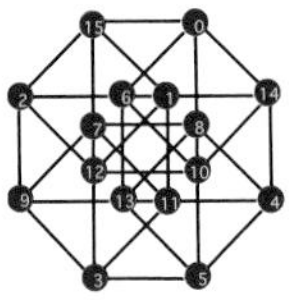

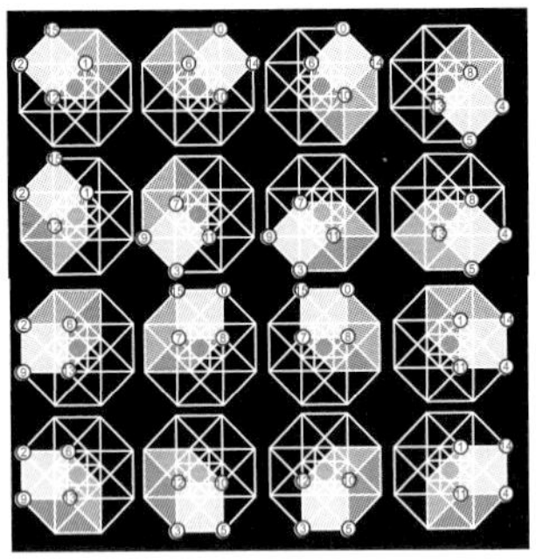

196

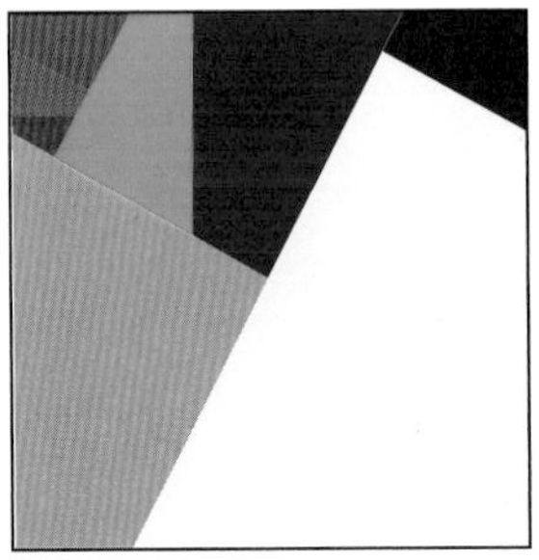

197 There are 773 dots forming Marylin.

198

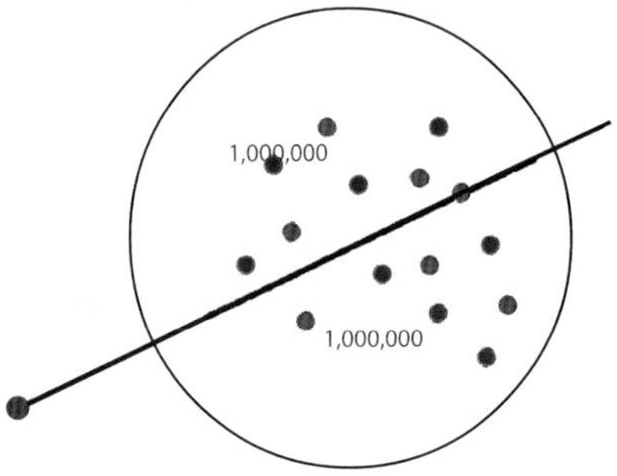

We can choose a point outside the circle from which a rotating line sweeps the circle in one direction while we count the dots that the line crosses until we count one million.

At this position of the line it divides the circle with one million dots on each side of the line. If, by bad luck, the line sweeps over two dots and jumps from 999,999 to 1,000,001, we choose another point outside the crcle and try again. The process always works eventually. This is a simplified version of a proof known as the pancake theorem.

199

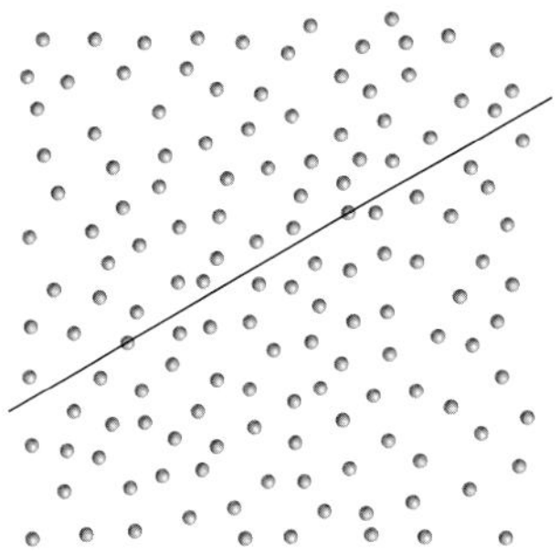

200-201 The sum of all 19 numbers is 190, which is divisible by five, which is the number of parallel rows in one direction. Thus the magic constant is 38. In general, is it possible to arrange a set of positive integers from 1 to n in a hexagonal honeycomb array of n cells, so that all straight rows have a constant sum (like in magic squares)? Or, in other words, is a magic hexagon of order-n possible?

An order-2 magic hexagon is clearly impossible. The simplest proof for this is that 28 divided by three is not an even number. The same proof tells us that a magic hexagon of order-3 is possible, but it was not an easy task to find one (which is our puzzle). Only after a long search was this hexagonal array of numbers discovered in 1910. At the beginning it was assumed that it was only one of the many possible order-3 magic hexagon patterns, but an extremely complicated proof conveyed the astonishing and surprising fact that ours is the only possible case, and the even more astonishing fact that no other magic hexagon of any size is possible!

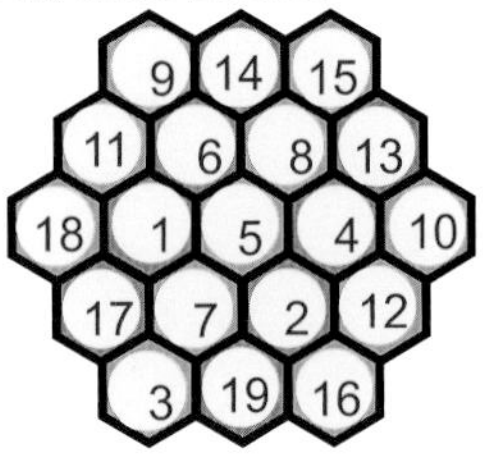

202 It is interesting to note that each 2-pire touches all the other colors, proving that 12 colors are necessary.

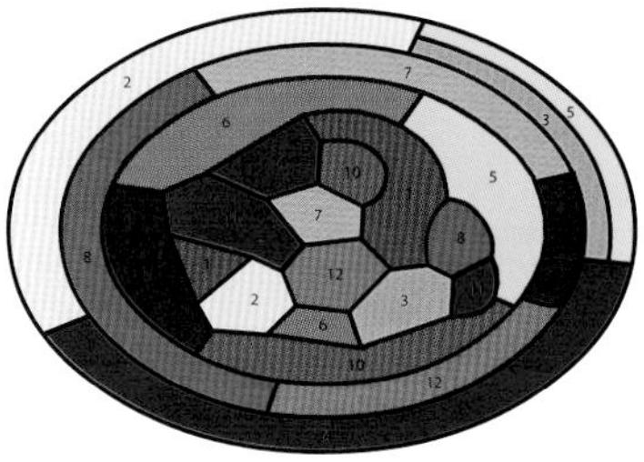

203 1- 24 sets of color triangles
2- 24 sets of color squares
3- 24 sets of color hexagons
4- 24 positions of color cubes
5- 30 two-, three- and six-color cubes
6- 16 sets of cubes and prisms

204 All cubes get the color violet at the bottom face. The front sides get one of the six colors in each line. The sides underneath get the third suitable color in each line. The remaining three sides get all permutations of the remaining three colors in a line.

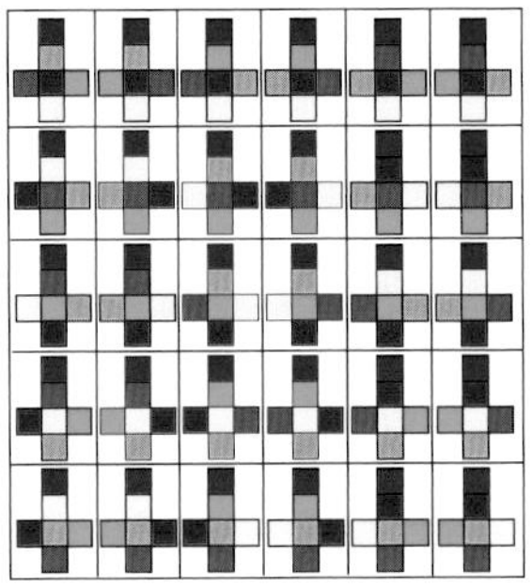

205 The hinged triangle-to-square four pieces transformation of Henry Ernest Dudeney is a real gem of recreational geometry. It has the curious property of folding continuously around the hinges from one position to the other as shown. Hinges marked in color connect the parts to each other.

When you rotate the part around the red shape counterclockwise you will end up with a square. In his book *Canterbury Puzzles* in 1907 Dudeney introduced his novel design variation of a four-piece solution for the dissection of a square into an equilateral triangle and vice -versa, with the parts hinged together. Thus he created a new puzzle category, that of hinged dissections and tessellations. Greg Frederickson's fantastic book *Hinged Dissections: Swinging & Twisting,* Cambridge University Press, is a treasure trove of this new category of recreational math puzzles.

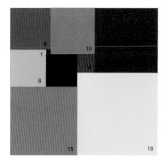

fixed

206 Altogether in 65 different ways as shown above.

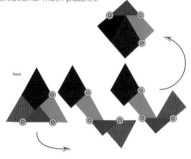

3 ways
6 ways
28 ways
14 ways
14 ways

207 One of the classic puzzles of Dudeney from 1931. An endless number of variations of the problem followed.

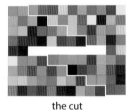
the cut

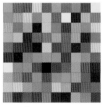
the new quilt

208

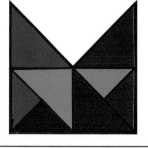

209 The result is a 32-by-33 rectangle.

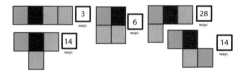

9
10
1
8
15
18

210 Can you imagine a shape that has an infinite length yet encloses only a finite area? Sounds impossible, but surprisingly enough, such figures exist. One of them is the beautiful snowflake curve and its opposite the anti-snowflake curve. These curves are basically growth patterns, created as sequences of polygons. The Koch snowflake of Helge von Koch (1870–1924) is one of the first fractals

The limit of this sequence of polygons is a remarkable curve. Its length is infinite, yet the area it encloses is finite. The snowflake curve is a good visual way to demonstrate the concept of the limit and fractals. It is not possible to draw the limiting curve. We can only create the polygons for the next sequence, and the ultimate curve is left to the imagination. As the curve grows, the size of its area will eventually enclose a finite area, 8/5 times the area of the original equilateral triangle. It is easy to prove that the area of the curve is finite.

The fact that the curve appears to remain within the page of our book is a good indication. Further than that, at no stage of the development will the curve extend beyond the circle circumscribed about the initial triangle. The limit of this infinite construction enclosed an area 8/5 that of the original triangle.

Now, concerning the length of this curve. If we suppose that the side of the original triangle is one unit long, then the perimeter is three units. In the construction of the second polygon, each segment is replaced by two line segments that together are equal to 4/3 its length. Thus, at each stage, the total length is increased by a factor of 4/3. Clearly this is not bounded. A curve of infinite length is the result. An important principle shown by the snowflake and similar so-called pathological curves is that complex shapes can result from repeated applications of very simple rules. These shapes are today called fractals.

There can be three-dimensional analogs of the snowflake and similar curves. For example, if tetrahedrons are constructed on the faces of tetrahedrons, according to similar rules, the limiting solid will have an infinite surface area and a finite volume.

211 For our simulation game of a simple linear random walk, probability theory says that after n flips, you will be on average a distance √n away from the starting point at the middle point. For 36 flips this distance would be six marks left or right from the middle. Despite this, the chances of eventually returning to the start are a certainty (1), though this may take a long time to happen. The most interesting aspect of the one-dimensional random walk appears if there are no barriers at all. The question that arises here is how often the walker is likely to change sides?

Because of the walk's symmetry, one would expect that in a long random walk, the walker would spend about half of his time on each side of the starting point. Exactly the opposite is true! The most probable number of changes from one side to the other is 0.

212 We really cannot say where the drunkard will be at the end of his walk, but we can answer the question about his most probable distance from the lampost after a given number of flips. The most probable distance D from the lampost after a certain large number of irregular turns is equal to the average length of each straight track of the walk L, times the square root of their total number N:D = L x √N. Amazingly, the drunkard will eventually get back to the safety of the lampost with certainty on a two-dimensional finite square grid as in our game, as the number of steps approaches infinity. When there are no barriers and the random walk is not finite the situation becomes quite complex, giving rise to many unsolved problems and theories. On an infinite three-dimensional lattice, a random walk has less than unit probability of reaching any point, including the starting point as the steps are approaching infinity. The probability of this is about 34 percent. The big surprise is that in a finite three-dimensional lattice a random walker is practically certain to reach any intersection in a finite time. In practical terms, if you are inside a large building or a maze with a very complex network of corridors and passages, you can be sure of reaching an exit in a finite time by walking randomly through the structure. However, this is not the case if the lattice is infinite.

213 Five pentominoes as shown

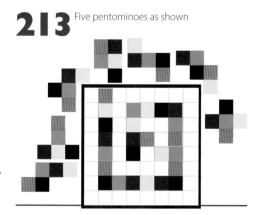

214

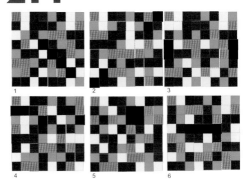

1   2   3   4   5   6

215

216

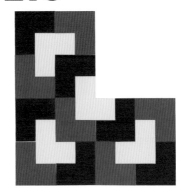

217 Divide any of the squares into four equal squares. The missing unit square belongs to one of the four squares. The other three squares form a right tromino, which is a rep - 4 tile, which can be tiled into smaller copies of themselves. You can treat bigger chessboards in the same way.

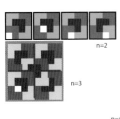

n=2

n=3

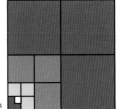

n=4

218–220

REP CATS

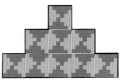

REP FISHNET

REP MONUMENT

221 You may not throw a six in six throws of a single die, so obviously, the probability is not one or 100 percent.

In fact, you have to calculate the probaility of not throwing a six, six times in a row.

The probability of not throwing a six in a throw is 5/6, and in six throws:

5/6 x 5/6 x 5/6 x 5/6 x 5/6 x 5/6=0.33

Therefore, the probability of throwing a six in six throws is high: 1 - 0.33 = 0.67 or 67 percent

222 The Cantor comb is a way to visualize the famous Cantor set. Georg Cantor was interested in this set because it is infinite and completely disconnected (its points are disconnected from each other), even though it is built by looking at line segments. In fractal terms such disconnected sets are called "fractal dust."

The total length after the nth stage: $(2/3)^n$. As n tends to infinity, Cantor's set tends to 0.

223 If the process is repeated infinitely the total area of the gold squares increases until it becomes the area of the initial square, a surprising and counterintuitive outcome – but such outcomes are not uncommon when dealing with infinity.

In the first stage: one gold square of area 1/9 = total area 0.111; Second stage: eight more gold squares of area $(1/9)^2$ = total area 0.209; Third stage: 64 more gold squares of area $(1/9)^3$ = total area 0.297; Fourth stage: 512 more gold squares of area $(1/9)^4$ = total area 0.375.

The pattern becomes clear; the total sum of gold areas is the infinite sum:

$1/9 + 8 \times (1/9)^2 + 64 \times (1/9)^3 + 512 \times (1/9)^4 + \ldots$

If we follow this series to the 25th stage, the total area of the gold squares is 0.947, and it becomes clear that this sum is steadily approaching one, which is the initial area of the blue square.

224

fourth generation

The fourth generation of the Sierpinski Triangle.

The proportions of the black triangles to whole triangle:

First generation: 25%

Second generation: ~ 44%

Third generation: ~ 58%

Fourth generation: ~ 68%

If you go on dividing the white triangles according to the rules the white area will constantly decrease as it approaches zero.

225 The total number of students participating in the three activities is 73, as shown above. The way to solve the puzzle is to place first the number of students who participate in all three activities. This makes it easy to determine the number of students in other groups.

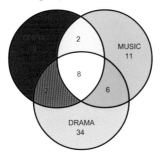

226 It takes five points to guarantee a convex quadrilatral. This was elegantly proven by the Erdos-Szekeres theorem. If you surround the given points by a rubber band (like lassoing the points), there can only be three possibilities:

1- the band forms a convex quadrilateral (with the fifth point inside);

2- the band forms a pentagon – connecting two vertices will always result in a convex quadrilateral;

3- the band will form a triangle, with two points inside. Draw a line through the two interior points – on one side of the line will be one vertex, and on the other will be two.

Take the latter two vertices and the two interior points – these will make a convex quadrilateral. It has been proven that nine points will unavoidably create a convex pentagon in any configuration of nine randomly placed points. Eight points can still be placed avoiding a convex pentagon. Any additional point will unavoidably create a convex pentagon.

227-229

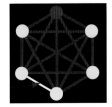

- ● you
- ○ your friends
- red lines - love
- blue lines - hate

No matter how you try to color the graph, you can't avoid forming a triangle of one of the colors. Having three in a group of six will either love or hate each other. As you can see, no matter which color you use for the last uncolored line, you will be forced to create either a red or blue triangle of a solid color. This is one of the applications of Ramsey theory: there are many others. Twenty different triangles can be colored. A maximum of 14 lines can be colored before the 15th line forces a red or blue triangle. There is no best first move, and no best strategy found so far. The second player has an advantage. The total number of games which can be played is 15. Some game will last 5 moves, others longer.

230
Connection is impossible without at least one intersection, or tunnelling under one of the houses.

In graph theory, as a graph, the classical problem is known as K3,3 , called a bipartite graph on six points, meaning that it makes all possible connections between two groups of three nodes.

231

232-234
The crossing number (k) is the minimal number of edge intersections in a planar representation of a graph. A graph is planar if its crossing number is zero. Not much is yet known about crossing numbers, since the work of Kazimierz Kuratowski from 1930. The Kuratowski theorem specifies graphs as planar with crossing number zero, if the graph does not contain a subgraph of (k5) a complete graph on five vertices, or a subgraph of (k3,3) a complete bipartite graph on six vertices, in two groups of three.

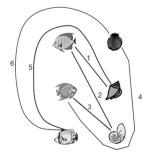

Multipartite puzzle 1: Of the total number of 6 connections, all were drawn without any intersections.

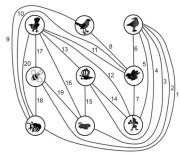

Multipartite puzzle 2:
Of the total number of connections, 20 were drawn without any intersections. Can you do better?

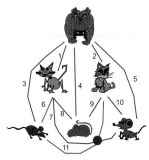

Multipartite puzzle 3:
Of the total number of 11 connections, 10 were drawn without any intersections. Crossing number is 1 as shown. Can you do better?

235
Six and 20 different ways.

236
Starting with 7, it takes a little longer and the sequence rises up to 52, after which it crashes to the endless loop of 1-4-2:7-22-11-34-17-52-26-13-40-20-10-5-16-8-4--2-1-4-2-

The answer to this new problem, named the "Hailstone Problem" is not yet known. None of the numbers from 1 to 26 survive long, until we reach the starting number 27. This one will take us on a long ride. It goes up to 9232 at the 77-th step, when the crash starts, until finally at the 111-th step the 142-142 loop is reached.

Malcolm E. Lines, a physicist, in his exciting book *Think of a Number*, mentions that all numbers up to one trillion have been tested by the University of Tokyo, and every one of these collapses to the 142142 loop. The most astonishing thing about these numbers is that none of them contains the same number twice.

237
An astonishing mathematical theorem called Benford's Law (or "first digit law") is a powerful and simple tool for pointing suspicion at frauds.

In our coin-tossing experiment a surprising probability was revealed by Benford's Law. There are overwhelming odds that at some point during the series of 200 tosses, either heads or tails will come up six or more times in a row. Most fakers don't know this, and while faking the results, are quite reluctant to include strings of more than four or five heads (or tails) in a row, trying to avoid such "non-random" occurences, assuming they are improbable.

Thus it is rather easy to discover simulated sequences. If you take a look at the test results, you can see that in test 1 there are no series of six consecutive heads or tails, so this test is probably fake.

238

The sample game ended in the sixth generation. Remember: the sequence of generations is representing the same length of land divided into an increasing number of equal divisions.

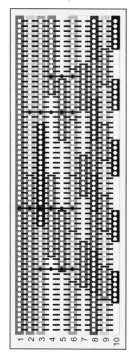

239

An opening move in 4A, 3B, 2C and 2D will give a win to the first player in seven moves as shown.

The first player can win on a five-by-fiv gameboard if he plays in the middle cell, on his or her move.

On larger gameboards the situation becomes more complex, and on a 11-by-11 gameboard reaches an enormous number of possible game situations.

It is interesting to note that although there is no specific procedure found to ensure a win, there exists a proof that a winning strategy exists for the first player on a board of any size.

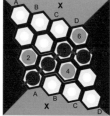

240

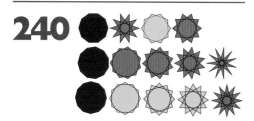

241

Two-unit length ruler with three markers – is an Imperfect Golomb ruler since distance one can be measured in two ways.

Three-unit length ruler with three markers – is the first Perfect Golomb ruler with all one, two and three distances measured.

Four-unit length ruler with three markers – is an Optimal Golomb ruler, in this placing of the markers with distance two missing.

Five-unit length ruler with four markers – is an Imperfect Golomb ruler, in this placing of the markers, distance two can be measured in two ways.

Six-unit length ruler with four markers – is the second Perfect Golomb ruler and no other perfect Golomb ruler exists.

Eleven-unit length ruler with five markers – is an Optimal Golomb ruler, in this spacing of the markers with distance six units missing.

Seventeen-unit length ruler with 6 markers – Optimal Golomb ruler.

In this solution 14 and 15 units are missing.

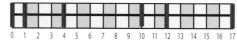

242

Length n=17, 6 marks. One solution is visualized in which lengths 14 and 15 cannot be measured. There can be other solutions.

243

The paradox lies in the fact that the diagonal line passes slightly below the lower left corner of the square at the upper right corner of the checkerboard. The addition of 1/7 of a unit to the height is not noticeable, but when it is taken into account, the rectangle will have the expected area of 64 square units. When the small square units are shown, a close inspection will reveal their inaccurate fitting along the diagonal cut.

244-245

Only these paradoxes are not true and there must be some mistakes in the resulting rectangles.

Zooming in on the first rectangle, the diagonal is not a line but a long thin parallelogram of a unit square area. In the second rectangle, a similar parallelogram is formed by the overlapping of the upper and lower parts, reducing the area by one unit square.

246

Fido can cover only 88 percent of the area of the big circle because of the obstruction caused by the shed. Regretfully he cannot reach his bone, which is in the 12 percent of the area that is not covered by Fido.

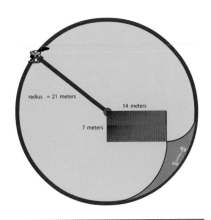

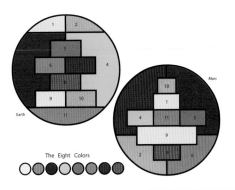

The Eight Colors

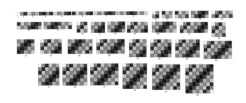

247
The horizontal and vertical sides of the octagon are all six units long, but the slanted sides are all longer than six units (7.07), thus it is not a regular octagon. (Pythagorean theorem can help).

248
"It's always too early to quit." I loved grooks, read them all, and many times told Piet so.

A grook ("gruk" in Danish) is a form of short aphoristic poem. It was invented by the Danish poet and scientist Piet Hein. He wrote over 7000 of them, most in Danish or English, published in 20 volumes. Some say that the name is short for "GRin & sUK" ("laugh & sigh" in Danish), but Piet Hein said he felt that the word had come out of thin air.

His grooks (consistency) first started to appear in the daily newspaper *Politiken* shortly after the Nazi Occupation in April 1940 under the signature Kumbel Kumbell. The poems were meant as a spirit-building, yet slightly coded form of passive resistance against Nazi occupation during World War II. The grooks are characterized by irony, paradox, brevity, precise use of language, sophisticated rhythms and rhymes, and are often satirical in nature.

249
This is a 2-pire problem, but with the added restriction that the two parts of the 2-pire lie on two disjoint regions (Earth and Mars). The number of colors needed is eight as illustrated.

250
There are 204 squares of different sizes from unit side and up to side 8, forming the sequence: $8^2 + 7^2 + 6^2 + 5^2 + 4^2 + 3^2 + 2^2 + 1^2 = 204$

The total number of different squares on a square matrix with n units on a side is the sum of the squares of the first n integers.

251
There are 8 x 8 x 8 individual cubes. There are 7 x 7 x 7 cubes of 2 x 2 x 2. There are 6 x 6 x 6 cubes of 3 x 3 x 3. And so on until one large cube of 8 x 8 x 8.

Total of $8^3 + 7^3 + 6^3 + 5^3 + 4^3 + 3^3 + 2^3 + 1^3 = 1296$. There is an alternative formula that can be used to get to the same answer: sum of cubes from 1 to n = $[n/2 \times (n + 1)]^2 = 1296$ when n = 8.

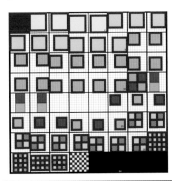

252
Puzzle 1: The number of rectangles (including squares) on square lattices from L(2) to L(8).
L(2) = 9 L(3) = 36 L(4) = 100 L(5) = 225 L(6) = 441 L(7) = 784 L(8) = 1296,
which number is the sum of the cubes from 2 to 8.

In general, for an n x n square lattice, the lattice numbers are expressed by the general formula; L(n) = $\{n (n + 1)/2\}^2$
Puzzle 2: In an L(8) square lattice (chessboard) there are 36 different sizes of squares and rectangles.

253
Many claim that there is not enough information provided to solve this problem. But that is because they have taken too narrow a view. The key lies in understanding what a lightbulb does: it produces not only light, but heat as well, and it remains warm many minutes after it has been switched off. With that in mind you can find the solution to both problems fairly easily

Puzzle 1 – First, turn on switch one and leave it on for several minutes so that the bulb will get hot. Next, turn off switch one and turn on switch two, and then go quickly to the attic. If the light is on, then switch two works the lamp; if the bulb is dark but warm switch one worked the lamp, if the bulb is both dark and cold, switch three – the one that has not been used – works the lamp.

Puzzle 2 – You do exactly as before. The hot lamp is turned on by switch one, the burning lamp by switch two, while the cold lamp by switch three, which has not been used.

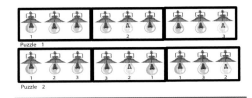

Puzzle 1

Puzzle 2

254
There are exactly eight possibilities for the product of three ages to be 36 Since Ivan could not solve the problem when he knew the product of their ages and the date of the encounter, this meant the sum must have been 13, for which there are two possibilities. The added information about the youngest son meant that one of the possibilities, a nine-year-old and two two-year-olds, can be ruled out, since there is no single youngest son in that case. That left Ivan with one solution: 1, 6 and 6.

Son 1	Son 2	Son 3	Product	Sum
1	1	36	36	38
1	2	18	36	21
1	3	12	36	16
1	4	9	36	14
1	6	6	36	13
2	2	9	36	13

255
1- Surprising procession
2- Surprising procession
3- Not surprising – because there are two instances of D=2, between a red egg and blue egg.
4- Not surprising – because there are two instances of D=4 between a red and yellow egg.
5- Surprising.
6- Not surprising – because there are two instances of a distance D=1 between a red and a blue egg.

256
Puzzle 1 – Four teams, two in each team
A unique solution for four pairs of Langford's problem.

Puzzle 2 – Nine teams, three in each team
For triplets n=9 Langford and his colleagues also found one unique solution shown below.
New and challenging mathematical problems can be created when mathematicians watch their offspring playing with colored balls.

Dudley Langford, a Scottish mathematician, watched his little son playing with three pairs of colored blocks. Finally the boy piled them in a stack in such a way that one block was between the red pair, two blocks were between the blue pair, and three blocks were between the yellow pair, as shown by the stack above.
John E. Miller found other variations of this principle:
For five and six pairs there are no solutions.
For seven pairs there are 26 different solutions.
For eight pairs there are 150 solutions.

Martin Gardner explained in 1967 that no one knows how to determine the number of solutions for a given number of pairs, except by an exhaustive trial-and-error method. If n is the number of pairs, the problem has a solution only if n is a multiple of four, or one less than such a multiple.

For triplets one solution was found which is our puzzle. Dudley's Son's colored stack of three pairs of color cubes.

257-258
Puzzle 1 –
There are six possible ways for the three hats to be mixed as shown. In four of these, one of the men gets his own hat. Therefore the probability of at least one man getting his own hat is four in six, or, 0.66 – a good bet.
Puzzle 2 – The number of ways that n hats can be permuted is n!, in our case 6! = 720.
How many of these permutations give each man a wrong hat.

A simple method of finding this number involves the transcendental number e = 2.718…

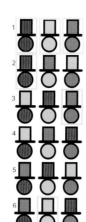

The number of all wrong permutations of n objects is the number that is closest to n! divided by e. In our case 720/2.718 = 265 or the probability that no man gets back his hat is 0.368055.
Subtracting this result from 1 (certainty) we obtain 0.6321, the probability of at least one man getting back his hat.

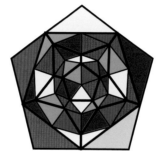

259
There are three different solutions of the quintomino dodecahedron puzzle, first published by John Horton Conway, the inventor of the problem; one of them is shown.

260
All the prisoners will be set free if they line up correctly.

The first prisoner starts the line. The others insert themselves into the line behind the last red hat they can see (or in front of the first black hat they can see). This will produce a line with all the red hats in front and all the black hats in the rear. Since the new prisoner is always in the middle (between red and black), he will know his color the moment the next prisoner joins the line. If the new prisoner joins the line in front of him, then he has a black hat. This takes care of 99 prisoners. So when the last prisoner joins the line, the one standing in the front simply leaves his position and reinserts himself between red and black. All the 100 prisoners are saved!

261
The best solution so far is shown, without the 7-by-7 square.

It is still not known whether a better of a complete packing is possible.

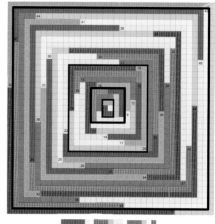

262
1- There are many solutions, two are shown that are not spirals.
2- After the smallest square formed by the first eight consecutive polyominoes (incorporating the smallest two-by-three rectangle), the next rectangle is a 14-by-15 rectangle, formed by the first 20 consecutive polyominoes (both shown in black outlines).
The area of this rectangle is 210 square units (the 20th triangular number – representing the sum of the first 20 integers.
3- The square is a 35-by-35 square, formed by the first 49 consecutive polyominoes. The area of this square is 1225 square units (the 49th triangular number – representing the sum of the first 49 integers).
You have to look a long way until the next square, which is the 41616 triangular number.
4- The formation of polyomino spiral can continue infinitely tessellating the plane.

263
To ensure I have a pair of socks of any color I must draw four socks in the dark.
To ensure I have a pair of each color, in the worst scenario, I must draw all the socks of two colors (12 socks) and then two socks more, altogether 14 socks.

264
To answer the problem we have to consider the worst scenario in which you may be unlucky enough to draw all the left-handed or all the right-handed gloves, and there are 14 of each. In such a case only one more, the 15th drawn glove will form a complete pair.

But you can do better than that, since, though it is completely dark, you can distinguish between left and right-handed gloves. In this case, the worst scenario would be to to choose 13 right or left-handed gloves and then choose one more, of the opposite handed group of gloves, the 14th chosen glove to have a perfect pair.

265 The best-case scenario – that the two missing socks make a pair, leaving you with four matched pairs – can happen in only five different ways.

If the socks can be labeled A1, A2, B1, B2, C1, C2, D1, D2, E1 and E2, then the best-case scenario occurs only when the missing socks are A1-A2, B1-B2, C1-C2, D1-D2 or E1-E2.

The worst-case scenario – that the two missing socks do not make a pair, leaving you with only three matched pairs and two orphan socks – occurs when the missing socks are

A1-B1, A1-B2, A2-B1, A2-B2, A1-C1, A1-C2, A2-C1, A2-C2, A1-D1, A1-D2, A2-D1,A2-D2, A1-E1, A1-E2, A2-E1, A2-E2, B1-C1, B1-C2,B2-C1, B2-C2, B1-D1, B1-D2, B2-D1,B2-D2, B1-E1,B1-E2, B2-E1, B2-E2, Ci-D1, C1-D2, C2-D1 C2-D2,C1-E1, C1-E2, C2-E1, C2-E2, D1-E1, D1-E2, D2-E1,D2-E2.

That's 40 different ways to get the worst-case scenario. As you can see, the worst-case scenario is eight times more likely to occur than the best-case scenario.

266

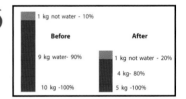

1 kg not water - 10%	
Before	**After**
	1 kg not water - 20%
9 kg water- 90%	4 kg- 80%
10 kg -100%	5 kg -100%

267 The weight of the melons is 1, 3, 5, 7, 9, 11 and 13 kg.

268 Through a systematic procedure you can fold Flexi-Twist into half of its area to get all the solutions shown:

For example to get the solution shown:

1- Down-Up – holding the square in your right hand, press down at the corner with the black point and turn it around 360 degrees. After that hold the folded left half and with your right hand fold up the corner with the white point up and like before turn it around 360 degrees .

You will get the pattern below left. Flexi-Twist was produced in 2012 by Fat Brain Toys as part of the folding puzzles compendium named "FOLD."

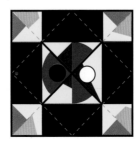

269 The other solution:

In move 2, hiker 2 is crossing back. The four hikers just made it.

They crossed in 17 minutes, just before the bridge collapsed.

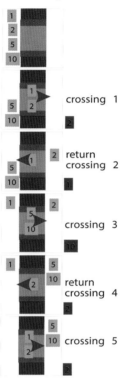

crossing 1

return crossing 2

crossing 3

return crossing 4

crossing 5

270 Sequence of n=11 and n=13:

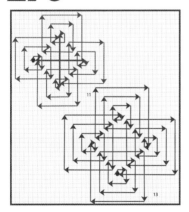

271 Folding a strip of three stamps:

You can achieve the complete set of six permutations by folding a strip of three labeled stamps, while there are only two unlabeled and symmetric folds.

Folding a strip of four stamps:

For a labeled strip of four stamps you can achieve 16 folds out of the total of 24 possibilities, five different unlabeled folds, and four different symmetric folds. For longer strips the number of different folds increases significantly.

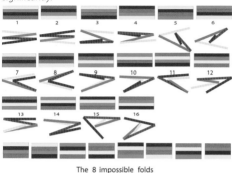

The 8 impossible folds

Folding a square of four stamps:

The four possible folds.

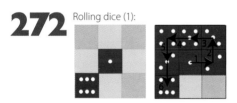

Folding a rectangle of six stamps:

Fold three is impossible. In general, it is not possible to fold the strip to make the diagonally appearing colors adjacent in the final stack.

Folding a rectangle of eight stamps:

Fold the right half onto the left so that 5 goes on 2, 6 on 3 and 7 on 8. Now fold the bottom half up so that 4 goes on 5 and 7 on 6. Then tuck 4 and 5 between 6 and 3 and fold 1 and 2 under the stack.

272 Rolling dice (1):

Rolling dice (2):

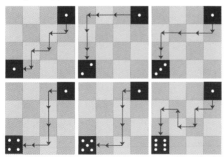

273
Puzzle 1 solution   Puzzle 2 solution

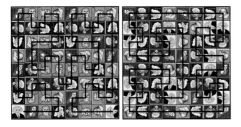

274
The problem is another example of the fixed point theorem.

Modern geometry simply explains it as follows: if two points are on either side of another line, then a line drawn through the two points intersects the other line (as long as one can prevent the line going around the end of the other line). This is one of the interesting problems whose visual representation leads to a fairly obvious solution, presenting two monks for the two journeys, one walking up and the other walking down. Regardless of the speed of ascent and descent of the monk during his journeys, or how long he rested, or even traveled backward at times, the two paths must intersect at a point somewhere along the route as shown in the diagram in which the two paths are superimposed, and therefore occupying the same spot at the same time of the day.

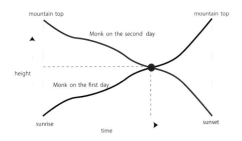

275
The secret of the transformation is easy to remember. All you have to do is exchange the two triangles in each of the four sides in one arrangement to go to the other.

Fitting in the rest of the pieces is fairly obvious in both arrangments. The two big squares seem to be identical, but as we learned, there are no miracles in Geometry – the areas of the two squares can't be identical. One of the squares is certainly smaller, but not much. It is of course smaller by exactly as much as the area of the small "superfluous" square, which as the result of the ingenious transformation becomes an irregular square ring of negligible thickness, so it is not surprising that the smaller size of the big square is not so obvious. One thing certainly to be learned from such vanishing puzzles is that disguise has to be subtle. The space of the extra piece which has to be concealed is effectively "lost" by sharing it out as thinly as possible all over

the perimeter of the big square, so as to be rendered all but invisible.

276
The set of dice convincingly demonstrates the probability paradox violating transitivity – or non-transitive dice, as they are known. Die A beats die B. Die B beats die C. Die C beats die D, and finally, die D beats die A. The game has a circular winning arrangement, which can best be worked out by creating score charts of all encounters between two dice in the set.

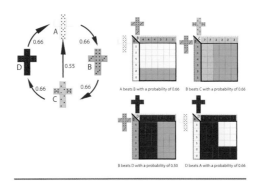

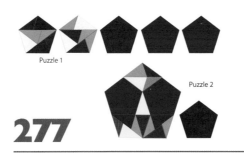

Puzzle 1

Puzzle 2

277

278
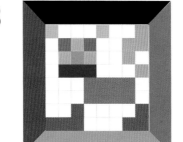

279
Trying first to pack the big blocks, the solution becomes elusive.

The secret is to place first the three small cubes, which must be positioned along one of the diagonals of the cube.

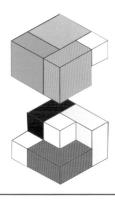

280
Like in the 3x3x3 cube, the secret is to place first the three 1x1x3 blocks as

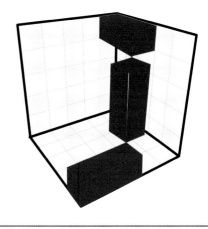

281-282
Here is how they solved the problem involving triangulation: triangulate the whole layout and color the vertices of each triangle in three different colors. The same three colors should be used for each triangle. The camera should be placed at the points that have the color that appears the fewest times.

For a gallery with n walls, this method guarantees a solution with n/3 or fewer cameras. If the shape of the gallery is a convex polygon then one camera would be enough placed anywhere in the gallery. A simple solution would be a circular shape or a polygon of 24 sides. But another solution would be a star-shaped gallery, which would require a minimum floor area.

283

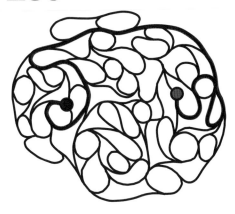

284

A shorter fence cannot consist of a single line, because the U-shape is the shortest fence that contains all four corners. The logical solution to the opaque square problem should be the so-called Steiner tree (minimal surface) spanning the four corners of a square (or any four points in general), creating a fence 2.732 units long, but it is still not the shortest opaque square fence.

A fence of two disconnected parts can be reduced to a length of about 2.639 units. Note that although this is believed to be the best solution, it is not yet proven. Some mathematicians even doubt whether a shortest opaque square fence can exist at all.

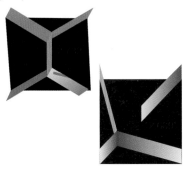

285

Puzzle 1- 1 x horizontal-clockwise, 4 x vertical-counterclockwise and 3 x horizontal-clockwise will replace the red marbles with blues in 8 single position moves.
Puzzle 2- I have found no solution with the yellow disks in their initial postions.

286

The water level will fall. When in the boat the mermaid pushed the boat down (thereby raising the water up around the boat) by an amount of water equal to the weight of the object. Once on the bottom, and not attached to the boat, the mermaid will cause the water to rise by an amount equal to the volume of water displaced. Bronze is extremely dense, so the weight of the sculpture displaces more water than the volume.

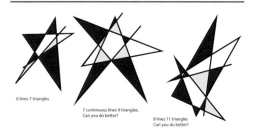

6 lines 7 triangles

7 continuous lines 9 triangles. Can you do better?

8 lines 11 triangles Can you do better?

287

With the requirement of continuous broken lines, solutions for 6, 7 and 8 lines are shown. Can you do better?
Solutions for n = 3, 4, 5 and 9 lines are already continuous closed broken lines. For seven lines the problem is no longer easy. The problem of finding a formula for the maximum number of triangles as a function of the number of lines appears to be difficult and is still unsolved.

288

In general, n people shake hands of n-1 people (one doesn't shake hands with oneself). Since two people share a handshake, this result must be halved to get the number of handshakes:
$H = n \times (n-1)/2 = n^2 - n/2$

The 17 members of the board were supposed to shake hands with 16 people, which would amount to 136 handshakes. But four people did not shake hands, so there were six handshakes less, i.e. 130 handshakes. A 10-point graph can help.

289

Five as shown below.

290

Assume A had the maximum number of eight handshakes.
J was left without a handshake, so it must be the wife of A.
B had seven handshakes, and I must be the wife of B.
C had six handshakes, and H must be the wife of C.
D had five handshakes, and G must be the wife of D. E had four handshakes, and F must be the wife of E, also having four handshakes. It follows that E must be me, and F my wife with four handshakes as well.

The number of handshakes of the other nine attendees were: 8 - 7 - 6 - 5 - 4 - 3 - 2 - 1 - 0. The two people (say, A and J) who answered "0" and "8" must have been a couple. The same argument is valid for the pair who answered "1" and "7," etc. From the graph we can see that this leaves me and my wife with four handshakes each.

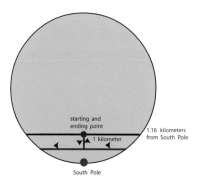

starting and ending point

1 kilometer

1.16 kilometers from South Pole

South Pole

291

The explorer can start from any point on a circle drawn around the South Pole at a distance slightly more than $1 + 1/2\pi$ kilometers (or about 1.16 kilometers taking into account the earth's curvature).

After walking a kilometer south, his next walk will take him on a complete tour around the Pole, and the walk one kilometer north from there will take him back to the circle where he started.

292

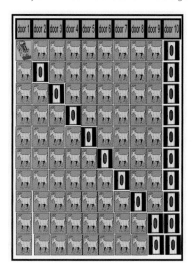

293

If we have a Monty Hall problem with 10 doors, it will be easier to get rid of the mental block associated with the original problem having only three doors. As before, behind one of the doors is the luxury car and behind the other nine are goats.

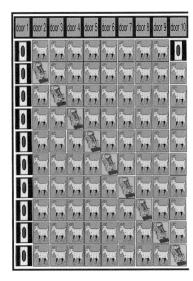

You are allowed to choose one door that stays closed. The host always opens eight doors, behind all of which are goats. He leaves one door closed, and you again have the choice to switch. As Marilyn said the correct method is always to switch in order to increase your chances of winning.

In problem 2 it is easier that there is only a one in 10 chance of the car being behind the door we picked. So switching is clearly the right strategy, which will result in increasing your chance to nine in 10.

One of the reasons it is so difficult to grasp the problem is the fact that the role of the host somehow goes unnoticed. But he is the one fixing the game. His role is made obvious if we have the problem having more than three doors, more, for example 10 or 100.

You choose door one, but now you have a probability of 1 in 10 of being right. Meanwhile the chance of the car being behind the other doors is nine in 10. But after the intervention of the host, there is only one door left representing nine of those other doors, and so the probability that your car is behind that remaining door is nine out of 10!

294 No, there is no mistake. Can you believe the blue lid fits the red coffin, and the red lid the blue one. Dracula's coffin was inspired by the "Tables Turned" optical illusion of Roger N. Shepard.

295

n = 11
r = 16

296

n = 12
r = 7

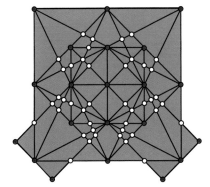

297 Twenty-one dots are required.

298

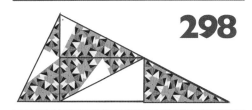

299 The parity of the glasses is odd. It cannot be changed to an even parity by an even number of moves.

300

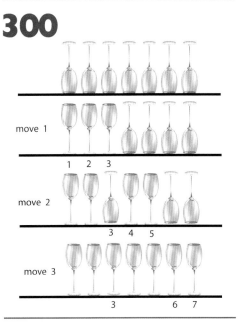

move 1

1 2 3

move 2

3 4 5

move 3

3   6 7

301 The parity of the glasses is odd. It cannot be changed to an even parity by an even number of moves.

302 If we start systematically, entering 9 as its first digit and filling in the other numbers as needed, we can see it won't work, since nine zeros cannot be placed; 8 and 7 have similar results, as shown below. Entering 6 as the first digit quickly gives the correct solution, which is unique according to Martin Gardner.

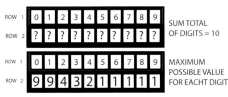

	0	1	2	3	4	5	6	7	8	9	
ROW 1	0	1	2	3	4	5	6	7	8	9	SUM TOTAL
ROW 2	?	?	?	?	?	?	?	?	?	?	OF DIGITS = 10
ROW 1	0	1	2	3	4	5	6	7	8	9	MAXIMUM
ROW 2	9	9	4	3	2	1	1	1	1	1	POSSIBLE VALUE FOR EACHT DIGIT

PREREQUISITE INSIGHTS TO THE PROBLEM

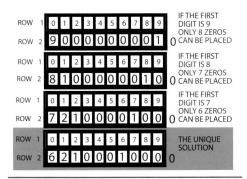

	0	1	2	3	4	5	6	7	8	9		
ROW 1	0	1	2	3	4	5	6	7	8	9		IF THE FIRST DIGIT IS 9
ROW 2	9	0	0	0	0	0	0	0	0	1	0	ONLY 8 ZEROS CAN BE PLACED
ROW 1	0	1	2	3	4	5	6	7	8	9		IF THE FIRST DIGIT IS 8
ROW 2	8	1	0	0	0	0	0	0	1	0	0	ONLY 7 ZEROS CAN BE PLACED
ROW 1	0	1	2	3	4	5	6	7	8	9		IF THE FIRST DIGIT IS 7
ROW 2	7	2	1	0	0	0	0	1	0	0	0	ONLY 6 ZEROS CAN BE PLACED
ROW 1	0	1	2	3	4	5	6	7	8	9		THE UNIQUE SOLUTION
ROW 2	6	2	1	0	0	0	1	0	0	0	0	

303 The 10 digits can be permutated in 10! or 3,628,800 ways. But since all the numbers starting with 0 don't count and must be dropped, the actual answer is 362,880 lower (or 9!), for a total of 3,265,920 10-digit numbers.

304 Yes, the man with the pipe will see the lit match reflected from two walls, no matter where he is standing in the room, so the whole room is illuminated by just one single match. Thus our room is illuminable.

305

306 Ask the man: "Please, point to the road leading to the city you are from?"

If the man is from Truthteller city, he will point to its road.

If he is from Liars city, he will also point in the same direction.

The interesting thing about his answer is that, though you have obtained the direction you asked for, you don't really know whether the man told you the truth or lied.

(The puzzles on this page were inspired by the mind-boggling and fascinating lecture of Raymond Smullyan at the Gathering for Gardner in Atlanta, 2000.)

307 The young man should ask one of the daughters: "Are you married?"

Regardless which daughter is asked the question, a "yes" answer means Amelia must be married, while a "no" means Leila is the one who must be married.

For example, if Amelia is the one the question is directed to, her answer "yes" is the truth – therefore she is married. Her answer of "no," again the truth, means she is not married, so Leila must be.

On the other hand, if the question is asked from Leila, her answer of "yes" is a lie, so she is not married, and Amelia must be.

Her answer "no" again is a lie, so she is married.

308 Ask the man the same question twice: "Are you one of those who alternately lie and tell the truth?"

Two "no's" would be a truthteller. Two "yes's" would be a liar.

A "no" and a "yes" would be one who alternately lies and tells the truth.

309 Your friend's answers were: YES - NO - YES - YES.

We have to note which colors of the four questions get YES as an answer. Then, starting from the blue D node in the lower right, we traverse those same colored edges (in any order) to arrive at a node.

The letter on this node will be the chosen letter. If it is blue, the answers were truthful, if red, the answers were lies. If he lied, his answers would be: NO - YES - NO – NO.

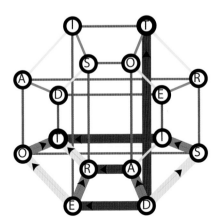

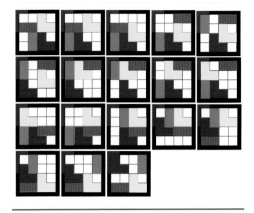

GRASSHOPPING
Solutions for the first 40 lengths

310 1- The next three lengths with solutions: n= 9, 13, 16.

2- For the first n=40, there are 16 solutions as shown below

3- It is believed that for all n: 0 or 1 mod 4 are possible for n=20 and higher, so the grasshop number sequence can be extended accordingly.

311 The probability that the woman has two girls is about 33 percent, whereas the probability that the man has two girls is about 50 percent.

Marylin also provided statistical evidence which supported this answer. For the woman with one girl there are three possibilties: BG -GB -GG - all equally likely at 1/3.

For the man there are only two possibilities:

BG and GG - 1/2

(The third possibility BG falls since the girl is older).

312 1- In analyzing the problem consider all the possible outcomes of the sexes of the children in their birth order: girl - girl, girl - boy, boy - girl, boy - boy.

It can be seen that the chance of both children being girls is 25 percent. It is a problem very similar to the coin-toss problem or, in mathematical language isomorphic to it.

Pascal's Triangle is usually used to give the answer to this type of problem.

2- Girl - Girl, Girl - Boy, Boy - Girl, Boy - Boy. It can be seen that the chance of at least one of the babies being a girl is 75 percent.

3- Girl - Girl , Girl - Boy, Boy - Girl, Boy - Boy.

The question eliminates the "Boy-Boy" possible outcome. It leaves "Girl-Girl" as the only favorable outcome at 33 percent.

313 The surprising answer is that the probability of a girl in a family of three children is near certainty. It is 7/8, or 87 percent.

314 This is a typical problem for which Pascal's triangle will easily provide the answer. The eighth row of the Pascal's triangle gives us the probability of getting four heads (girls or boys) of 70/256 or about 27 percent. While the probability of getting eight heads (girls or boys) is 1/256 or less than one percent. Note: it may be that more than mere chance is involved in producing families with six or more children of only sons and no daughters, which occur twice as often as chance says they should.

315 The 18 move solution.

ACKNOWLEDGMENT

First and foremost, I would like to thank Martin Gardner for everything.

His work, his personality and great friendship have been my inspiration since the mid-'50s, when I first read his "Mathematical Games" column in the *Scientific American*, a moment that changed the course of my life.

I miss him very much. I miss his near-monthly or bimonthly typewritten hand-corrected letters that, since 1957, became part of my life.

His immense contribution to the popularization of recreational mathematics (and mathematics in general) has created a world-wide environment of creativity. Without him there would have been fewer Puzzle Parties, and certainly no Gatherings for Gardner, an event like no other.

Over the last 20 years or so, these conventions of like-minded souls have allowed me to meet "Martin's People," at the enjoyable biannual "Gathering for Gardner" conferences in Atlanta, a wonderful and diverse group of mathematicians, scientists, puzzle collectors, magicians, inventors and publishers unified by a fascination with mind games and a love of recreational mathematics and, of course, Martin Gardner.

They have provided me with endless hours of enjoyment, intellectual enrichment and, often, precious friendships.

My appreciation and thanks to all of them, mentioning just a few: Paul Erdös, my famous relative, who provided the first sparks; David Singmaster, with whom we dreamed of a very special puzzle museum; Ian Stewart for his early help; Piet Hein, John Horton Conway, Solomon Golomb, Frank Harary, Raymond Smullyan, Edward de Bono, Richard Gregory, Tom Rodgers, Victor Serebriakoff, Jeremiah Farrell, Edward Hordern, Nick Baxter, Jerry Slocum, Nob Yoshigahara, Lee Sallows, Greg Frederickson, Hal Robinson, James Dalgety, Mel Stover, Mark Setteducati, Bob Neale, Tim Rowett, Scott Morris, Will Shortz, Bill Ritchie, Richard Hess, Cliff Pickover, Colin Wright, Dick Esterle, Mark Carson, Erik Quam and many, many others.

In a way *The Puzzle Universe* is a visual synthesis of the whole of recreational mathematics.

– Ivan Moscovich, 2015, Nijmegen

ABOUT IVAN MOSCOVICH

Ivan Moscovich is a mechanical engineer born in the former Yugoslavia. He was the founder and former director of the Lasky Museum of Science and Technology & Planetarium, and the author of many books on recreational math and puzzles.

Over many years he has invented many puzzles, games and toys, of which over 100 have been commercially produced, highly acclaimed worldwide for their originality receiving many awards and prizes.